Sustainable Development and Quality of Life

K. Muralidharan

# Sustainable Development and Quality of Life

## Through Lean, Green and Clean Concepts

 Springer

K. Muralidharan
Faculty of Science
Maharaja Sayajirao University of Baroda
Vadodara, Gujarat, India

ISBN 978-981-16-1834-5      ISBN 978-981-16-1835-2   (eBook)
https://doi.org/10.1007/978-981-16-1835-2

This Springer imprint is published by the registered company Springer Nature Singapore Pte Ltd.
The registered company address is: 152 Beach Road, #21-01/04 Gateway East, Singapore 189721, Singapore

*I dedicate this book to all those who eat-less, sleep-less and use-less*

# Preface

This book builds on simple concepts like Lean and Green concepts, which becomes an integral part of clean sciences and technologies. This author believes that everything in excess is 'wastes,' whether it is energy, food, infrastructure, machinery, comforts, facilities, techniques, and so on. The environment, air, land, ecosystem, and climate are for all. Climate change issues and environmental problems are prevalent these days and pose various kinds of problems sustainable in our day-to-day lives. The lean principle advocates waste elimination and improve quality in process or product, and the Green concepts facilitate sustainable practices and promotions. Therefore, there is an immediate requirement to educate people to act positively towards the ill-effects of waste and pollution. I firmly believe that the Lean, Green and Clean concepts can facilitate a positive approach for a better quality of life.

The concept of writing this book was started when I was asked to design a course for an undergraduate foundational level program offered under Choice based credit system (CBCS) of our university, where students from any discipline can join and acquire credits. Foundational level courses are generally popular topics and offered to students for enriching their tastes and preference for skill development and personal excellence. I found a strong demand for teaching the students about the quality concepts to be practiced in their daily life, as other papers offered at the undergraduate program are not catering to these needs. The book can be used for any foundation or regular courses offered at any university irrespective of faculties and disciplines. The necessary ingredients are incorporated into this book. The book can attract a wider audience, including students, teachers, Quality professionals, management consultants, Lean, and Six Sigma professionals. Emphasis is laid on understanding and applying the concepts of quality through project management and measurement-based assessment methods.

There are plenty of books on Lean concepts available in the literature. Even Lean ideas are discussed as a part of Lean Six Sigma, Lean manufacturing, Six Sigma kind of philosophies as well. The author's book entitled *Six Sigma for Organizational Excellence: A Statistical Approach* also covers lean concepts to some extent. Most of the time, the discussion is centered around the Toyota Production System (TPS) only, as they are responsible for introducing Lean concepts way back in the 1980s. The author has also referred to some books written on Green ideas, discussing the importance of being sustainable in business and trade. The books by Keivan Zokaei A. (*Creating a Lean and Green Business System: Techniques for Improving Profits and Sustainability*, CRC, 2013) and Walter Crinnion (*Clean, Green, and Lean: Get Rid of the Toxins That Make You Fat*, Wiley, 2010) are closely related to the title of this book. The first book discusses the benefits of using Lean and green concepts through TPS, as mentioned above. The second book is written from a medical dietary point of view, where some lean ideas are used to reduce food intake and its usages. Almost all the literature offers how to improve productivity and efficiency in their service processes. None of the books attempts to make any recommendation for sustaining quality in personal or professional life. Hence, I do not find any reason to believe that there is any competition for this book. The proposed text is absolutely a deviation from many other books in terms of its concepts, methodology, and presentation. Along with improving business profits and growth, this book also offers productivity and quality improvement through sustainable and clean business practices.

Chapter 1 is an introduction, which will explore why we need to have a sustainable quality of life. A green and clean environment facilitates the best science/technology/business practices to optimize the resources and availabilities, which in turn help us to address the issues concerning sustainable and climate change issues. I feel that, for sustainable quality life, developing a quality culture is essential for universal peace and happiness. Chapter 2 presents the concepts of LGC. Note that, for improving customer relationships, improved profitability, and service/industrial/environment growth, the organizations should strive for best business practices in tandem with environmentally sustainable notions and practices. This is achieved through the development of LGC concepts. To increase the knowledge of the process, it is necessary to quantify the process variation inherent in the process. For quantitatively understanding the process variations, we discussed various measures of accuracy and precision in this chapter. The importance of process performance and process capability measures are also discussed in detail in this chapter.

The scientific and technological notions of LGC are presented in Chap. 3. Here, we understand the importance of Green energy, Green chemistry, Green supply chains, Green biology, etc., as they are the core components for managing products and services subject to climate and environmental change issues. We feel that the Learn, Green, and clean methods can potentially recover energy, manage solid and liquid wastes, and facilitate renewable energy requirements for future generations.

A detailed discussion on various quality improvement methods like Total Quality Management (TQM), Total Productive Maintenance (TPM), Kaizen, 5S+ Safety and Security Techniques, 3M Techniques, Poka Yoke, Set up Reduction, Just-in-Time Approach, Kanban, Six Sigma concepts, Capability Maturity Model Integration (CMMI), etc. are presented in Chap. 4. The improvement tools are critically examined for their relevance in pursuit of LGC concepts. Among all improvement methods, Six Sigma supports various tenets of quality improvement by reducing variation and waste. Hence, it is considered to be one of the powerful tools for recovering Return on Investment (ROI) and maximizing the process quality.

The aspects of LGC quality assessment models are presented in Chap. 5. The relevance of ISO 9000 series, Environmental management system (EMS), Eco-Management and Audit Scheme (EMAS), Corporate sustainable business practices, Occupational Health and Safety Assessment Systems (OHSAS), Emission auditing, Carbon Management, Climate models and Life cycle assessment is critically examined in this chapter.

The essence of LGC is presented through the concept of Green Statistics (this author first studied the idea) in Chap. 6. It is stressed that any quality concept creates constancy of purpose toward improving product and service to become competitive and to stay in business and provide jobs. Towards this, a new concept called "Green Statistics" is introduced by the authors Muralidharan and Ramanathan (2012) based on measurement-based evaluation and assessment of the product, process, and service quality. Green Statistics are used to evaluate the economic and financial performance of business activities and improve the environmental concerns which impact climate change. It is recommended that there is a need for tools to promote the development and improvement of green performance measures and supply chain management and data and information use relating to green supply chain management (GSCM). This requirement is studied in detail, connecting GSCMs relevance with E-commerce, Big Data Analytics, Industry 4.0, Reverse logistics, Reverse Engineering, and Reliability engineering. The chapter also provides the quality guidelines underlying the Green statistics concepts of sustainable products and services.

The methods of achieving and promoting LGC concepts in daily life are studied in Chap. 7. It is believed that a business that is carbon-positive, wastes nothing, is regenerative to environments, and has a positive effect on people's lives. Hence, there should include a flexible quality audit and policies for producing sustainable products and services for the generations to come. It is also found that plans for legislative, social, and political pressures are also imposed on manufacturers to reduce their global warming contribution during production and use. We recommend that quality policy shall be consistent with the company's strategy, and the top management must share all the necessary measures to ensure that its corporate quality policy is clearly understood, implemented, and maintained. This will ensure the extent of the commitment of top management to quality. To resist the modern-day challenges in business, it is essential to carry out Green quality policy and quality audits when new products and services are introduced in the organization.

We provide an elaborate study on the control, monitoring, and deployment of LGC principles for a sustainable quality life in Chap. 8. Although quality concepts are perceptional and notional, monitoring and control of quality are equally essential as we do it in the case of processes and products for understanding assignable and chance causes of variations. To make a product defect-free, it is natural to minimize the variation of a process. Hence quantifying the amount of variation in a process is the first and the critical step towards improvement. Therefore, we study and show the relevance of various statistical quality tools of monitoring variation and control of the quality of a process. We strongly recommend that visual presentations better communicate with the audience faster and effectively, and convincing quality aspects become easy. Some detailed discussion on correlation and regression analysis, Design of Experiments (DOE), and Cause and Effect (C&E) diagram is presented in this chapter. The necessity of organizing company-wide Quality Function Deployment (QFD) is also stressed in this chapter. QFD promotes cross-functional teamwork and gets the right people together, early, to work efficiently and effectively to meet customers' needs. QFD is a structured methodology to identify and translate customer needs and wants into technical requirements and measurable features and character-istics. We feel that developing performance metrics and track them every day to monitor for quality is the best things one should do for realizing the LGC benefits. To identify those metrics, one should get into the root cause of the problem. Root cause analysis (RCA) is the process of identifying causal factors using a structured approach with techniques designed to provide a focus for identifying and resolving problems and risks. The chapter provides a comprehensive discussion on Dashboards and prioritization concepts too.

Chapter 9 provides some case studies based on LGC philosophy. The case studies are prepared based on organizational problems and contexts. Chapter 10 is the concluding chapter. We have pointed out that quality life addresses all concerns of society and environment suitably incorporating all types of externalities expe-rienced by the stakeholders of an organization. Establishing a close watch on the externalities should be the primary concern of all human beings. In this process, human beings should forgo their egos and priorities for the betterment of other living people, if it matters the livelihood of others. The need for increasing value-added activities is necessary for understanding the right product with the right capabilities. The identification of non-value added activities is essential to ascertain the COQ and COPQ associated with any process or organizational events. For new generation business ventures and startup initiatives, the development of Lean culture is a must to get speedy results quickly. With this aim, we offer many quality guidelines and recommendations for increasing quality in everyone's life.

Vadodara, India                                                                K. Muralidharan

# Acknowledgements

Writing this kind of book was not as easy as I initially thought. Firstly, the book was developed on a philosophical note incorporating a new generation thought process on livelihood, business, and sustainable practice. Secondly, since the materials are all dispersed and integrating three different concepts (Lean, Green, and Clean) in a single platform was a difficult task altogether. Along with many text and reference books, I was forced to read much literature based on magazines and general books. Close monitoring of news items appearing in daily newspapers and contemporary articles also helped me a lot in this process. Notably, the *Speaking Tree* edition of Time of India was beneficial for changing my attitude towards life and approach, which further penetrated my interest in writing this kind of book to its completion. Some of my lecture notes, conference presentations, and articles published in various journals and magazines also helped me to realize this task.

I am thankful to several people for their suggestions and personal feelings on this topic of discussion. I thank The Editor of Indian Association for Productivity, Quality, and Reliability for permitting me to use my articles published in their short communication "*Communiqué*" in various issues. I also consulted and used Wikipedia and other online site materials for understanding the concepts and current developments happening in Lean, Green, and Clean concepts. I sincerely acknowledge the online library and web sites for this. I am grateful to my faculty colleagues Dr. Aruna Joshi, Professor of Botany, and Dr. Bhavna Trivedi, Professor of Chemistry, for contributing some materials on Green energy and Green Chemistry, respectively. I also thank my friend, Mr. Venkatesh, a professor of English at my university, to go through the manuscript for language corrections and grammar.

It is my duty to acknowledge Mr. Dhiraj K. Patel of my department and Dr. V. O. Thomas of Mathematics department for their lovable affection and encouragement conferred on me. Thanks are also due to Dr. Khimya Tinani, Dr. Pratima Bhavagosai, and Mr. Agniva Das (Research Scholar) of my department and Linta Paulson of Zoology department for their valuable help rendered for proof-reading and other assistance.

I have a dual mixed feeling of happiness and sadness to state that the lockdown due to COVID-19 was a blessing in disguise for me, as I got enough time to spare for revising the chapters and its contents. At the same time, I am very much concerned about the increasing incidence of infection and deaths due to the virus pandemic. This incident has sensitized my thought process to some extent. The pandemic is giving us many lessons to learn from nature and the environment. I acknowledge all those scientists and doctors close to our ecosystem and biodiversity for defeating the virus menace.

I must also submit that this book took many excruciatingly small steps and exacted a heavy toll on my evenings, weekends, and vacations. Therefore, it is a matter of pride for me to express my feelings to my wife, Mrs. Lathika, and sons Vivek and Varun for their complete silence throughout the writing of this book.

I also take this opportunity to thank the editor and the team of production of Springer Nature for their valuable service rendered to me in this endeavor.

Vadodara, India                                                                    K. Muralidharan

# Contents

**1 Introduction** .................................................. 1
References ..................................................... 9

**2 Lean, Green, and Clean Quality Concepts** ...................... 11
2.1 Introduction ............................................ 11
2.2 Lean, Green, and Clean Concepts ......................... 14
2.2.1 Developing Lean Culture ......................... 16
2.2.2 Developing Green Culture ........................ 21
2.2.3 Developing Clean Culture ........................ 23
2.3 Process Variations and Characteristics .................. 25
2.3.1 Process Quality and Histogram ................... 27
2.3.2 Process Accuracy Measures ....................... 29
2.3.3 Process Precision Measures ...................... 32
2.3.4 Process Capability Measures ..................... 38
2.4 Lean, Green, and Clean Quality Walk .................... 49
2.5 Exercises ............................................... 52
References ..................................................... 56

**3 Lean, Green, Clean Sciences and Technologies** .................. 57
3.1 Introduction ............................................ 57
3.2 Green Energy ............................................ 58
3.3 Green Biology ........................................... 60
3.4 Green Chemistry ......................................... 63
3.5 Green Supply Chain Management ........................... 65
3.6 Green Technologies ...................................... 70
3.7 Exercises ............................................... 73
References ..................................................... 74

**4 Lean, Green, and Clean Quality Improvement Models** ............ 77
4.1 Introduction ............................................ 77
4.2 Uncertainty and Probability Models ...................... 82
4.2.1 Probability Model Development ................... 83
4.2.2 Normal Distribution ............................. 90

| | 4.2.3 | Central Limit Theorem Explained | 95 |
| 4.3 | | Total Quality Management | 98 |
| 4.4 | | Total Productive Maintenance | 101 |
| 4.5 | | Kaizen | 103 |
| 4.6 | | 5S + Safety and Security Techniques | 105 |
| 4.7 | | 3Ms Technique | 106 |
| 4.8 | | Poka-Yoke | 106 |
| 4.9 | | Setup Reduction | 108 |
| 4.10 | | Just-in-Time Approach | 108 |
| 4.11 | | Kanban | 109 |
| 4.12 | | Six Sigma Concepts | 109 |
| | 4.12.1 | Six Sigma Project Essentials | 118 |
| 4.13 | | Capability Maturity Model Integration | 127 |
| 4.14 | | Exercises | 129 |
| | | References | 132 |

**5 Lean, Green, and Clean Quality Assessment Models** ............... **135**
| 5.1 | Introduction | 135 |
| 5.2 | ISO 9000 Series | 136 |
| 5.3 | Environmental Management System | 140 |
| 5.4 | Eco-Management and Audit Scheme | 141 |
| 5.5 | Corporate Sustainable Business Practices | 142 |
| 5.6 | Occupational Health and Safety Assessment Systems | 144 |
| 5.7 | Emission Auditing | 145 |
| 5.8 | Carbon Management | 145 |
| 5.9 | Environmental Stress Screening | 146 |
| 5.10 | Climate Models | 148 |
| 5.11 | Life-Cycle Assessment | 149 |
| 5.12 | Cleaner Production Assessment | 150 |
| 5.13 | Other Assessment Models | 151 |
| 5.14 | Exercises | 152 |
| | References | 153 |

**6 Green Statistics: Essence of Lean, Green, and Clean Sciences** ...... **155**
| 6.1 | | Introduction | 155 |
| 6.2 | | Green Statistics | 156 |
| 6.3 | | Green Statistics Supply Chain Dynamics | 158 |
| 6.4 | | Key Performance Measures Associated with Green Statistics | 162 |
| 6.5 | | Green Methods of Evaluation | 165 |
| 6.6 | | Green Statistics for Emerging Aspects of Business | 165 |
| | 6.6.1 | E-Commerce and E-Governance | 165 |
| | 6.6.2 | E-Governance in Information and Communication Technology | 169 |
| | 6.6.3 | E-Governance in Education | 170 |
| | 6.6.4 | Quality in e-Governance | 171 |

6.7 Big Data Analytics .......................................... 174
6.8 Industry 4.0 ............................................... 178
6.9 Reverse Logistics .......................................... 179
6.10 Benchmarking and Reverse Engineering .................... 180
6.11 Reliability Engineering ..................................... 182
   6.11.1 Exponential Distribution ........................... 184
   6.11.2 Weibull Distribution .............................. 186
   6.11.3 Software Reliability ............................... 189
6.12 Simulation and Process Model Generation ................... 190
   6.12.1 Simulation of the Probability Model ................ 191
   6.12.2 Simulation of an Engineering Model ................ 191
   6.12.3 Simulation of Reliability Model .................... 193
6.13 Green Statistics and Quality Certifications ................... 196
6.14 Exercises ................................................. 198
References ..................................................... 199

7 Achieving and Promoting Lean, Green, and Clean Quality
   Objectives ................................................... 203
7.1 Introduction ............................................... 203
7.2 Achieving Lean Quality .................................... 204
7.3 Green Quality Policy ....................................... 208
7.4 Green Quality Audit ....................................... 209
7.5 Promoting Lean, Green, and Clean Thinking ................. 210
7.6 Benefits of Being Lean, Green, and Clean ................... 211
7.7 Exercises ................................................. 213
References ..................................................... 213

8 Control, Monitoring, and Deployment of Lean, Green,
   and Clean Activities ......................................... 215
8.1 Introduction ............................................... 215
8.2 Quality Assurance ......................................... 217
8.3 Quality Control ........................................... 219
   8.3.1 Control Chart ..................................... 221
   8.3.2 Time Series Plot .................................. 231
   8.3.3 Pareto Chart ..................................... 232
   8.3.4 Scatter Plots ..................................... 235
   8.3.5 Cause and Effect Analysis ......................... 247
   8.3.6 Testing and Confirmatory Analysis ................. 250
   8.3.7 Design of Experiments ............................. 262
8.4 Quality Function Deployment .............................. 272
   8.4.1 Dashboards ....................................... 276
   8.4.2 Prioritization Method ............................. 278
   8.4.3 Root Cause Analysis .............................. 279
8.5 Exercises ................................................. 283
References ..................................................... 289

**9    Lean, Green, and Clean: Some Case Studies** ...................... 291
    9.1    Introduction ............................................... 291
    9.2    Case Study Based on Lean Six Sigma Project ................ 291
    9.3    Case Study Based on 5S and Sustainable Practices ........... 295
    9.4    Case Study Based on the Quality Walk ...................... 298
    Reference ....................................................... 303

**10   Moving Toward Sustainable Quality Life** ....................... 305
    10.1   Introduction ............................................... 305
    10.2   Identifying Value-Added Activities ......................... 306
           10.2.1   Value Stream Mapping ............................ 309
           10.2.2   Cost of Quality/Cost of Poor Quality ............... 313
    10.3   Leadership and People Engagement ......................... 315
    10.4   Moving Toward Sustainable Quality Life .................... 318
    References ....................................................... 322

**Appendix** ......................................................... 323
**Glossary of Terms** ................................................ 365

# About the Author

**K. Muralidharan** is Professor of Statistics at the Faculty of Science, Maharaja Sayajirao University of Baroda, Gujarat, India. He completed his post-doctoral fellowship from the Institute of Statistical Science at Academia Sinica, Taiwan, in 2003, and Ph.D. in Statistics from Sardar Patel University, India, in 1997. He is an internationally certified Six Sigma Master Black Belt from Indian Statistical Institute, Bangalore. His research interests are statistical inference, stochastic point process models, reliability and life testing, artificial intelligence, quality and Six Sigma.

He has authored 3 books, out of which *Six Sigma for Organizational Excellence: A Statistical Approach* was published in 2015 by Springer Nature, and over a 135 publications in international and national journals of repute. He has completed 5 major research projects funded by the UGC and DST, New Delhi, India; guided many student projects on statistical methods and Six Sigma and offered consultancy services for many industries. He has about 30 years of teaching and research experience and has guided 7 research students for the doctoral degree.

# Acronyms

| | |
|---|---|
| ALT | Accelerated life test |
| ANOVA | Analysis of variance |
| AR | Augmented reality |
| ARL | Average run length |
| ASCM | Agile Supply Chain Management |
| ASQ | American Society for Quality |
| ATS | Average time to signal |
| BB | Black belts |
| BDA | Big Data Analytics |
| BIBD | Balanced incomplete block design |
| BIC | Bayesian information criterion |
| BIS | Bureau of Indian Standards |
| BSC | Balanced score card |
| C&E | Cause and effect |
| CBA | Cost-benefit analysis |
| CC | Cloud Computing |
| CED | Cause and effect diagram |
| CL | Control limit |
| CLT | Central limit theorem |
| CMM | Capability maturity model |
| CMMI | Capability maturity model integration |
| CNG | Compressed natural gas |
| COPQ | Cost of poor quality |
| COQ | Cost of quality |
| COV | Covariance |
| CPA | Cleaner production Assessment |
| CPM | Critical path method |
| CRD | Completely randomized design |
| CSF | Critical success factors |
| CSR | Corporate social responsibility |
| CTC | Critical to cost |
| CTD | Critical to delivery |

| CTP | Critical to process |
|---|---|
| CTQ | Critical to quality |
| CTS | Critical to safety |
| CV | Coefficient of variation |
| CVM | Contingent Valuation Method |
| DEIT | Department of Electronics and Information Technology |
| DFM | Demand Flow Manufacturing |
| DFR | Design for reliability |
| DFSS | Designed for Six Sigma |
| DIDOV | Define-Identify-Design-Optimize-Verify |
| DMADV | Define-Measure-Analyze-Design-Validate |
| DMAIC | Define-Measure-Analyze-Improve-Control |
| DMEDI | Define-Measure-Explore-Develop-Implement |
| DOE | Design of experiments |
| DPMO | Defects per million opportunities |
| DPO | Defects per opportunity |
| DPU | Defect per unit |
| EDA | Exploratory data analysis |
| EMAS | Eco-Management and Audit Scheme |
| EMS | Environmental management standards |
| ERP | Enterprise Resource Planning |
| ESS | Environmental Stress Screening |
| EWMA | Exponentially weighted moving average |
| FMEA | Failure mode effect analysis |
| FTA | Fault tree analysis |
| GB | Green belts |
| GDP | Gross Domestic Product |
| GHG | Greenhouse Gas |
| GNP | Gross National Product |
| GRPI | Goals-roles and responsibilities-processes-procedures |
| GS | Green Statistics |
| GSCM | Green supply chain management |
| GSS | Green Six Sigma |
| GSSCM | Green Statistics Supply Chain Management |
| ICT | Information and Communication Technology |
| ID | Interrelationship diagram |
| IoT | Internet of Things |
| IPO | Input-Process-Output |
| IPR | Intellectual Property Right |
| ISO | International Organization for standardization |
| IT | Information technology |
| ITES | Information technology enabled services |
| JUSE | Union of Japanese Scientists and Engineers |
| KMI | Key marketing indicators |
| KPI | Key performance indicators |

| KPIV | Key process input variables |
| KPOV | Key process output variables |
| LCA | Life-cycle assessment |
| LCL | Lower control limit |
| LEED | Leadership in Energy and Environmental Design |
| LGC | Lean, Green, and Clean |
| LPG | Liquefied petroleum gas |
| LSCM | Lean supply chain management |
| LSD | Latin square design |
| LSL | Lower specification limit |
| LSS | Lean Six Sigma |
| MBO | Management by objectives |
| MDG | Millennium Development Goals |
| MIS | Management information system |
| MRL | Mean residual life |
| MSA | Measurement system analysis |
| MSE | Mean square error |
| MTBF | Mean time between failures |
| MTTF | Mean-time-to-failure |
| MTTR | Mean-time-to-repair |
| NAPCC | National Action Plan on Climate Change |
| NeGP | National e-Governance Plan |
| NVA | Non value added |
| OEE | Overall Equipment Effectiveness |
| PBIBD | Partially balanced incomplete block design |
| PDCA | Plan–Do–Check–Act |
| PM | Project management |
| PPM | Parts per million |
| PPP | Public-Private Partnership |
| QC | Quality control |
| QCI | Quality council of India |
| QD | Quartile deviation |
| QE | Quality engineering |
| QFD | Quality function deployment |
| QM | Quality management |
| QMS | Quality Management System |
| QPM | Quantitative process management |
| R&R | Repeatability and reproducibility |
| RBD | Randomized block design |
| RCA | Root cause analysis |
| RE | Reverse engineering |
| RL | Reverse logistics |
| ROI | Return of investment |
| RP | Recursive partitioning |
| RPN | Risk priority number |

| | |
|---|---|
| SCM | Supply chain management |
| SD | Standard deviation |
| SDG | Sustainable Development Goals |
| SIPOC | Supplier-Input-Process-Output-Customer |
| SMART | Specific-Measurable-Achievable-Relevant-Timely |
| SNG | Synthetic Natural Gas |
| SOP | Standard operating procedures |
| SOW | Statement of work |
| SPC | Statistical process control |
| SQC | Statistical quality control |
| SS | Six Sigma |
| SSM | Six Sigma Marketing |
| SWOT | Strength-Weakness-Opportunities-Threats |
| TCE | Total Customer Experience |
| TFM | Total Flow Management |
| THRM | Total Human Resource Management |
| TOC | Theory of Constraints |
| TOI | Times of India |
| TOP | Total opportunity |
| TPM | Total productivity maintenance |
| TPS | Total Performance Scorecard |
| TQC | Total quality control |
| TQM | Total quality management |
| TRIZ | Theory of inventive problem solving |
| TSM | Total Service Management |
| UCL | Upper control limit |
| UNDP | United Nations Development Program |
| UNFCC | United Nations Framework Convention on Climate Change |
| USL | Upper specification limit |
| VA | Value added |
| VAM | Value analysis mapping |
| VIF | Variance inflation factor |
| VOC | Voice of customer |
| VOP | Voice of process |
| VR | Virtual Reality |
| VSM | Value stream mapping |
| WHO | World Health Organization |
| WIP | Work in progress |
| WIQ | Work in queue |
| WTA | Willingness to accept |
| WTP | Willingness to pay |

# List of Figures

Fig. 2.1    Typical organizational process ........................... 12
Fig. 2.2    Process flow diagram .................................... 13
Fig. 2.3    Quality dimensions of components of LGC ................. 16
Fig. 2.4    Typical lean process ................................... 17
Fig. 2.5    Perceptions of a process. *Source* Muralidharan &
            Syamsundar (2012) ..................................... 20
Fig. 2.6    Histogram of endurance level in minutes .................. 28
Fig. 2.7    Skewed distributions .................................... 28
Fig. 2.8    Symmetric or bell-shaped distributions ................... 29
Fig. 2.9    Accuracy and precision explained ....................... 36
Fig. 2.10   Summary of thickness measurement of piston rings ......... 38
Fig. 2.11   Graphical summary of the center thickness data ............ 40
Fig. 2.12   Incapable process ...................................... 41
Fig. 2.13   Perfect capable process ................................. 41
Fig. 2.14   High capable process ................................... 42
Fig. 2.15   Process capability of center thickness ................... 44
Fig. 2.16   Two normal processes with the same variance .............. 45
Fig. 2.17   Normal processes with a different mean and standard
            deviation ............................................. 46
Fig. 2.18   The process capability of time to failure of electronic chips .... 48
Fig. 3.1    Sector-wise GHG emissions in India. *Source* GHG Platform ... 59
Fig. 3.2    Deming's PDCA cycle ................................... 67
Fig. 3.3    Lean green supply chain management diagram ............. 68
Fig. 4.1    Quality flow diagram .................................. 80
Fig. 4.2    Graph of binomial distribution .......................... 85
Fig. 4.3    Graph of Poisson distribution ........................... 87
Fig. 4.4    Graph of normal distributions $X$ and $Z$ .................... 90
Fig. 4.5    The area correspond to $P(Z < -1)$ ........................ 91
Fig. 4.6    The area correspond to $P(Z > 3)$ ......................... 92
Fig. 4.7    The area correspond to $P(-2 < Z < 1)$ .................... 92
Fig. 4.8    The area correspond to $P(Z1) = 0.70$ .................... 93
Fig. 4.9    Area under normal curve ............................... 94

| | | |
|---|---|---|
| Fig. 4.10 | Graphical summary of the amount of radiation . . . . . . . . . . . . . . | 95 |
| Fig. 4.11 | Plot of sample statistics from binomial distribution . . . . . . . . . . | 97 |
| Fig. 4.12 | Plot of sample statistics from an exponential distribution . . . . . . | 98 |
| Fig. 4.13 | TQM process diagram . . . . . . . . . . . . . . . . . . . . . . . . . . . . . . . . . | 99 |
| Fig. 4.14 | Six Sigma problem-solving approaches . . . . . . . . . . . . . . . . . . . . | 111 |
| Fig. 4.15 | Six Sigma process with $1.5\sigma$ shift . . . . . . . . . . . . . . . . . . . . . . . . | 112 |
| Fig. 4.16 | Normal probability plot of wheel dimension . . . . . . . . . . . . . . . . | 114 |
| Fig. 4.17 | Process capability of wheel diameter . . . . . . . . . . . . . . . . . . . . . . | 116 |
| Fig. 4.18 | Value of Six Sigma . . . . . . . . . . . . . . . . . . . . . . . . . . . . . . . . . . . | 118 |
| Fig. 4.19 | A high-level Six Sigma process map. *Source* Muralidharan (2015) . . . . . . . . . . . . . . . . . . . . . . . . . . . . . . . . . . . . . . . . . | 124 |
| Fig. 4.20 | Gantt chart . . . . . . . . . . . . . . . . . . . . . . . . . . . . . . . . . . . . . . . . . | 125 |
| Fig. 6.1 | Green Statistics Supply Chain Management system . . . . . . . . . . | 159 |
| Fig. 6.2 | Lean-enabled process uncertainty model . . . . . . . . . . . . . . . . . . . | 161 |
| Fig. 6.3 | Plot of $f(x)$, $R(x)$, and $h(x)$ of an exponential distribution . . . . | 185 |
| Fig. 6.4 | Plot of $f(x)$, $R(x)$, and $h(x)$ of Weibull distribution . . . . . . . . . | 187 |
| Fig. 6.5 | Simulated t-distribution . . . . . . . . . . . . . . . . . . . . . . . . . . . . . . . . | 192 |
| Fig. 6.6 | Graph of all variables . . . . . . . . . . . . . . . . . . . . . . . . . . . . . . . . . | 192 |
| Fig. 6.7 | Process capability of $y$ . . . . . . . . . . . . . . . . . . . . . . . . . . . . . . . . | 193 |
| Fig. 6.8 | Distribution overview plot of the pressure . . . . . . . . . . . . . . . . . . | 194 |
| Fig. 6.9 | Six-pack capability of pressure measurement . . . . . . . . . . . . . . . | 195 |
| Fig. 7.1 | Shingo principles . . . . . . . . . . . . . . . . . . . . . . . . . . . . . . . . . . . . | 205 |
| Fig. 8.1 | A normal (controlled) process . . . . . . . . . . . . . . . . . . . . . . . . . . . | 220 |
| Fig. 8.2 | Box plot of the sample values . . . . . . . . . . . . . . . . . . . . . . . . . . . | 224 |
| Fig. 8.3 | Control chart corresponds to piston diameter . . . . . . . . . . . . . . . | 224 |
| Fig. 8.4 | Control chart of imperfections in rolls production . . . . . . . . . . . | 228 |
| Fig. 8.5 | p-chart for non-conformities . . . . . . . . . . . . . . . . . . . . . . . . . . . . | 230 |
| Fig. 8.6 | Time series chart of $x_1$ . . . . . . . . . . . . . . . . . . . . . . . . . . . . . . . . | 231 |
| Fig. 8.7 | Time series plot of accidental deaths in India . . . . . . . . . . . . . . . | 233 |
| Fig. 8.8 | Pareto chart of lost time . . . . . . . . . . . . . . . . . . . . . . . . . . . . . . . | 234 |
| Fig. 8.9 | Scatter plot of anxiety versus marks and vice versa . . . . . . . . . . | 236 |
| Fig. 8.10 | Fitted line plot . . . . . . . . . . . . . . . . . . . . . . . . . . . . . . . . . . . . . . . | 237 |
| Fig. 8.11 | Cause and effect diagram for high petrol consumption. (*Source* Muralidharan, 2015) . . . . . . . . . . . . . . . . . . . . . . . . . . | 248 |
| Fig. 8.12 | Pareto chart of material causes . . . . . . . . . . . . . . . . . . . . . . . . . . | 249 |
| Fig. 8.13 | Types of statistical testing errors . . . . . . . . . . . . . . . . . . . . . . . . | 252 |
| Fig. 8.14 | Rejection region of various types of test . . . . . . . . . . . . . . . . . . . | 255 |
| Fig. 8.15 | Test statistic and critical region . . . . . . . . . . . . . . . . . . . . . . . . . | 256 |
| Fig. 8.16 | Z-test statistic and critical region . . . . . . . . . . . . . . . . . . . . . . . . | 257 |
| Fig. 8.17 | $F$-test statistic and critical region . . . . . . . . . . . . . . . . . . . . . . . | 259 |
| Fig. 8.18 | $Z$ test statistic and critical region . . . . . . . . . . . . . . . . . . . . . . . . | 260 |
| Fig. 8.19 | $t$ test statistic and critical region . . . . . . . . . . . . . . . . . . . . . . . . | 261 |
| Fig. 8.20 | $x^2$ test statistic and critical region . . . . . . . . . . . . . . . . . . . . . . | 262 |
| Fig. 8.21 | Main effect plots of efficiency parameters . . . . . . . . . . . . . . . . . . | 269 |
| Fig. 8.22 | Interaction plot of efficiency parameters . . . . . . . . . . . . . . . . . . . | 270 |

Fig. 8.23    Main effect plots ........................................ 271
Fig. 8.24    Pareto chart of main effects ............................. 272
Fig. 8.25    Structure of house of quality ........................... 274
Fig. 8.26    A QFD matrix ........................................... 276
Fig. 8.27    House of quality: relationship building .................... 277
Fig. 8.28    Vendor selection method ............................... 280
Fig. 8.29    5-Why drilled down .................................... 282
Fig. 9.1    Timeline chart ......................................... 295
Fig. 9.2    C&E diagram for excess utilization of wire drawing area ...... 296
Fig. 10.1    Comparison of VA/NVA activities and eight deadly wastes.
         ( *Source* Harry & Mann, 2010) ........................... 308
Fig. 10.2    Typical value stream map ............................... 310

# List of Tables

| | | |
|---|---|---|
| Table 2.1 | Metrics: lean versus Six Sigma | 19 |
| Table 2.2 | Quality dimensions of products as per LGC | 26 |
| Table 2.3 | Time in minutes to complete a one-minute game | 27 |
| Table 2.4 | Process accuracy measures | 31 |
| Table 2.5 | Process precision measures | 33 |
| Table 2.6 | Calculation of dispersion measures | 33 |
| Table 2.7 | Calculation of CV | 35 |
| Table 2.8 | Measures of skewness and kurtosis | 37 |
| Table 2.9 | Thickness measurements of piston rings | 37 |
| Table 2.10 | Center thickness data | 39 |
| Table 2.11 | Process capability indices | 40 |
| Table 2.12 | Time to failure of electronic chips | 47 |
| Table 2.13 | Quality Walk checklist form | 48 |
| Table 4.1 | Summary of some frequently used probability distributions | 83 |
| Table 4.2 | Distribution of four tosses of an unbiased coin | 84 |
| Table 4.3 | Binomial samples | 96 |
| Table 4.4 | Exponential samples | 97 |
| Table 4.5 | Area under the normal curve and the corresponding PPM | 110 |
| Table 4.6 | Sigma level versus DPMO (includes a 1.5 Sigma shift) | 113 |
| Table 4.7 | Data on wheel dimension | 113 |
| Table 4.8 | Project identification process | 120 |
| Table 4.9 | A project charter for optimization of the cycle time of pump generation process | 121 |
| Table 4.10 | Essential questions to ask in defining the problem | 122 |
| Table 4.11 | Project balanced score card | 123 |
| Table 4.12 | Lean versus Six Sigma metrics | 125 |
| Table 6.1 | Critical quality parameters for a select list of services | 172 |
| Table 8.1 | Control limit for variable control charts | 222 |
| Table 8.2 | Measurement of piston diameter | 223 |

Table 8.3    Control limit for attributes control charts . . . . . . . . . . . . . . . . . .    226
Table 8.4    Data on imperfections in rolls production . . . . . . . . . . . . . . . . .    227
Table 8.5    Data of non-conforming units . . . . . . . . . . . . . . . . . . . . . . . . . .    229
Table 8.6    Accidental data due to natural causes . . . . . . . . . . . . . . . . . . . .    232
Table 8.7    Data on causes of lost time of production . . . . . . . . . . . . . . . . .    234
Table 8.8    $x$ versus $y$ . . . . . . . . . . . . . . . . . . . . . . . . . . . . . . . . . . . . . . . . .    235
Table 8.9    Data on players and age group . . . . . . . . . . . . . . . . . . . . . . . . . .    235
Table 8.10   Anxiety scores and marks . . . . . . . . . . . . . . . . . . . . . . . . . . . . . .    236
Table 8.11   ANOVA for linear regression . . . . . . . . . . . . . . . . . . . . . . . . . . .    242
Table 8.12   Measurements on moisture content and relative humidity . . . . .    243
Table 8.13   ANOVA of linear regression . . . . . . . . . . . . . . . . . . . . . . . . . . .    244
Table 8.14   Data on psychological measurements . . . . . . . . . . . . . . . . . . . .    245
Table 8.15   ANOVA table . . . . . . . . . . . . . . . . . . . . . . . . . . . . . . . . . . . . . .    245
Table 8.16   Coefficients' table and their summary statistics . . . . . . . . . . . . .    246
Table 8.17   Formulation of hypothesis . . . . . . . . . . . . . . . . . . . . . . . . . . . . .    251
Table 8.18   Process of testing of hypothesis . . . . . . . . . . . . . . . . . . . . . . . . .    254
Table 8.19   Critical values of z . . . . . . . . . . . . . . . . . . . . . . . . . . . . . . . . . . .    254
Table 8.20   Classification of designs . . . . . . . . . . . . . . . . . . . . . . . . . . . . . .    264
Table 8.21   Models and their ranking . . . . . . . . . . . . . . . . . . . . . . . . . . . . . .    265
Table 8.22   ANOVA table for overall significance of the model . . . . . . . . . .    265
Table 8.23   Yield of corn . . . . . . . . . . . . . . . . . . . . . . . . . . . . . . . . . . . . . . .    266
Table 8.24   ANOVA table for yields . . . . . . . . . . . . . . . . . . . . . . . . . . . . . .    267
Table 8.25   Time is taken to cover a particular distance . . . . . . . . . . . . . . . .    268
Table 8.26   ANOVA for the efficiency of cars . . . . . . . . . . . . . . . . . . . . . . .    269
Table 8.27   Measurement of current (in microamps) . . . . . . . . . . . . . . . . . . .    270
Table 8.28   ANOVA table of measurement of current . . . . . . . . . . . . . . . . . .    271
Table 8.29   A prioritization matrix . . . . . . . . . . . . . . . . . . . . . . . . . . . . . . . .    279
Table 9.1    Prioritization of causes . . . . . . . . . . . . . . . . . . . . . . . . . . . . . . . .    292
Table 9.2    Variable association and failure effects . . . . . . . . . . . . . . . . . . . .    293
Table 9.3    Control plans . . . . . . . . . . . . . . . . . . . . . . . . . . . . . . . . . . . . . . .    294
Table 9.4    Sigma level of the processes . . . . . . . . . . . . . . . . . . . . . . . . . . . .    294
Table 9.5    Processes and 5S scores . . . . . . . . . . . . . . . . . . . . . . . . . . . . . . .    297
Table 9.6    Job profile versus improvement components . . . . . . . . . . . . . . . .    299
Table 9.7    Quality checklist . . . . . . . . . . . . . . . . . . . . . . . . . . . . . . . . . . . . .    300
Table 9.8    RACI matrix . . . . . . . . . . . . . . . . . . . . . . . . . . . . . . . . . . . . . . . .    302
Table 9.9    Process versus LGC components . . . . . . . . . . . . . . . . . . . . . . . . .    303
Table 10.1   Sigma level versus quality costs . . . . . . . . . . . . . . . . . . . . . . . . .    314
Table 10.2   Comparison of COQ/COPQ against industry sectors . . . . . . . .    314
Table 10.3   Comparison of improvement methods concerning
             the cost of quality . . . . . . . . . . . . . . . . . . . . . . . . . . . . . . . . . . . .    316
Table A.1    Tables of binomial probability sums . . . . . . . . . . . . . . . . . . . . . .    324
Table A.2    Cumulative Poisson distribution tables . . . . . . . . . . . . . . . . . . . .    338
Table A.3    Cumulative standard normal distribution . . . . . . . . . . . . . . . . . .    340

Table A.4    Sigma value versus DPMO ............................    345
Table A.5    Values of t for a specified right tail area ..................    346
Table A.6    Chi-squared values for a specified right tail area ...........    349
Table A.7    Values of F for a specified right tail area $F_{0.01(v_1, v_2)}$ ..........    353
Table A.8    Control chart coefficients for variables ...................    363

# Chapter 1
# Introduction

*The best way to find yourself is to lose yourself in the service of others.*
—Mahatma Gandhi

*"Vasudhaiva Kudumbakam"*—"The world is one family" is India's core slogan to bring universal peace, tranquility, and welfare to people. Our nation strives for establishing core values, traditions, and cultural cohesion among the people, and it is the best way of safeguarding every living being's (human, animal, plants, and planet) right to existence. Everybody wants to live a quality life. For some, it is material satisfaction, and for some, it is a way of living a long and healthy life. According to the Gandhian principle, being humble and straightforward (a lean philosophy) can lead to a quality life. According to Rabindranath Tagore, *"quality of life"* is a relative term, and there is no absolute standard of living that is universally accepted. Quality of life varies in time, space, and culture. Objective as well as subjective criteria are used for the evaluation of the quality of life. Material wealth, health condition, a social and political position, and mortality may be considered a basis for assessing the quality of life (Adhikary, 2003).

The concept of "quality of life" was popularized by the world bodies like UNDP and WHO, in measuring improvement in the status of the living condition of people belonging to the developing countries as against those of the developed countries. However, it has been found that the prescribed human development indices are not adequate to meaningfully appraise the quality of life. The concept of the useful life of particular individuals or societies does not always conform to the quality of life defined by the elite. It involves, among others, the wants, the desires, the expectations, the social and cultural norms as well as the perceptions of self, society, and nature of the concerned people. Quality of life is broadly defined as an individual's happiness or satisfaction with life and environment, including needs and desires, aspirations, lifestyle preferences, and other tangible and intangible factors that determine his overall well-being. When an individual's quality of life is aggregated to the community level, the concept is linked to existing social and environmental conditions such as economic activity, climate, or the quality of cultural institutions (Cutter, 1985).

K. Muralidharan, *Sustainable Development and Quality of Life*,
https://doi.org/10.1007/978-981-16-1835-2_1

1

As a matter of concern, it is recommended to include the targets and actions for protecting access to daylight and sunshine be included in the UN 2030 Agenda for Sustainable Development and the WHO's "health for all" policy (The TOI dated 30.04.2019). Daylight will become a luxury in future, as high-rise buildings and skyscrapers are on the rise across the world. Therefore, the direction is sought from city authorities, urban planners, and policymakers to prioritize good "daylighting," means sufficient access to natural light to improve public health and quality of light. Access to natural light must be integrated into the early stages of urban planning. It is recommended that businesses and schools could encourage employees and children to take longer lunch breaks outside to have a healthy life.

According to Mukherjee (2003), the quality of life has two distinct dimensions, one personal and the other institutional. The first reflects the attitude of an individual toward life, determines his/her needs and desires, and delineates his/her adjustment to the environment that encompasses physical resources and facilities available to the individual. The second is provided to the individual in terms of goods and services made available to the individual by the system he/she represents. Quality of life also depends on many other factors, including environment and climatic changes. Quality life should be free of tensions, chaos, conflicts, errors, variations, and wastes. In the mad rush to make life beautiful and livable, we have compromised our value systems to a great extent, where there is no immediate return possible. Attempts to produce more goods and services of better and more diverse quality and to ensure access by many more people to such products and services through more even and effective distribution mechanisms will continue and, in the process, more energy will be consumed and more significant depletion of non-renewable resources will take place, our shared environment will get more polluted posing more significant problems to our health and eventually leading to a deterioration of our quality of life (Mukherjee, 2003). This cycle will continue without end, aggravating a pathetic situation. We have earned money, fame, and comforts all at the cost of harming our ecosystem, natural resources, climate, and environment. Poor law and order, uncivil attitude, corruption, terrorism, etc. became a part of our life, leaving it full of uncertainties. There is no level playing ground, as the authorities (administrators, politicians, business people, etc.) are interested only in selfish motives and gains. Everyone washes off their hands when it comes to taking responsibility for all the wrongdoings. The problem will be aggravated further in future resulting in all kinds of unrest, apathy, and dissatisfaction among living beings. The novel coronavirus pandemic is the latest to join the bandwagon of human-made biological problems, threatening the whole world alike. Millions of people have already been affected, and lakhs have already lost their lives, leaving the future completely uncertain and doomed.

Science can provide a better framework for analyzing the quality of life, identifying its determinants and correlates, and suggesting avenues for improving it. We believe that lean, green, and clean science concepts can facilitate increased awareness among our people for leading a sustainable quality life. However, it is expected that at a desirable level, everyone should:

- discharge their moral duties
- have a good civic sense
- practice the best part of civil rules and regulations
- promote positive thinking
- be adaptable to attitudinal changes for the better.

A green and clean environment facilitates the best science/technology/business practices to optimize resources and availabilities, which will help us address the issues concerning sustainable and climate change issues (Muralidharan, 2018). It is a matter of concern for us to see that our natural habitat is under constant threat because of various kinds of wastes and contaminations, causing a massive scale depletion of natural resources like safe drinking water, energy, and materials. Shrugging off responsibilities and accountabilities during such emergencies can create a whole array of problems for the government and civil society. Time has come to work together to achieve a win-win situation for all living beings.

Lean, as the name suggests, is to make the process thin (simple) and flexible. Lean is a methodology that relies on a collaborative team effort to improve performance by systematically removing waste and variations. The perceived wastes could be anything like errors (natural or unnatural); defects; excess inventory of raw materials, finished goods; long wait of information, content, and people; underutilized talent, ideas, and resources; inefficient layouts or poor ergonomics at workstations or in offices; extra-processing of files and documents; the excessive movement of people, workstations and transportations, etc. Thus, lean principles advocate value creation and mapping all processes from the customer's point of view. As value is specified, value streams and gaps are identified, wasted steps are removed, and flow and pull are introduced, and then begin the process again and continue it until a state of perfection is reached in which absolute value is created with no waste (Chap. 2, Sect. 2.2 offers a very detailed discussion on lean concepts).

From the discussions above, it is clear that the lean, green, and clean (LGC) philosophies together can achieve incredible results in terms of quality outputs and world-class products and services. A name that is very close to quality is reliability, where *reliability* is defined as the probability of an item to perform its required function under specific conditions for a specified time. If quality is an observable characteristic, then reliability is an experimental characteristic. Therefore, reliability incorporates the passage of time, whereas quality is a static descriptor of an item. For instance: consider two transistors of equal quality, where one is used in a television set and another in a cannon launch environment. Since the working environment is different, they have different reliabilities, although qualities are the same. Consider another example: There are two automobile tires produced for the same purpose: high quality. One was built in 1960, the other in 2000. Both the tires have different reliabilities due to improved design (e.g., tread or steel belts), components (e.g., rubber), or processes (e.g., manufacturing advances). Some quality improvements (e.g., improved tread design) improve the reliability of the tire, while others (e.g., enhanced white wall design) will not (Leemis, 2000).

According to Feigenbaum (1983), reliability is one of the more critical "qualities" of the product; it cannot be operationally or systematically separated from other product-quality considerations. We buy products based upon the quality and come back and buy again based upon reliability. There are many ways to improve reliability in the way we improve quality. *Redundancy* is a highly effective method of improving the reliability of a system, if components are organized in an independent way (for technical details and modeling concepts on reliability engineering, see Chap. 6, Sect. 6.11).

Rapid advances in technology, the development of highly sophisticated products, intense global competition, and increasing customer expectations have put new pressures on manufacturers to produce high-quality and reliable products. Customers expect purchased products to be safe and dependable, be it vehicles, machines, devices, or systems useful in our day-to-day lives. Everything is scheduled to perform its intended function with high probability under usual operating conditions, for some specified time. In statistical studies, reliability is the consistency of a set of measurements or measuring instruments, often used to describe a test. Reliability is inversely related to random error and does not imply validity. A reliable measure is measuring something consistently, but not necessarily what it is supposed to be regulating. For example, while there are many reliable tests of specific abilities, not all of them would be valid for predicting, say, job performance. In terms of accuracy and precision, reliability is analogous to precision, while validity is analogous to accuracy (Muralidharan & Syamsundar, 2012). For sustainable (green) products and services, the reliability assessment is, therefore, mandatory for tangible and intangible benefits (see Chap. 2 for details).

Industrialization and urbanization have already damaged our environment and climate a lot. The changing patterns of human life, insurgence, and overuse of scientific technologies have multiplied the lousy effect on climate change and global warming issues. Global warming has been generally agreed to be caused primarily by the emission of greenhouse gases (such as carbon dioxide ($CO_2$), methane ($CH_4$) and nitrous oxide ($N_2O$), chlorofluorocarbons) and other chemicals into the atmosphere. The greenhouse gas (GHG) effect is a natural effect, which helps prevent excessive heat loss from the earth's surface. Without that effect, the universe would have been colder and might have been less habitable for humans, animals, and plants. The aggravated emission of greenhouse gases is linked to human efforts toward industrialization. Machines of any type (reliable or sustainable), materials of any brand, and infrastructures of any size are bound to produce variations and wastes. According to the World Bank's "What a Waste 2.0" report (2018), waste disposal is causing 1.6b tones of $CO_2$ emissions globally. The report articulates that 2.01 billion tones of municipal waste generated worldwide every year, and 34% of all waste is made by just 16% of the world's population, mainly from high-income countries. In this, more than a tenth of the world's waste is generated in India, which is a matter of concern for our people. In 30 years from now, India tipped to double the amount of garbage it makes now (The TOI dated Mar 7th, 2020). Food waste is the major contributor to this menace. Let us admit that wastes (or scraps or residues) and variations are part and parcel of our life. To mitigate the menace of wastes and problems needs, commitment, and focus along with people's participation.

Over the years, air pollution has become a concern in most of the urban and rural centers of the country. Contributing to this is the increasing population, leading to an increase in the number of vehicles, and growth in the industrial and power sectors are exerting tremendous pressures on the atmosphere. According to TEDDY (2010), the factors attributable to problems linked with the transport sector are the types of engines, the age of vehicles, poor road conditions, and congested traffic, etc. The other contributing factor is air and noise pollutions. It is proved that high levels of air pollution lead to severe health issues. India is one of the developing countries with the most significant burden of diseases due to the use of household fuels. Many Indian cities (Delhi being the highest) are reeling under air and noise pollution. Although interventions from government and civil administrations are passively made to improve the air quality and reduce noise pollutions, the regulations are not fully adopted by the industrial organizations to control the menace.

The problem is further aggravated due to the climatic changes in the environment. Climate change can be described as the persistent change in the weather pattern engendered by anthropogenic activities mostly linked to industrialization. It manifests in a long-term shift in the statistics of the weather (including its averages). One of the significant drivers of climate change is global warming. Global warming refers to an average increase in the temperature of the atmosphere near the earth's surface. As per record, the last decade of the twentieth century and the beginning of the twenty-first century have been the warmest period in the entire world. Global warming contributes to the melting of ice leading to wastewater accumulation and the rise of sea levels. However, this water may not be useful for consumption and irrigation usages because of its arsenic and fluoride nature. The primary sources of water pollution are municipal sewage, effluent generated from industrial processes, and agricultural runoff contaminated with fertilizers and pesticides.

The effect of global warming and other climate change issues have already taken their toll and are seen frequently on the earth. The year 2019 has seen the deadliest meteorological disasters in the world, including India. Inundated rain and landslide are frequent in most of the states in India. Countries across Asia and Africa suffered the most significant human toll from typhoons, heat waves, and floods. This includes: Cyclone Idai happened in March, the heat wave in India and Japan happened in June and July, Typhoon Hagibis occurred in Japan in October, and Typhoon Lekima in China occurred in August 2019. According to The Guardian International weekly, a new study has revealed that intolerable bouts of extreme humidity and heat, which could threaten human survival, are on the rise across the world, suggesting that worst-case scenario warnings about the consequences of global heating are already occurring. The report says that humidity is more dangerous than dry heat alone because it impairs sweating—the body's life-saving natural cooling system (https://www.theguardian.com/environment/2020/may/08/climate-change-global-heating-extreme-heat-humidity, last accessed on 09.05.2020). Natural calamities, including wildfires, are widespread in European and African countries. Many unusual occurrences of natural disasters have already impacted massively on human lives, along with the extinction of many species of animals and birds. The extent of damage caused to natural habitat is unimaginable and is still increasing. The year 2020 has opened with the COVID-19 virus across the world. As of July 4, 2020, the total numbers of confirmed cases are more than 11 million, and numbers of deaths have crossed more than 5 lakhs worldwide and are still counting. The figure for India is 625,544 confirmed cases and 18,213 deaths, respectively, as of 04.07.2020. In the absence of medicine and vaccination, the virus spread will be a challenge to contain the disease. What is in store for the future is still unknown to scientists and policymakers?

The best way to manage climate change is to prevent it. To improve and prevent detrimental climate change effects, one should look for result-oriented and efficiency-enhancing methodologies like lean or Lean Six Sigma philosophies (see Sects. 2.2 and 4.12 for the detailed discussion). Lean is considered to be a practical methodology for improving the speed and execution of a process, along with the identification of wastes. According to lean philosophy, wastes are non-value-added activities of a process and should not be allowed to recur. Lean enables smooth flow in the process by identifying value-added activities in the process. Because of its tangible benefits, people are using lean methods in every field of organizational activities. Lean provides tools to eliminate waste from the manufacturing process resulting in improved efficiency, thereby reducing GHG emissions. Less energy needed to operate product manufacturing leads to fewer carbon emissions and less dependency on non-renewable energy. As per some study, about 960 MT per year of solid waste generated in India from industrial mining, municipal, and agricultural sectors and other sources and is increasing over the years due to changing lifestyles and growing consumerism resulting from rapid urbanization and economic growth (Pappu et al., 2007; TEDDY, 2010).

According to Toyota's "lean manufacturing" movement (Taiichi Ohno), the elimination of transportation waste directly leads to reduced in-process inventory levels and reduces $CO_2$ concentrations at a manageable level. Practicing lean manufacturing over long periods translates into the ever-improving quality of goods. As manufacturing guru, Edwards Deming was always quoted to say, "Quality always costs less." As counterintuitive as that sounds, it is a fact. The implication is that by eliminating transportation waste and leaning out production, you create far more efficient systems and produce far higher quality goods for less. In this way, one can vastly reduce $CO_2$ emissions and create more profitable businesses (see Honeycutt, 2013). Green Statistics (Muralidharan, 2015) is a relatively new concept to account for the systematic collection and analysis of data that could lead to assessing environmental impacts of the operations of a firm (see Chap. 7 for more details). The philosophy is also beneficial for green practices, and its implementation can help reduce global warming to some extent (see also Shoeb and Javaid (2016)). We are always reminded of the *Reduce, Reuse*, and *Recycle* practice with an emphasis on the first one. Let us add two more concepts like *Recover* and *Resist* to grab the end-effects of the first 3Rs. Lean manufacturing is started with manufacturing industries, one of the best solutions for speedy results and cost-effectiveness in the current scenario.

Implementations of these technologies reduce the cost of poor quality (COPQ) and product cost and improve product quality.

A lean environment that is green is also clean (Muralidharan, 2018). A clean quality life promotes the well-being of the person and society. Toward this, the government, industrial organizations, and civil society are supposed to work in close association with each other to implement and monitor the environmental regulations in its true spirit and sense. For any improvements to sustain, one needs to change the attitude and perception. This is realized through the *Change Management* philosophy, which is all about anticipating and planning for human reaction to improvements, in addition to the technical and business components of the change. It begins with understanding the *values* and *habits* of the population that must change. It develops the means to promote that change among the target population members, including communication; changes in reward and recognition; and adaptation of details of the change in ways that are consistent with the population values (see Chap. 2 and many other sessions of this book).

According to Gryna et al. (2007), value is quality divided by price. The reality is that people do not separate quality from price; they consider both parameters simultaneously. Improvements in quality that can be provided to customers without an increase in price result in better "*value.*" As customers compare sources of supply, organizations must evaluate the value they provide relative to the competition. Quality, productivity, costs, cycle time, and value are interrelated. Quality activities must try to detect quality problems early enough to permit action without requiring a compromise in cost, schedule, or quality. The emphasis must be prevention rather than just correction of quality problems (see Chap. 4 for a detailed discussion on these issues).

For any changes to happen in the system, we need positive thinking and the right attitude. Mere perception is not enough to make changes. Attitude and belief are the first steps to get you out of the rut and into the groove. Along with the attitude and perception, we also need an influential quality culture imbibed by the eternal quest for universal peace and happiness. *Quality culture* is the organizational capabilities and habits and beliefs that enable an organization to design and deliver products and services that can meet a customer's needs and succeed in the marketplace over the long term. According to Juran Institute, Inc., the new quality culture should focus on

- Moving from cost-based quality to value-based quality
- The relentless pursuit of process efficiency and effectiveness
- A better understanding of the universal managing for quality to integrate new generation quality improvement programs like Six Sigma, Lean Six Sigma, etc.
- Process optimization through multi-functional improvements
- Enterprise-wide assurance systems to maintain daily control
- Multi-tasking employees trained in quality assurance and control, not just improvement
- Annual and bi-annual renewal of programs aimed at performance improvement to avoid improvement programs surges and collapses.

In this modern, developed, and technologically sound society, it is the people's right to receive and the administration's duty to provide objective information about people's lives, economic, and social conditions. This is possible through good governance and transparent organizational setup. Because of a lack of transparency in the system, the information is either not appropriately transferred to the stakeholders or the people at the receiving end are not interested in because of delays and obstacles. Unfortunately, the lack of information is creating a lot of confusion among ordinary people, and because of this, there is less acceptability and no confidence in government policies and programs. Leading a quality life under such a non-transparent situation will be detrimental for both society and man.

Information is the core of strong governance. The government should play an active role in promoting valuable services and transparent piece of information to the public. The rule of law should be the same for all. At the same time, the citizen in a democracy should have an obligation to understand the powers and duties of the government. Such information can be broadly classified into three categories: strategic, management, and operational (Kumar and Srivastava (1998), Muralidharan (2000)). Strategic information is required by planners and policymakers to set objectives, mission statements, and to frame policies. Information needed in this context is in broad and relevant areas. Management information is required for monitoring and surveillance. The information requested is at a much more exceptional level of aggregation but with limited perspectives. Operational data is needed by the person who is at the scene of action. This kind of information is minimal in the aggregate but high in accuracy and should be available in short periods.

To some extent, modernization can facilitate this. This is what exactly is happening now. There is no shortage of operational information. Unlike strategic and management, operational or functional information is timely and more frequent. Having advanced scientifically and technologically to a great extent, we are still not able to cash in on the full potential of the quality use of information to address the societal requirements and environmental concerns adequately (see Chap. 6 for enhanced study on information-based quality improvement).

To bring any changes in the system or society, we also need to discipline the human mind, to set right the priorities, and to encourage modernization in the existing order. The role of leadership plays a crucial role in taking up the responsibility to a new height. The kind of leadership is related to the tasks and the levels of maturity. At an introductory level, the jobs have to be structured for each individual so that each person knows what is expected of him/her. As the maturity level of the associate increases, owning of responsibility and sense of belongingness also increases. Working with cross-functional teams and assignments makes one identify one's strengths and weaknesses. Thus, a climate for success can easily be created once the dictation leaves to the delegation.

Modernization is a study of the social, political, and cultural consequences of economic growth and the necessary conditions for industrialization and economic growth (Eisenstadt, 1966; Hountondji, 1996). According to Chulu (2016), modernization is a pathway to achieving sustainable economic growth and investments. As far as modernization is concerned, it can be either institutional or attitudinal (Sharma,

1998). Institutional modernization refers to the modernization of the institutional structures of society. Attitudinal modernization is concerned with the modernization of the minds of people. Once the modernization at the institutional level is done, attitudinal modernization can be achieved automatically, as people tend to get along with the system's changes. These two are the potential elements in society, where changes influence the system immediately. Quality of life also depends on the extent we adopt these two changes. The increasing promotion of digital technology in every form has changed the quality of everyone's life. Lean (or simple), shorter, and reliable is the norm of life now. All these changes are creating a massive impetus for being up to date and modern.

In the remaining chapters of this book, I will dwell upon the significance of the concepts of lean, green, and clean sciences and technologies for a leading quality life. The ideas, methods, tools, and techniques are described constructively. The importance of being responsible to the environment and society is articulated through various qualitative and quantitative methods. Wherever possible, the guidelines for quality and sustainability are prioritized and classified. Thus, this book will try to nourish the positive and negative aspects of our livelihood by portraying the nuances of green, lean, and fresh scientific concepts.

# References

Adhikary, A. K. (2003). *Tagore's view on quality life*, an edited volume of selected papers presented at the National seminar of Quality of life, Edited by Das and Bhattacharya, 2003, (pp. 29–31).

Chulu, J. (2016). Modernization: Is it a pathway to achieving sustainable economic growth and investments. https://doi.org/10.2139/ssrn.2717083.

Cutter, S. L. (1985). *Rating places: A geographer's view on quality of life.* Association of American Geographers.

Eisenstadt, S. N. (1966). *Modernization: Protest and change.* Prentice-Hall.

Feigenbaum, A. V. (1983). *Total quality control* (3rd ed.). McGraw-Hill.

Gryna, F. M., Chua, R. C. H., & Defeo, J. (2007). *Juran's quality planning & analysis for enterprise quality.* Tata McGraw-Hill.

Honeycutt, R. (2013). Lean manufacturing: Addressing climate change through reductions in waste, 3p contributor on Tuesday, Jan 8.

Hountondji, P. J. (1996). *African philosophy: Myth and reality.* Bloomington and Indianapolis: Indiana University Press. Hunt.

Kumar, A., & Srivastava, T. R. (1998). *Statistics and information. An analysis of the present and perspectives for the future* (pp. 116–120). Indian Statistical System.

Leemis. (2000). *Reliability: Probabilistic models and statistical methods* (1st ed.). Prentice-Hall. 1995.

Mukherjee, S. P. (2003). *Quality of life*, an edited volume of selected papers presented at the National Seminar on Quality of life, Edited by Das and Bhattacharya, 2003 (pp. 29–31).

Muralidharan, K. (2000). Statistics and information sciences: A higher education perspectives. *Perspectives in Education, 16*(1), 61–64.

Muralidharan, K. (2015). *Six Sigma for organizational excellence: A statistical approach.* India: Springer.

Muralidharan, K. (2018). Lean, green and clean: Moving towards sustainable life, Communiqué. *Indian Association for Productivity, Quality and Reliability, 40*(4), 3–4.

Muralidharan, K., & Syamsundar, A. (2012). *Statistical methods for quality, reliability and maintainability*, India: PHI Learning in PVT.

Pappu, A., Saxena, M., & Asolekar, S. R. (2007). Solid waste generation in India and their recycling potential in building materials. *Building and Environment, 42*(6), 2311–2320.

QuEST Management Solutions (www.questmanagementsolutions.in).

Sharma, S. L. (1998). Role of education in India's modernization. In *Professional competency in higher education* (pp. 80–86).

Shoeb, M., & Javaid, M. (2016). Lean manufacturing beneficial for green supply chain management to reduce global warming. *Ijates, 4*(6), 386–390.

TEDDY. (2010). *TERI energy data directory and yearbook*, New Delhi.

# Chapter 2
# Lean, Green, and Clean Quality Concepts

*People who lean on logic and philosophy and rational*
*exposition end by starving the best part of the mind.*
—William Butler Yeats

## 2.1 Introduction

An organizational process is a series of interconnected sub-processes. A process can be transactional as well as non-transactional. Transactional processes such as order processing, inventory control, order preparation/generation, credit checking/rating, mortgage application processing, admission process, examination process, project scheduling, medical testing, communication engineering, registration process, etc. are processes having many activities working simultaneously or in isolation. They are all measured in terms of *cycle time* (the total time required to convert raw materials into finished goods), transparency of transactions, stakeholder involvement, and the quality and credibility of the management, and so on. Naturally, every management is interested in improving the quality of products and services, reducing cycle time and speed of delivery, reducing waste and variations in the process activities, etc. This is best achieved through the practice of lean, green, and clean (LGC) concepts.

A process is a set of linked or related activities that take inputs ($x$-variables) and transform them into outputs ($y$-variables) to produce a product, information, or service for external or internal customers (see Fig. 2.1 for a general process diagram). Technically, a process is represented as $y = f(x)$. As seen in the picture, a process can be seen as a "value chain" or "value stream." By its contribution to the creation or delivery of a product or service, each step in a process should add value to the preceding steps. It is generally believed that the quality of inputs decides the quality of outputs. But managing functional activities efficiently and effectively in the process is also vital for quality products and services. A critical step in understanding processes is the development of *process maps* and *flowcharts*. The format of these documents will vary with the situation, but in general, flowcharts show each step

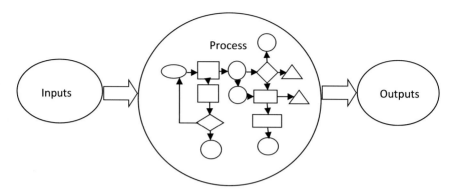

**Fig. 2.1**  Typical organizational process

in a process, decision points, inputs, and outputs. Process maps usually contain additional information about the steps, including data and outputs, costs, setup time, inventory, types of defects, probability of errors, and other useful information (Kubiak & Benbow, 2010). Thus, before initiating any improvement activities, it is desirable to prepare a process map, where processes map:

- Graphically outlines the sequence of a process
- Shows how steps in a process related to each other
- Identifies bottlenecks
- Pinpoints redundancies
- Locates defects (or wastes) and errors in the process.

The best way to document the activities of a process is done through a *process flow diagram* or *flowcharts*. A flow chart depicts the information relating to the actual or potential problems. The study of the flow chart, in conjunction with other information, enables one to develop a list of product characteristics and critical process parameters. A process flow diagram involves a series of tasks and decisions connected by arrows to show the workflow. The flow chart of a delivery process with incomplete address details is explained in the flow diagram shown in Fig. 2.2. When a process is documented and validated through a process flow diagram, one can analyze it for some of the following specific problem areas (Muralidharan, 2015):

- *Disconnects*. Points where handoffs from one group to another are poorly handled, or where a supplier and customer have not communicated clearly on one another's requirements.
- *Bottlenecks*. Points in the process where volume overwhelms capacity, slowing the entire flow of work. Bottlenecks are the "weak link" in getting products and services to customers on time and in adequate quantities.
- *Redundancies*. Activities that are repeated at two points in the process also can be parallel activities that duplicate the same result.
- *Rework loops*. Places where a high volume of work is passed "back" up the process to be fixed, corrected, or repaired.

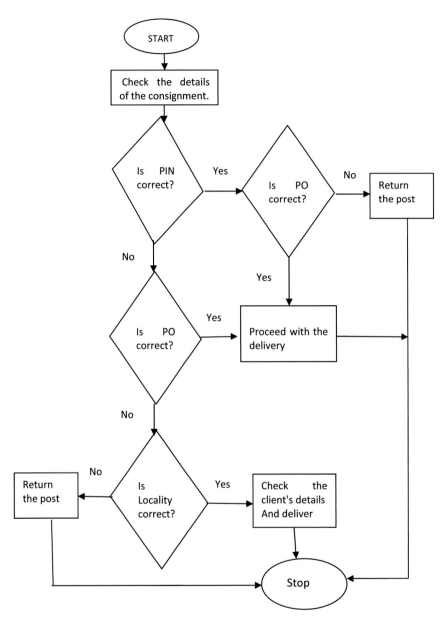

**Fig. 2.2**  Process flow diagram

- *Decisions/Inspections.* Points in the process where choices, evaluation, checks, or appraisal intervene-creating potential delays. These activities tend to multiply over the life of a business and process.

The Six Sigma professionals often use the *Supplier–Input–Process–Output–Customer* (SIPOC) diagram for mapping their process activities. This is a useful tool for identifying the critical process input (KPI) variables and critical process out (KPO) variables and their interrelationships for optimizing the process. A detailed discussion on SIPOC in the Six Sigma concept is provided in Chap. 6 of this book. The value stream or value chain mentioned above consists of all activities required to bring a product from conception to commercialization. It includes detailed design, order taking, scheduling, production, and delivery. Understanding value stream allows one to see value-added steps, non-value but needed steps, and non-value-added measures (see Chap. 10 for details). Value-added activities transform material or information to meet customer requirements. Non-value-added events take time or resources but do not add value to the customer's specification (Gryna et al., 2007).

The analytical process designed to enhance the benefits of a value delivery system while reducing or eliminating all non-value-added costs associated with value delivery is called *Value Stream Analysis* (VSA). Similarly, the technique for following the production path for product or service from beginning to end while drawing a visual representation of every process in the material and information flow is called the *value stream mapping* (VSP). Both VSA and VSP facilitate the identification of critical process and product characteristics (Allen, 2006; Kubiak & Benbow, 2009). Some more discussion on VSM is given in Chap. 10, Sect. 10.2.1.

Apart from logistical and technical issues, the quality of a process depends on many other factors, including environmental factors. Global warming, climate change issues, and geographical imbalance are also creating all kinds of hazards and unrest in the organizational structure. The risks can be anything like: physical, chemical, biological, psychological, and behavioral. It affects all people in some way or another. Keeping environmental impact low and improving efficiency in every organizational process and providing high-quality results every time to increase total customer experience must be the core of any business. This will increase attention to the organization's environmental performance and inspire people to become sensitive toward climatic conditions and harmful emissions affecting our society. Organizations work with the goal of zero waste, zero harmful emissions, and zero use of non-renewable resources outperform their competitors. A lean, green, and clean environment is everyone's right. Hence, addressing lean, green, and fresh thinking in our day-to-day activities can increase value addition in the organization, which is the need of time.

## 2.2 Lean, Green, and Clean Concepts

The modern business is quite often attributed to high-quality products and services with fast and on-time performance. In the process of establishing opportunities to develop market share, increased global coverage, and competition, people often forget about business ethics, climatic conditions, and environmental compliance. To sustain competitions and rapid technological advancement, the greening of the environment and business is equally important. Exploring ideas and innovations is continuous and sporadic, as data (information) is getting bigger and bigger every day. For improving customer relationships, improved profitability, and service/industrial/environment growth, the organizations should strive for best business practices in coordination with environmentally sustainable notions and practices (Muralidharan, 2018). This should start with the articulation of the following questions:

- What/where is the problem with sustainability? Is it in the process, product, or service?
- What exactly is causing the problem? Is it the general features of the problem? Or the suitability and adaptability of the problem?
- How/why are the environmental issues impacted on the output of the process?
- How does the service mechanism address green issues?
- How frequently are innovations required in the process?
- Is the organization responding the first time to the environmental issues? If not, how often?
- At what frequency do the customers demand green improvements in the product and service?
- How transparent (clean) is the administrative, research, and marketing system of your organization? For instance: the opportunity to learn and acquire new skills and knowledge; opportunity to create new and more productive working practices; the extent of supportive management and so on.

Some of the above questions may have optimistic answers, and some have organization-specific answers. Still, for a practical implication, one should find a more data-based decision-making opportunity to support the issues. This is where the importance of LGC concepts relies on. LGC could be a rewarding experience for some and a learning experience for some. However, the necessity of quality permeates every sphere of human endeavor. According to Juran and De Feo (2010), quality means *fitness for purpose*. To be fit for purpose, every goods and service must have the right features to satisfy customer needs and be delivered with few failures. It must be sufficient to meet customer requirements and efficient for superior business performance. We recommend LGC principles in every aspect of products and services for improved quality and customer satisfaction. LGC features can influence the customer's decision and choice if quality dimensions are identified and incorporated separately for every product and service. Figure 2.3 shows the quality dimensions identified for each component of LGC.

Below, we briefly discuss LGC issues of managerial importance, and then we embark on the importance, necessity, and benefits of LGC.

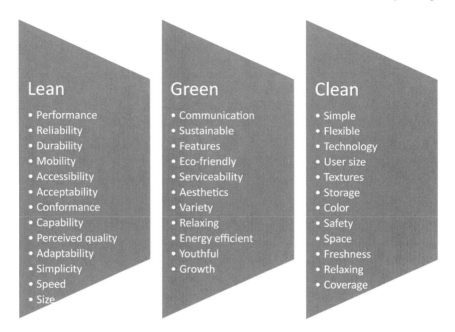

**Fig. 2.3**  Quality dimensions of components of LGC

## 2.2.1  Developing Lean Culture

Lean is a quality improvement philosophy, began with Japanese automobile manufacturing in the 1960s and was popularized by Womack et al. (1990) in *"The Machine that Changed the World."* The Machine that Changed the World is essentially the story of the Toyota way of manufacturing automobiles. The characteristics of lean, sometimes referred to as Toyotaism, are that materials flow "like water" from the supplier through the production process onto the customer with little if any stock of raw materials or components in warehouses, no buffer stocks of materials and part-finished goods between stages of the manufacturing process and no output stock of finished products. True lean, manifested in the Toyota Production System (Sobek et al., 1998), is concerned with doing the right things from a whole system point of view and enhancing value for all people proximate or distant, in the present or future. In the Toyota sense, lean and green are part of the same approach. According to the authors, Toyota integrates product development using (i) "social processes" among engineers to foster in-depth technical knowledge and efficient communication and (ii) design standards to speed development, allow flexibility, and build the company's base of knowledge.

According to the Chinese language, lean means the relentless pursuit of perfection. Toyota, as well as many other Japanese industries, learned from Dr. Deming to focus on doing the right things rather than doing the wrong things righter. Deming was a systems thinker. He spoke about the 94/6 rule, where 94% of problems can be traced to the process and only 6% to the person. But it is the person that is often measured and even blamed for problems. Deming said that any meaningful improvement comes from the action on the system, the responsibility of the management, and that wishing, pleading, begging, or even threatening employees to do better is entirely futile until we have fixed the system. It is the system of work that we need to act upon to improve, not the output, and certainly not blaming the people.

The focus of lean is to fix processes (any meaningful activities) and improve quality (Gryna et al., 2007; Rampersad & El-Homsi, 2008). The philosophy can be summed up as: "reduce the time it takes to deliver a defect-free product or service to the customer with a minimum amount of waste and complexity" (Muir, 2006; Muralidharan, 2015). It is not the question of whether the customer wants it right or quickly, but both. The lean philosophy develops with the identification of the problems and ends with the retention of the benefits of the program. Thus, lean is the methodology of eliminating waste through continuous improvement, thereby increasing value-added activities within an organization (Womack et al., 1990). It identifies value from the client's perspective and keeps the flow of events moving progressively to improve the process. Any activity or task that transforms the "deliverables" of a process in such a way that the client is both aware of it and willing to pay for it is value-added (Chap. 10 provides a detailed discussion on these aspects). But, most of the time, the processes involve more no-value-added activities than value-added activities. A typical lean process is shown in Fig. 2.4.

In comparison to Fig. 2.1, here, the process activities are simplified, and continuous flow is established. Working with this kind of process is secure, and hence, execution can be made faster. Quality output is sure to happen with the least amount of defectives leading to higher customer satisfaction in lean processes.

According to lean philosophy, any activities that add no value to the client are, by definition, "waste." For any strategic improvements to occur, it is necessary to have full control over the elimination of the waste, whatever type it may be. This is possible if we root out the symptoms of wastes when it started becoming an issue of quality. A symptom is an observable phenomenon arising from and accompanying a defect. A defect will have multiple symptoms. For example, the defect, "insufficient torque," may include the symptoms of vibration, overheating, erratic function, etc.

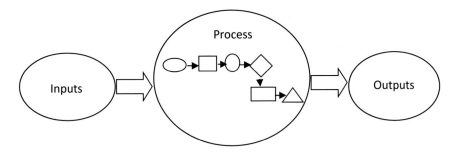

**Fig. 2.4** Typical lean process

The number of wastes will grow as per the size and occasion of symptoms. The symptoms of waste can be anything like:

- Machine breakdowns
- Transportation issues
- Poor scheduling
- Poor time management
- Poor housekeeping
- Poor vendor delivery
- Long setup times
- Absenteeism
- Quality defects
- Communication problems
- Uncertain behavior of climate and environment
- Delays of all nature.

Hence, the first stage for improving quality is to avoid the recurrence of waste symptoms generated through non-value-added activities. This control can further escalate to a more significant reduction of variation and abnormalities from the process, thereby achieving waste-free (high quality) products and services. The wastes can be anything like:

- *Defects*: All types of errors, scrap materials, rework, non-conformance items, etc.
- *Overproduction*: Making or doing more than is required or earlier than needed. For example, keeping extra copies of documents, purchasing items before its indented need, or accumulating additional consignments, etc.
- *Waiting*: For information, materials, people, maintenance, etc., for example, waiting to be served, waiting for repair, resolving downtime, etc.
- *Underutilized talent, ideas, resources,* etc.: Ideas that are not listened to, skills that are not utilized, mismatch of people, expertise and work, etc.
- *Transportation*: Moving people or goods around or between sites without any control. Examples include poor warehouse management, poor workplace management, etc.
- *Inventory*: Raw materials, work in progress, finished goods, papers, electronic files, etc. Examples include non-responding to customer queries, not realizing customer requests, not responding to customer emails, etc.
- *Motion*: Inefficient layouts or poor ergonomics at workstations or in offices. Some of the examples are improper handling of resources, files and raw materials, and office layouts, etc.
- *Extra-Processing*: Too many/too few steps, non-standardization, inspection rather than prevention, etc. Repeated manual entry of information on a computer, not following standard operating procedures (SOP), multiple formats of data, etc. are examples.

One can go to the administration or shop floor and observe operations to identify and feel each of these types of waste. What needs to be done to eliminate the waste may be neither obvious nor straightforward, but the directions to focus our energy become clear. Therefore, implementing lean tools will address waste and its root causes. lean also focuses on the efficient use of equipment and people and minimizes issues by standardizing work. A lean cost model says that decreased cost always leads to increased profit. The tools and techniques specific to lean are value stream mapping (VSM), 5S, workplace management, 3 M (Mude, Mura, Muri), one-piece flow, Kaizen (continuous improvement) approach, visual management, Kanban or workplace management (pull and push), etc. Among the tools discussed above, one of the flexible and straightforward productivity improvement management tools is the 5S concept. 5S (a Japanese idea) is a lean practice used to keep production workspace or service environment orderly and keep the workforce committed to maintaining order (see Chap. 4 for a detailed discussion on the tools as mentioned above).

Waste points us to problems within the system. So eliminating waste from processes is the core concept of lean. A more technically sound and strategically placed methodology for waste identification and elimination (both are essential for quality improvement) is the Lean Six Sigma concept. It is also a process improvement program that combines two ideas: *Lean*—a collection of techniques for reducing the time needed to provide products or services. *Six Sigma*—a collection of statistical methods for improving quality by reducing the variation. Lean Six Sigma (LSS) is a proven business management strategy that helps organizations operate more efficiently. According to Muralidharan (2015), a Six Sigma initiative is a customer-focused problem-solving approach with reactive and proactive improvements of a process leading to sustainable business practices. Sustainable business practices include innovation, development, competition, environmental compliance, customer satisfaction, and organization growth. It works with the DMAIC philosophy: *Define–Measure–Analyze–Improve–Control* the process (see Sect. 4.11 for detailed technical discussion). Hence, combining the two concepts can yield pleasurable experience for both organization and its stakeholders. The metrics associated with lean and Six Sigma are presented in Table 2.1.

The success of lean or LSS depends on many aspects of project management. An essential component of project management is the definition of the process $(y = f(x))$ itself and its characterizations. The quality of any operation depends on the scope of the process and its deliverables. This is facilitated through strong process

**Table 2.1** Metrics: lean versus Six Sigma

| Lean metrics | Six Sigma metrics |
| --- | --- |
| – Process flow | – DPMO |
| – Customer value | – Goal value |
| – Waste (all forms) | – Initial Sigma level |
| – VA-NVA activities | – Throughput yield |
| – First-time quality | – Rolled throughput yield |
| – Takt time | – Cost of poor quality |
| – Lead time | – Process capability indices |
| – Cycle time | – Process performance indices |
| – Change over time | – Final Sigma level |

management. *Process management* is the totality of defined and documented procedures that are monitored on an ongoing basis, which ensures that measures provide feedback on the flow and function of a process (Muralidharan & Syamsundar, 2012). The adequacy of process quality is assessed through the process capability study. *Process capability* is the ability of a process to produce items consistently within the specified functional measurement limits. For quality planning and prediction, it is essential to know the inherent variation in operations. This is achieved through the evaluation of process capability, which is a quantitative measureand becomes a significant part of statistical process control (SPC) or statistical quality control (SQC) of industrial applications (see Sect. 2.3.4 of this chapter for a comprehensive discussion on process capability indices).

Both SPC and SQC articulate the importance of process mapping, analysis, and monitoring of variations inherent in the process. They increase knowledge about the process and steer the process in the desired way. A detailed process map (or flow charts) and process analysis facilitate a deeper understanding of what happens in a process giving an idea about the value-added and non-value-added activities (see Chap. 10 for more discussions on these concepts). They often capture decision points, rework loops, complexity, etc. in the process model. To understand a process with too many circuits and decision points may be tedious, but to get along with the processing activity, one needs to simplify the process somehow. Lean does this simplification to some extent. Interestingly, everyone likes a process to be a model of the reality away from perceptions, so that reinventing the process parameters and their performance may not be warranted anytime and every time. Figure 2.5 describes a diagrammatic presentation of a perceived process.

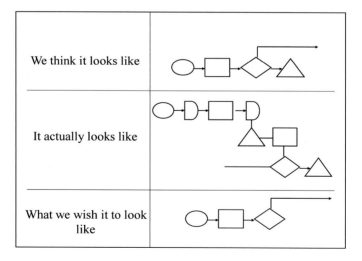

**Fig. 2.5**  Perceptions of a process. *Source* Muralidharan & Syamsundar (2012)

According to many business analysts and quality improvement experts, LSS is the most popular business performance methodology in the history of corporate development of quality of the process, products, and services, substantially contributing to increased customer satisfaction and confidence (Muralidharan, 2015). The LSS characteristics can be summed up as

- Increased speed and flexibility
- Involves all employees and their participation
- Simplifies the process and their participation
- Positive results in short time frame
- Focused on smaller-scale projects
- Less scientific: Often trial and error.

It is apparent that lean methods simplify the whole process and guaranteed to provide the customer with the best possible quality, cost, delivery, and pleasurable service. Lean directly contributes to the process, service, and environment to become green. A waste-free environment is synonymous with green initiatives and practices. Any activity which does minimum damage to the environment is green. Any product made of environmentally friendly materials, flexible processes, sustainable approaches, safe handling, pleasure, and freshness in the product are all green. As per the current scenario, green is a necessity and essential for a safe and clean environment. To sustain the environmental threats and challenges, the organizations should build a strong confidence base among its stakeholders and transparency in their day-to-day activities. Greening all the business and service activities is, therefore, a necessity rather than a mere requirement. Let us understand the concept of green or greenness a bit more.

## 2.2.2  Developing Green Culture

The term green business indicates that the item being described is more environmentally friendly or incorporates preservation of the environment and sustainable methods. Green business or sustainable business is an enterprise to be that, which has a minimal negative impact on the global or local environment, community, society, or economy. The green business environment facilitates less harmful, pollution-free products and services. It eliminates all those detrimental effects of our biodiversity, culture, and climatic conditions. A sustainable business that participates in environmentally friendly or green business activities ensures that all processes, products, and manufacturing activities adequately address current environmental concerns while maintaining profit and organizational growth. Green business has been seen as a possible mediator of economic-environmental relations, and if increased, would serve to diversify our economy, even if it has a negligible effect at lowering atmospheric $CO_2$ levels.

According to a TOI report dated 23.01.2019, India has about 25,940-ton plastic waste problem. Out of this, 40% remains uncollected causing choking of drainage and river systems, littering of the marine ecosystem, soil and water pollution, ingestion by stray animals, and open-air burning leading to adverse impact on human health and environment. Among the top ten cities in India, Delhi produces about 689 tons/day and remains at the top followed by Chennai (429 tons/day), Kolkata (425 tons/day), and Mumbai (408 tons/day). This is growing with the urbanization and modernization every day. There is a strong demand for India to ban all single-use plastic materials from human use and all other sources, including domestic applications.

Respecting the ecosystem and reducing the usage of resources, time, and consumption is everyone's responsibility. Toward this, green procurement programs are contributing to the popularity of green products and services. Green procurement involves identifying and giving priority to green products and services in corporate and government purchasing decisions. In particular, governments tend to be large volume purchasers and provide preferential treatment to green products and services, assisting their market penetration. Because products and services designed on sustainability principles are still new and exceptional, it can often be challenging to find those pumps, motors or fans, which are most energy efficient or manufactured without using environmentally damaging processes. Green procurement programs facilitate such information gathering and ensure that the purchases made do the job.

A term often used for green business is the corporate social responsibility or sustainability itself (D'Amato et al., 2009). The idea is to address topics of business ethics; organizational social performance, global corporate citizenship, and stakeholder management (see Sect. 5.5 for more details). Management education can be an essential source of new ideas about shifting toward an integrated rather than fractured knowledge economy. This also means that the role and meaning of a socially responsible leadership need to be updated. With the growing industrialization, the organization is supposed to change society at large. Changing culture for a better living involves plenty of opportunities as well as challenges. The challenges include coping with higher temperatures, extreme weather conditions, changing human lifestyles, and changing philosophies.

From a spiritual perspective, a green initiative represents those activities that are safe, flexible, and fresh to make, use, and consumption. It shows the abundance, renewal, and growth of people, processes, and the planet. All green activities closely associate nature as responsible for all environmental and climate change issues. Therefore, a close harmony with people, processes, and the world should be maintained to improve the foundations of the environmental management system. This is the need of the hour.

As discussed above, there are plenty of opportunities for improvement in every aspect of our life. The lean and green principles can be very effectively applied in various sectors like education, public utility systems, agricultural productivity, production and manufacturing systems, transportation, tourism, etc. where waste generation is very high. According to Woodall (2006), and Montgomery and Woodall (2008), the healthcare service is one of the potential areas, where LGC concepts can play a significant role in improving the services and delivery. Through proper understanding and identification of problems, analysis-based decision making, and close monitoring and control, one can bring effective communication and transparency in the system. We can quickly make people aware of the importance of natural resources, utilization, and sustainability. Thus, a LGC culture facilitates a positive approach to creating the best business practices in the organization.

### 2.2.3 Developing Clean Culture

Any lean business activity that is green is also clean. A clean business impacted the lean and green concept is guaranteed to generate less wastage and increase quality and productivity. Companies that sought to improve their condition by adopting green business practices often found that they had to change their entire culture. To convince all the stakeholders, the company should regularly promote confidence-building programs and moral development training in every aspect of the job profile. To build a better future, one should be ready to sacrifice comforts and facilities with a minimum occupancy of space, infrastructure, and time.

A clean initiative facilitates:

- Best business practices
- Cost-effective products and processes
- Sustainable thinking and practice
- Collaborative and mutual participation of people and processes
- Waste treatment and disposal a common practice
- Increased concern for future generations and livelihood.

As we know, a better and improved quality product always contribute less rework, fewer mistakes, fewer delays, and better use of time and materials. A significant part of the organization's time will most often be utilized to find and correct mistakes. Hence, a quality conscious atmosphere is in demand for producing a quality product in any organization. Note that, the entire market and supply of quality has undergone a sea change from the last two decades with the advent of scientific innovations putting pressure on both customers and producers. Although reasons can be sighted from many angles, here are some of the potential causes:

- The growth of foreign markets and the shrinking nature of transportation
- The growth of information technology and its adaptability has increased over a while
- The discriminating attitude of consumers as they are offered more choices and preferences
- Consumer sophistication leads to the demand for new and better products and services
- Increasing global competition and free-market concepts due to the erosion of economic and political boundaries
- Innovative concepts over the non-conventional aspects of quality and its practice
- The surge of information technology and digital marketing has changed the buying habit of people

- The overall perception of the quality of products and services has changed exponentially.

The necessity of quality increases according to the organization's growth, output, and customer's requirements. Therefore, the dimensionality of the quality characteristics includes the cost, innovative products, time to market, lead time, employee morale, competitive environment, safety, engineering techniques including process and product specifications, etc. For a detailed discussion on these and other quality dimensions, see Garvin (1987), Montgomery, and Muralidharan and Syamsundar (2012) and the references contained therein.

As mentioned earlier, with the growth of technology and science, there come many opportunities and challenges. Therefore, to resist the positive and adverse impact of this need-based requirement, we need to bring in technically and scientifically sound topics like LGC concepts to the forefront for reaping the benefits. Making use of the essence of the three theories discussed so far, below we outline some of the easy to implement techniques for sustainable business:

- Maintain all types of personal hygiene
- Sanitize the surroundings and community places
- Reduce, reuse, and recycle, with an emphasis on the first one. Also, use recovery and resist policy to withstand the end-effect of the first 3Rs
- Promote e-business and e-commerce activities for all transactional and service business activities
- Have own sources of waste disposal and renewable methods
- Identify the use, necessity, and source of energy in each activity
- Measure and monitor the quantity of energy used, the minimum the best
- Safeguard flora and fauna of your ecological system
- Use environmentally friendly products for safe use, consumption, and practice
- Promote multi-user transportation methods like vehicle-pooling, public transport, and bicycling
- Incorporate renewable energy and energy efficient equipment in homes and organizations
- Use a holistic method of manufacturing methods. If necessary, go back to the old system, where sustainable products, machines, and buildings are used
- Monitor and practice measurement-based evaluations and interpretations of every consumption
- Monitor carbon management and GHG emissions
- Be content and Learn to compromise with the availability of natural resources
- Do "much more" with "much less."

A clean business process activity entails mutual participation and collaborative research practices for shaping a better organizational culture from spiritual perspectives. It encourages cost-effective methods of manufacturing and production, so that maximum throughput yield is generated. A clean business activity promotes enhanced relationships with nature and long-term sustainability of people, processes, and planet.

Conversely, a clean business is mandated for lean and green principles to work for long-term sustainability. Since the regulations and policies are getting tougher and tougher for the better, one should always pursue quality deliverance in every aspect of the business. A clean, green, and lean transaction still costs extra efforts and investment but does not lead to any kind of loss, chaos, or quality compromise. Tom Peters, the co-author of *In Search of Excellence*, asserts that "in pursuit of improved quality of service, the organization has found that generating measures, posting those measures, then creating an educational process around the measures, has led to the most extraordinary improvement increments." The author also recommends that along with an environmental plan, there is a need for radical decentralization for getting everyone involved for overall improvement. According to time and space, one should always allow some margin for growth in a continuous way. Interestingly, the new generation quality concepts include many new dimensions centered on features and consumption of the products and materials. Table 2.2 compiles the quality dimensions identified for some new generation favorite items.

It makes sense that companies that produce better quality products at lower prices with smaller shipping times will be more successful in the long run. However, keeping any business activity to that extent is a continuous process and involves a significant amount of strain and patience. Since uncertainties and unanticipated developments mar all organizational operations, it becomes a priority for the management to make an appropriate course of action without harming the corporate activities in entirety. Therefore, one should be prepared to take bold decisions to set higher goals to realize the best output from one's business without compromising the environmental and climate change issues.

Uncertainty or chance, as mentioned above, is the foundation of any data (information) based reasoning. This can lead to certain illusions, as relationships are rarely deterministic. Uncertainties lead to errors and variations in decision making. They can influence people, processes, and products in anyways. Statistical significance is, therefore, necessary for any scientific inference. Like any other studies, the exploration of LGC studies also warrant the planned intervention of procedural and conceptual ideation. To realize its full potential, in various chapters henceforth, we study the concepts of lean, green, and clean in detail through different business assessment models, improvement models, and deployment models by being sensitive toward our environment and climate.

## 2.3  Process Variations and Characteristics

Variation (or random variation) is inherent in every process in varying amounts. Variation is a fact of nature and a fact of organizational/industrial/corporate life as well. They usually appear as a result of chance (or uncertain) events (change of season, voltage fluctuations, breakdown of machines, traffic blocks, inconsistency in time, poor quality of products, unnecessary movements, measurement inconsistency, non-acceptability of a product in the market, the outcome of the launch of a satellite,

**Table 2.2** Quality dimensions of products as per LGC

| Products | Quality dimensions included | | |
|---|---|---|---|
| | Lean | Green | Clean |
| Mobiles | – Performance<br>– Reliability<br>– Durability<br>– Mobility<br>– Accessibility<br>– Acceptability<br>– Conformance<br>– Capability<br>– Perceived quality | – Size<br>– Communication<br>– Sustainable<br>– Features<br>– Eco-friendly<br>– Serviceability<br>– Aesthetics<br>– Variety<br>– Relaxing | – Simple<br>– Flexible<br>– Technology<br>– Large user size<br>– Textures<br>– Speed<br>– Storage<br>– Color<br>– Safety |
| LED/CFL bulbs | – Performance<br>– Reliability<br>– Durability<br>– Acceptability<br>– Conformance<br>– Aesthetics<br>– Capability<br>– Perceived quality | – Size<br>– Less harmful<br>– Minimum space<br>– Sustainable<br>– Eco-friendly<br>– Serviceability<br>– Energy efficient<br>– Variety | – Simple<br>– Flexible<br>– Large user size<br>– Textures<br>– Space<br>– trust<br>– Color<br>– Safety |
| Noodles | – Nutritional<br>– Acceptability<br>– Conformance<br>– Flavor<br>– Satisfaction<br>– Soothing | – Energy efficient<br>in preparation<br>and usage<br>– Sustainable<br>– Aesthetics<br>– Perceived<br>quality<br>– Youthful | – Healthy<br>– Large user size<br>– Textures<br>– Instant<br>– Relaxing<br>– Freshness |
| Drowns | – Performance<br>– Reliability<br>– Acceptability<br>– Adaptability<br>– Perceived quality<br>– Size<br>– Speed | – Sustainable<br>– Eco-friendly<br>– Serviceability<br>– Coverage<br>– Variety<br>– Youthful<br>– Growth | – Simple<br>– Flexible<br>– Technology<br>– Coverage<br>– Capability<br>– Safety<br>– Nature |

political instability, and so on). The impact of these uncertainties can sometimes be very nominal and occasionally severe. Disregarding the existence of variation can lead to incorrect decisions on significant problems. Whatever may be the type and source of change; the quality will be suffered and will lead to wastes and variations. There is no procedure to eliminate the variation form the process, but its effect can be minimized with a holistic approach. So, the question is not how much variation is present, but how to account for these variations in a process, so that, its effect can be nullified. In some statistical investigations, there are instances where one may be interested in the extent of allowable variation to maintain the quality of the process at a certain level. This is to create a win–win situation for all.

### 2.3.1 Process Quality and Histogram

Presenting data in a convenient form is the first step in understanding the quality of the data. Data presented in tables, diagrams, charts, and plots can convey unusual patterns in the situation. By doing so, one will be able to plan, revise, and modify the future course of action. One such diagram is the histogram and frequency distribution formed from them. A *histogram* is a graphical display of frequency of measurements taken from a process and shows how those data are distributed and centered over a measurement scale. A histogram gives us a picture of what a process is producing over a specified period and how much variation exists. This picture can then be compared with our expectations of the process or older histograms to see if it operates within specifications and to the desired plan for that particular process. Among the seven quality tools, a histogram is the most widely used tool for assessing quality pictorially.

Consider an example to illustrate the use of a histogram. The data given in Table 2.3 is about the time in seconds to complete a one-minute physical game given to kids during an endurance test. The kids are all in the age group 4–5 years. The purpose is to see whether there is any abnormality in the timings, which will speak about the weak (scores more than 65), moderate (scores between 55 and 65), and high (scores below 55) endurance of the kids. Also, by visualizing the number of students in each category, the coach can recommend the level of physical exercise necessary for kids in the early stage of their growth.

The histogram with the frequency curve is presented in Fig. 2.6. By visualizing the data, it may be concluded that the data is symmetric and hence normal (the complete diagnosis investigation of normality check and the normal plot will be discussed in Sect. 4.2). In the sections to follow, we will be discussing these issues descriptively. As per the diagram, it is possible to conclude that about 22 kids have high endurance (scores below 55), 14 kids have a weak endurance (scores more than 65) level. The percentage of kids at a moderate level is about 40% (24/60). So the quality of the data is not inferior in this case.

The histogram is a quick and easy to construct diagram and is typically one of the first tools you will use in looking at the output of a process. Using the histogram, an assessment can be done on the process's average (mean, median, and mode) and amount of dispersion (standard deviation, mean deviation, quartile deviation, etc.). If all process averages are the same (i.e., mean = median = mode), then we conclude that the data is *normal* and have controlled variation. If the information is skewed to

**Table 2.3** Time in seconds to complete a one-minute game

| | | | | | | | | | |
|---|---|---|---|---|---|---|---|---|---|
| 60.95 | 36.67 | 45.26 | 57.22 | 50.11 | 55.08 | 60.02 | 49.43 | 62.22 | 66.71 |
| 58.92 | 57.65 | 56.31 | 69.35 | 50.24 | 56.37 | 74.04 | 70.08 | 39.69 | 72.19 |
| 47.31 | 54.75 | 61.52 | 63.88 | 51.31 | 70.65 | 44.78 | 43.03 | 68.04 | 56.00 |
| 40.55 | 70.20 | 53.04 | 48.05 | 53.79 | 30.89 | 71.00 | 65.56 | 53.40 | 59.67 |
| 79.05 | 62.51 | 56.48 | 57.58 | 53.35 | 59.48 | 77.44 | 74.57 | 48.60 | 63.20 |
| 62.41 | 59.20 | 69.94 | 58.85 | 47.41 | 55.11 | 47.28 | 43.90 | 59.29 | 65.86 |

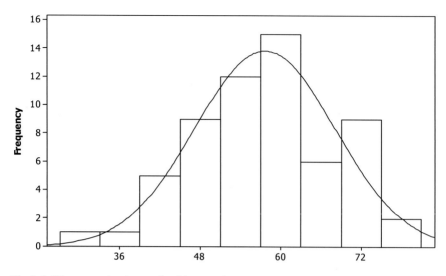

**Fig. 2.6**  Histogram of endurance level in seconds

the left, then the distribution will be left-skewed (mean < mode) distribution, and if it is skewed to the right, we say that the distribution is right-skewed (mean > mode) distribution.

There are many left-skewed and right-skewed distributions in the statistics literature, which are all used for modeling various practical situations (see Fig. 2.7 for skewed distributions). Data with symmetric or bell shape (or normal distribution) is always preferred over other cases for business decisions and forecasting (see

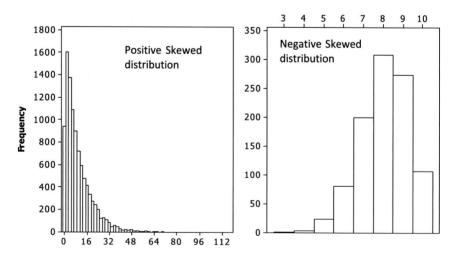

**Fig. 2.7**  Skewed distributions

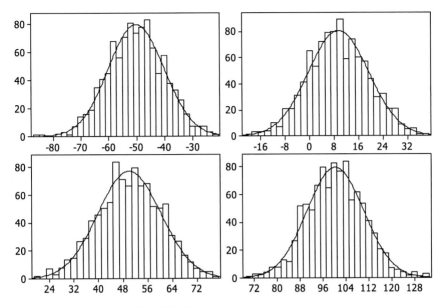

**Fig. 2.8** Symmetric or bell-shaped distributions

Fig. 2.8). The normal distribution is considered to be a distribution for quality assessment because of its numerous properties. At various chapters in this book, we will be using those properties. One can also check the normality of data using multiple statistical principles, which is beyond the scope of this book.

The objective of all statistical or empirical studies is to support the research of a small *sample* of items (a limited number of items) drawn from the *population* (also called *census* study) and then characterize it. This is facilitated through the sample measurements, from which we calculate some characteristics of the sample called a *statistic*. This is then used to describe the population under study and is called a *parameter*. It is usually assumed that the sample is random, i.e., each possible sample of *n* items has an equal chance of being selected from the population size say *N*. Characterizing population based on a sample of observations is the objective of all explorative data analysis, statistical inference, multivariate techniques, and prediction analysis. Therefore, the sample-based characteristics are the focus of discussion from here onwards.

## 2.3.2 Process Accuracy Measures

A customer typically relates quality to feature-based measures and its ability to satisfy its desires. They are usually determined by design and performance measures in conformance to some *ideal value* or *target value*. Obtaining an ideal value is pertinent to the project objectives, and for that, we need measurement to consolidate.

That ideal or perfect value depends on the analysis of process activity. Many concerns including: "When" is to be measured? "What" is to be measured? And in how much accuracy the measurement is desired?, and so on are appropriately mapped with the voice of process and customer. A group of measurements forms "data," which is the core of all decision-making process. So, the credibility of the results drawn from the application of data analysis depends on the genuineness of the information collected and included in the study. Hence *accuracy measures* or *average values* are essential for quality improvement and control.

The accuracy measures or location measures signify the central value (or target value) around a numerical data set. The objective of compressing the information into a single measure is to find a universal value that can be used to locate and summarize the entire set of data values. This one value can be used to make decisions concerning the whole set. The measures of central tendency also enable us to compare two or more similar sets of data to facilitate comparison. The commonly used measures of accuracy measures are mean (or average), median, mode, geometric mean, harmonic mean, and so on, and are all computed form a sample of $n$ observations $x_1, x_2, \ldots, x_n$ taken from a population under study. The summary of these statistics is shown in Table 2.4. Some examples based on descriptive measures are studied below.

***Example 2.1*** Compute the descriptive accuracy measures for the data: 12, 6, 8, 10, 8, 9, 5, 8, 10.

$$\text{Mean, } \bar{x} = \frac{x_1 + x_2 + \cdots + x_n}{n}$$

*Solution.*
$$= \frac{12 + 6 + \cdots + 10}{9}$$

$$= 8.44.$$

Median: Observations in increasing order: 5, 6, 8, 8, 8, 9, 10, 10, 12.
So the median is the middle value here, which is equal to 8.
Mode: The most frequently appearing item $= 8$

$$\text{GM} : \bar{x}_{\text{GM}} = \sqrt[n]{x_1 x_2 \ldots x_n}$$
$$= \sqrt[9]{12 * 6 * \ldots 10}$$
$$= 8.19$$

$$\text{HM} : \bar{x}_{\text{HM}} = \frac{n}{\frac{1}{x_1} + \frac{1}{x_2} \cdots + \frac{1}{x_n}}$$
$$= 7.93$$

It is evident from the computations that the accuracy measures are all centered on a standard value in the middle of the data. Note that the mean (arithmetic mean) is a simple measure of average and is based on all observations. The median is a positional measure, and it does not depend on extreme values. It divides the entire data set into two equal parts and hence called the second quartiles ($Q_2$). Although the mode is

**Table 2.4** Process accuracy measures

| Accuracy measure | Computational procedure | Importance of measure |
|---|---|---|
| Arithmetic mean | $\bar{x} = \frac{x_1 + x_2 + \cdots + x_n}{n}$ | – Universal measure<br>– Depends on all observations<br>– Uniquely defined |
| Median | $\bar{x}_{\mathrm{Med}} =$<br>$\begin{cases} \text{the value corresponds to } \left(\frac{n+1}{2}\right)\text{th item, } n \text{ is odd} \\ \text{the average corresponds to} \left(\frac{n}{2}\right)\text{th and } \left(\frac{n}{2}+1\right)\text{th item, } n \text{ is even} \end{cases}$ | – Based on ordering<br>– Not depends on extreme observations<br>– Affected by number of observations |
| Mode | $\bar{x}_{\mathrm{Mod}} = $ the most frequently repeated item | – Depends on repeated observations<br>– A mode may or may not exists |
| Geometric mean | $\bar{x}_{\mathrm{GM}} = \sqrt[n]{x_1 x_2, \ldots, x_n}$ | – Powerful measure<br>– Not easy to calculate<br>– Used for calculating various indices |
| Harmonic mean | $\bar{x}_{\mathrm{HM}} = \frac{n}{\frac{1}{x_1} + \frac{1}{x_2} \cdots + \frac{1}{x_n}}$ | – Good measure<br>– Restricted applications<br>– Used as average in relative terms |

easy to calculate, it depends on the frequency of observations and therefore, may not be unique. If the maximum frequency is at the extreme, then the model value can be the minimum value or the maximum value of the data. It is also possible that there can be more than one mode for given data. In such cases, a mode can be uniquely found using an alternative formula: mode = 3 Median – 2 Mean. Data with the same mean, median, and mode is considered quality data, and hence called symmetric (Normal) data. The distribution based on symmetric data is called *normal distribution*. This is a particularly a significant case, and therefore, a detailed discussion on normal distribution is provided in Chap. 4.

The other measures like geometric mean (GM) and harmonic mean (HM) are not computationally sound and hence not used frequently. However, they are used as accurate measures in many financial and management applications. Among all accuracy measures, a mean or simple average is considered a universal measure because it satisfies many interesting (statistical) properties, which can be proved theoretically. The formulas for various accuracy measures when data is organized in frequency distribution can also be attempted by organizing the data in some suitable form.

There are many other positional accuracy measures, as well. They are called *quartiles* ($Q_1, Q_2, \ Q_3, Q_4$), *deciles* ($D_1, \ D_2, \ldots, D_{10}$), and *percentiles* ($P_1, \ P_2, \ \ldots, P_{100}$). They are often used for assessing the quality of various segments (portions) of data in some particular order. For example, the data volume centered around lower 25% or middle 40–70% or upper 90%, etc. For competitive examinations and interviews, percentiles are frequently used for ranking the candidates for selection and placement. Note that the value corresponds to $Q_2$, $D_5$, and $P_{50}$ correspond to the median of the data values.

### 2.3.3   Process Precision Measures

We know that all goods and services are produced in various grades or levels of quality. These variations in grades or levels of quality are natural and are due to design or other quality measures. The extent of the degree to which numerical data tend to spread about an average value is called the dispersion or spread of the data. *Precision is the amount of accuracy you plan to accommodate in a process.* According to Montgomery (2003), quality is inversely proportional to variability. That is less the variability more the quality. Since statistical terms can only describe variability, statistical methods play a central role in quality improvement efforts. The precision measures are presented in Table 2.5.

***Example 2.2*** Compute the precision measures for the data: 22, 26, 32, 28, 18, 24, 25, 32, 30, 20.

*Solution.* The complete calculation is shown in Table 2.6.

    Mean $= \bar{x} = 25.7$

    Range $= 32 - 18 = 14$,

$$QD = \frac{Q_3 - Q_1}{2},$$

where the observations in increasing order are: 18, 20, 22, 24, 25, 26, 28, 30, 32, 32. Also

$$Q_1 = \text{the average correspond to} \left(\frac{n}{4}\right)\text{th and } \left(\frac{n}{4} + 1\right)\text{th item}$$

**Table 2.5** Process precision measures

| Precision measure | Computational procedure | Importance of measure |
|---|---|---|
| Range | $\text{Range} = H - L$ where $H$ = Highest observation, $L$ = Lowest observation | – Crude measure<br>– Applied in control charts |
| Quartile deviation | $QD = \frac{Q_3 - Q_1}{2}$ where $Q_1$ and $Q_3$ are the first and third quartiles | – Limited applications<br>– Used for quarterly variations |
| Standard deviation | $s = \sqrt{\sum_{i=1}^{n} \frac{(x_i - \bar{x})^2}{n}}$ where $\bar{x}$ is the mean | – Powerful measure of variation<br>– Based on all observations<br>– Plenty of applications |
| Mean deviation | $MD = \frac{1}{n}\sum_{i=1}^{n}|x_i - \bar{x}|$ where $\bar{x}$ is the mean | – Limited applications<br>– Gives variation in absolute terms |
| Coefficient of variation | $CV = \frac{\sigma}{\bar{x}}$ where $\bar{x}$ is the mean | – Relative measure of variation<br>– Used for checking consistency of data sets |

**Table 2.6** Calculation of dispersion measures

| Observations | $x$ | $x - \bar{x}$ | $(x - \bar{x})^2$ | $|x - \bar{x}|$ |
|---|---|---|---|---|
| 1 | 22 | −3.7 | 13.69 | 3.7 |
| 2 | 26 | 0.3 | 0.09 | 0.3 |
| 3 | 32 | 6.3 | 39.69 | 6.3 |
| 4 | 28 | 2.3 | 5.29 | 2.3 |
| 5 | 18 | −7.7 | 59.29 | 7.7 |
| 6 | 24 | −1.7 | 2.89 | 1.7 |
| 7 | 25 | −0.7 | 0.49 | 0.7 |
| 8 | 32 | 6.3 | 39.69 | 6.3 |
| 9 | 30 | 4.3 | 18.49 | 4.3 |
| 10 | 20 | −5.7 | 32.49 | 5.7 |
| Total | 257 | | 212.1 | 39 |

$= \text{the average correspond to} (2.5)\text{th and } (3.5)\text{th item}$

$= \text{the average correspond to } (20 + 0.5(20 - 18)) \text{ and } (22 + 0.5(24 - 22))$

$= \text{the average of } (21 \text{ and } 23)$

$= 22$

and

$Q_3 = \text{the average correspond to } \left(3\frac{n}{4}\right)\text{th and } \left(3\frac{n}{4} + 1\right)\text{th item}$

= the average correspond to (7.5)th and (8.5)th item

= the average correspond to $(28 + 0.5(30 - 28)$ and $(30 + 0.5(32 - 30)$

= the average of(29 and 31)

= 30

Hence,

$$QD = \frac{30 - 22}{2} = 4$$

$$s = \sqrt{\sum_{i=1}^{n} \frac{(x_i - \bar{x})^2}{n}}$$

$$= \sqrt{\frac{212.1}{10}}$$

$$= \sqrt{21.21} = 4.60$$

$$MD = \frac{1}{n} \sum_{i=1}^{n} |x_i - \bar{x}|$$

$$= \frac{39}{10} = 3.9$$

and

$$CV = \frac{\sigma}{\bar{x}} = 17.9$$

Note that the dispersion measures are all equal (= 4 roughly) except the value of range in the above example. As mentioned in Table 2.5, range is a crude measure of dispersion, and hence it is not widely accepted. When the data values are consistent for every measurement, only then it should be used. Also note that the standard deviation here is denoted as $s$, which is the sample standard deviation. The square of sample standard deviation is called *sample variance* and is denoted by $s^2$. Variance is always a positive quantity and will be zero if all data values are the same. According to statistical reasoning, the sample variance ($s^2$) is an unbiased estimator of the population variance $\sigma^2$, where the sample variance is defined as $s^2 = \frac{\sum_{i=1}^{n}(x_i - \bar{x})^2}{n-1}$. For large samples, dividing by $n$ or $n - 1$ does not make any difference to the sample and population estimates. So for practical convenience, the standard deviation is always denoted as $\sigma$, which is considered to be one of the best and widely used measures of precision for describing variation. Among many relative measures of dispersion, the coefficient of variation (CV) plays a vital role in deciding the consistency or stability of a data set. The CV is usually expressed in percentages and can be used to judge consistency across other data sets. This is illustrated in the following example.

***Example 2.3*** In a study of complaints of fatigue among men with brain injury, following depression scores from three samples of subjects of men are selected. There are three categories of subjects: (A) brain injured with a complaint of fatigue, (B) brain injured without complaint of fatigue, and (C) age-matched normal controls. The experiment is conducted of nine men of each category to understand the type of subjects consistent with the depression scores so that a suitable treatment plan can be thought of. The results were as follows:

| A | 46 | 61 | 51 | 36 | 51 | 45 | 54 | 51 | 69 |
|---|----|----|----|----|----|----|----|----|----|
| B | 39 | 44 | 58 | 29 | 40 | 48 | 65 | 41 | 46 |
| C | 36 | 34 | 41 | 29 | 31 | 26 | 33 | 56 | 54 |

*Solution.* To understand the category of subjects consistent with the depression scores, we will compare the coefficient of variation of all the three types of men. So we need to calculate the mean and standard deviation of all three groups. The calculation is shown in Table 2.7.

Here, the coefficient of variation in percentage is 17%, 22%, and 26%, respectively, for A, B, and C. Hence, it is concluded that category A (i.e., brain injured with a complaint of fatigue) subjects of men are more consistent than the other two.

Engineers and business professionals use mean (as a measure of target value) and Sigma (as a measure of variation) for their managerial decisions. These two measures characterize the strength and consistency of the data. See Fig. 2.9 for a better understanding of this fact.

**Table 2.7** Calculation of CV

| | A | | B | | C | |
|---|---|---|---|---|---|---|
| | $x_A$ | $(x_A - \overline{x_A})^2$ | $x_B$ | $(x_B - \overline{x_B})^2$ | $x_C$ | $(x_C - \overline{x_C})^2$ |
| | 46 | 30.8642 | 39 | 42.97531 | 36 | 3.160494 |
| | 61 | 89.19753 | 44 | 2.419753 | 34 | 14.2716 |
| | 51 | 0.308642 | 58 | 154.8642 | 41 | 10.38272 |
| | 36 | 241.9753 | 29 | 274.0864 | 29 | 77.04938 |
| | 51 | 0.308642 | 40 | 30.8642 | 31 | 45.93827 |
| | 45 | 42.97531 | 48 | 5.975309 | 26 | 138.716 |
| | 54 | 5.975309 | 65 | 378.0864 | 33 | 22.82716 |
| | 51 | 0.308642 | 41 | 20.75309 | 56 | 332.0494 |
| | 69 | 304.3086 | 46 | 0.197531 | 54 | 263.1605 |
| Total | 464 | 716.2222 | 410 | 910.2222 | 340 | 907.5556 |
| Mean | 51.55556 | | 45.55556 | | 37.77778 | |
| SD | 8.920776 | | 10.05663 | | 10.04189 | |
| CV | 0.173032 | | 0.220755 | | 0.265815 | |

| Measures | Alignments | Situations |
|---|---|---|
| Inaccurate, but precise | 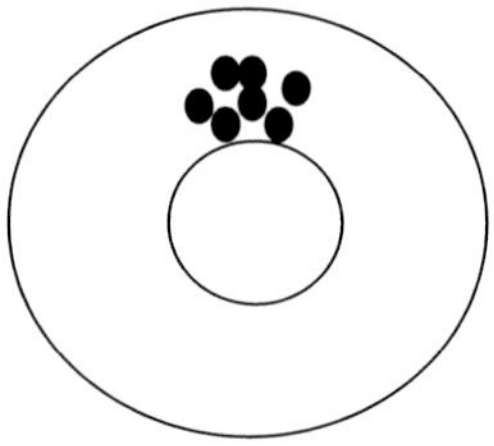 | - Inaccuracy is generally due to bias in measurements.<br>- Bias is the difference between the average measured value and a reference value.<br>- Precision is the amount of stability over some time. |
| Accurate, but imprecise | 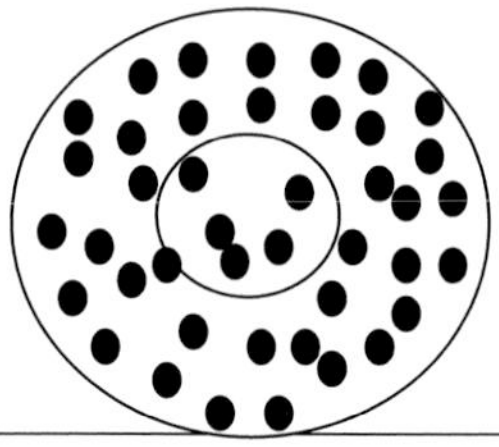 | - If the alignment of measurements is calibrated to a standard value, then accuracy can be improved.<br>- If repeatability and reproducibility don't match with the process, imprecise estimates will occur. |
| Accurate and precise | 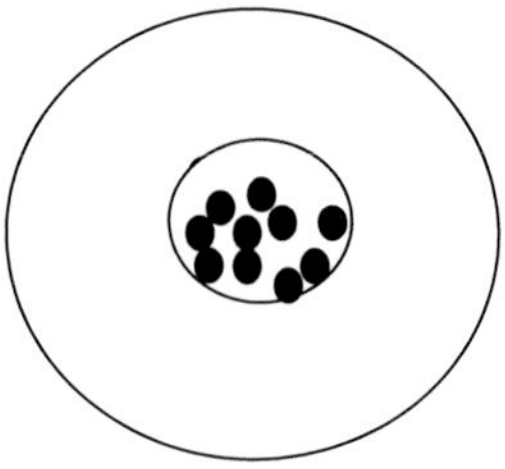 | - If repeatability, reproducibility, and stability in the process are only due to random causes, then accurate and precise estimates can be produced.<br>- Accurate and precise estimators produce sustainable quality products and processes. |
| Neither accurate nor precise | 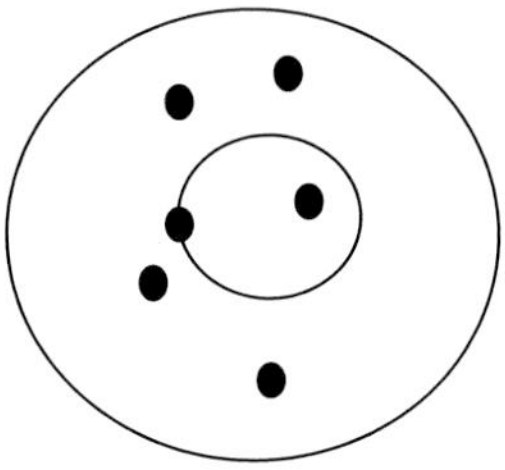 | - In this case, the estimators are most undesirable.<br>- Process variation cannot be controlled and will plague other processes also.<br>- Complete overhauling may be required to control the process. |

**Fig. 2.9**  Accuracy and precision explained

It is possible that two distributions may have the same mean and standard deviation but may differ widely in the shape of their distribution. So we need to understand the concepts like skewness (a measure of shape) and kurtosis (a measure of size or height) of the distribution along with accuracy and precision measure to characterize processed data completely. Skewness refers to the lack of symmetry in distribution, whereas kurtosis refers to the height (or peakedness) of the distribution. A distribution whose skewness is zero and kurtosis equal to 3 is called a moderate or bell-shaped distribution. A moderate distribution is also called a normal distribution and hence is

**Table 2.8**  Measures of skewness and kurtosis

| Measure | Method of calculation |
|---|---|
| Skewness | $\frac{\text{Mean}-\text{Mode}}{\sigma}$, based on averages |
| | $\frac{Q_1+Q_3-2Q_2}{Q_3-Q_1}$, based on quartiles |
| | $\frac{\left[\sum_{i=1}^{n}(x_i-\bar{x})^3\right]^2}{\left[\sum_{i=1}^{n}(x_i-\bar{x})^2\right]^3}$, for raw data |
| Kurtosis | $\frac{\left[\sum_{i=1}^{n}(x_i-\bar{x})^4\right]}{\left[\sum_{i=1}^{n}(x_i-\bar{x})^2\right]^2}$, for raw data |

**Table 2.9**  Thickness measurements of piston rings

| 189 | 181 | 183 | 191 | 180 | 182 | 187 | 188 | 189 | 189 |
|---|---|---|---|---|---|---|---|---|---|
| 177 | 191 | 192 | 180 | 184 | 190 | 179 | 198 | 197 | 179 |
| 171 | 188 | 165 | 191 | 176 | 172 | 199 | 185 | 183 | 183 |
| 187 | 189 | 186 | 191 | 185 | 172 | 182 | 184 | 185 | 201 |
| 189 | 187 | 186 | 183 | 178 | 173 | 172 | 193 | 184 | 183 |

an ideal distribution. An ideal distribution has controlled variation, and therefore, all business professionals focus their decision making on this perfect distribution. The measures of skewness and kurtosis are compiled in Table 2.8.

To understand the significance of the measure of shape and height, we consider an example here. Table 2.9 shows 50 piston rings used in automobiles coated with hard chrome plating. For a particular stage of production, the requirement of plating thickness is $195 \pm 20$ µm (µm). The objective here is to understand whether the plating process is normal or not. This will give a judgment about the quality problems the process owner is facing and the course of action required for improving the process.

Figure 2.9 shows the histogram corresponding to the thickness measurements of piston rings. The mean and median are respectively obtained as 184.58 and 185. But there are two modes: viz. 183 and 189 respectively, and hence, the distribution is bimodal. Although skewness ($= -0.25$) is very low, we cannot conclude that the distribution of this data is normal. But the data follows a normal because its $p$-value is high (see the right-side panel of the figure). The $p$-value is interpreted as the smallest level of significance at which a correct hypothesis is rejected. In this hypothesis, the true hypothesis is that the data follows normal is accepted with a probability of 0.639. However, there are still some quality issues with the data. To understand the presence of any spurious observations, we need to look at the box plot of the data shown in Fig. 2.10. It is evident from the plot that there is one observation, 165, which seems to be an inliers (lower outlier) observation. The presence of such observation can influence the results of the whole experiment. People often remove such observation from further analysis to make it normal and make it favorable to satisfy their objectives. If we remove the observation 165 from the data and draw the

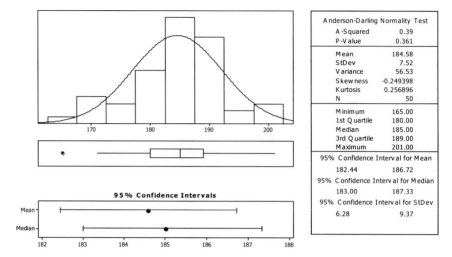

**Fig. 2.10** Summary of thickness measurement of piston rings

histogram again, there is all probability that the data will be normal. As a statistician, we do not recommend removing any outliers or inliers observations from the data.

### 2.3.4   Process Capability Measures

It is known that the amount of variation in a process tells us what that process is capable of achieving (tolerance). In contrast, specifications tell us what we want a process to be able to achieve. Improving the performance of any technical process is the first step toward quality control and quality assurance. This is possible if the accuracy and precision measures of the quality characteristics under study are stable and consistent level. A statistically in control process (or normal process) can be achieved through the implementation of statistical quality control (SQC) and statistical process control (SPC) techniques.

Statistical process control (SPC) is the method of data gathering and analysis to monitor processes, identify performance issues, and determine variability existing in the process. It is also a strategy for reducing variation in products, deliveries, processes, materials, attitudes, and equipment. The statistical quality control (SQC), on the other hand, is an improvement technique carried out on analyzed data for understanding the performance of the process. The issues related to capability analysis, yield analysis, Sigma level calculations are performed as a part of SQC. Both SPC and SQC techniques help us to identify the controllable variables which affect the quality of the process. Quite often, these techniques are discussed as a part of total quality management (TQM) to assess the process variations and improve process efficiency. See Chap. 4 for more details.

**Table 2.10** Center thickness data

| 6.90 | 6.73 | 6.61 | 6.81 | 6.80 | 6.80 | 6.83 | 6.79 | 6.67 | 6.77 |
|------|------|------|------|------|------|------|------|------|------|
| 6.84 | 6.71 | 6.80 | 6.71 | 6.77 | 6.83 | 6.96 | 6.70 | 6.83 | 6.77 |
| 6.77 | 6.83 | 6.98 | 6.71 | 6.83 | 6.64 | 6.85 | 6.64 | 6.92 | 6.76 |
| 6.83 | 6.76 | 6.83 | 6.78 | 6.77 | 6.71 | 6.68 | 6.73 | 6.89 | 6.77 |
| 6.78 | 6.92 | 6.67 | 6.76 | 6.98 | 6.75 | 6.62 | 6.82 | 6.84 | 6.76 |
| 6.79 | 6.81 | 6.76 | 6.77 | 6.80 | 6.91 | 6.72 | 6.83 | 6.94 | 6.78 |
| 6.74 | 6.88 | 6.82 | 6.86 | 6.85 | 6.80 | 6.84 | 6.80 | 6.83 | 6.73 |
| 6.91 | 6.83 | 6.78 | 6.74 | 7.00 | 6.86 | 6.80 | 6.80 | 6.81 | 6.87 |
| 6.91 | 6.79 | 6.81 | 6.87 | 6.73 | 6.84 | 6.77 | 6.82 | 6.90 | 6.77 |
| 6.77 | 6.75 | 6.67 | 6.87 | 6.80 | 6.78 | 6.72 | 6.90 | 6.86 | 6.67 |

Consider an example to understand the necessity of monitoring variation in a process: The Company OurVision manufactures various spherical concave mirrors for its clients across Asia. The specialty of the company is in providing overall imaging capability of the spherical mirror for contact lenses. Among various quality characteristics, the center thickness (CT) was monitored for capability study and quality assessment. One of the spherical concave mirrors produced by the company is of the type SM100A. The SM100A is a 100-mm radius-of-curvature mirror with a target center thickness of 6.8 mm and specification limits ±0.2 mm around the target. The data in Table 2.10 is about the center thickness data collected on 100 randomly chosen spherical mirrors.

The first step toward the process capability calculation is to check whether the data follows a normal distribution or not. If the process is statistically under control, only then the process capability ratios (PCR) or capability index make any sense. The PCRs are simply a means of indicating the variability of a process relative to the product specification tolerance. The graphical summary of the data is shown in Fig. 2.11. Although there are few outliers (inconsistent observations) present in the data, the data follows a normal (see p-value). This fact can be supported through the value of skewness as well, which is almost zero here. The p-value is interpreted as the smallest level of significance at which a correct hypothesis is rejected. In this hypothesis, the true hypothesis is that the data follows normal is accepted with probability 0.59.

The method of determining the extent to which an industrial process's long-term performance complies with engineering requirements or managerial goals is called *capability analysis* (Montgomery, 2003). A *process capability index* (Oakland, 2005) is a measure relating the actual performance of a process to its specified performance, where operations are considered to be a combination of the equipment, materials, people, plant, methods, and the environment. If LSL and USL are the lower specification limit and upper specification limit, the process capability is simply the ratio of USL-LSL (specified tolerance) to the $6\sigma$ (total tolerance). Here, the numerator is controlled by design engineering parameters, and the process engineering parameters

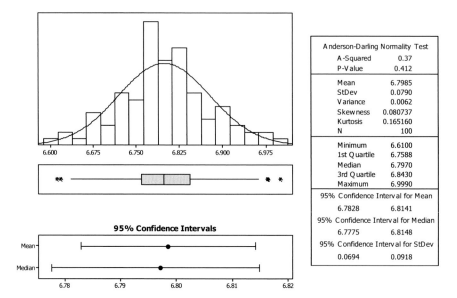

**Fig. 2.11** Graphical summary of the center thickness data

control the denominator. The percentage of the specification band that the process used up is obtained as $\hat{P} = \left(\frac{1}{C_p}\right)100\%$. The capability ratios under short-term and long-term variations are presented in Table 2.11.

**Table 2.11** Process capability indices

| Type | Capability | Index | Purposes |
|------|-----------|-------|----------|
| Short term or within | Process potential capability or process capability ratio (PCR) | $C_p = \frac{\text{USL}-\text{LSL}}{6\sigma}$ | – Predict whether rejections will Take place on the higher side or the lower side |
| | Process performance index or actual process capability | $C_{pk} = \min(C_{pl},\ C_{pu})$, where $C_{pl} = \frac{\mu-\text{LSL}}{3\sigma}$, and $C_{pu} = \frac{\text{USL}-\mu}{3\sigma}$ | – Take centering decisions – Decide whether to consider broadening of tolerances – Take decisions on whether to go in for new measurements |
| Long term or overall | Process potential capability or process capability ratio (PCR) | $P_p = \frac{\text{USL}-\text{LSL}}{6\sigma_{\text{LT}}}$ | – Predict whether rejections will Take place on the higher side or on the lower side |
| | Process performance index or actual (overall) Process capability | $P_{pk} = \min(P_{pl},\ P_{pu})$, where $P_{pl} = \frac{\mu-\text{LSL}}{3\sigma_{\text{LT}}}$, and $P_{pu} = \frac{\text{USL}-\mu}{3\sigma_{\text{LT}}}$ | – Take both centering and process variation into account – The long-term variability includes process drift – Decide on the level of inspection required |

Although $C_p$ and $C_{pk}$ are both used as short-term capability indices, $C_{pk}$ is widely used to communicate the process capability, as it accounts for both the process variation and the centering. Some of the $C_p$ index interpretations are shown in Figs. 2.12, 2.13, and 2.14.

If $C_{pk} < 1$, then there will be a situation in which the producer is not capable, and there will inevitably be nonconforming output from the process. In this case, the process uses up more than 100% of the tolerance band.

Figure 2.12 shows a competent process. In this case, the process uses up all tolerance band. In this case, the process performance will be according to normal distribution yielding about 2700 parts per million nonconforming units. Figure 2.13 is corresponding to a highly capable process. In this case, the process makes use of a less tolerance band. If $C_p = 2$, the process is excellent. There is a high level

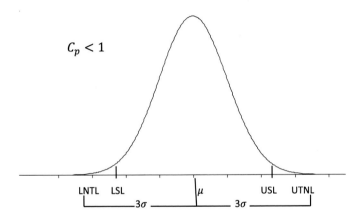

**Fig. 2.12** Incapable process

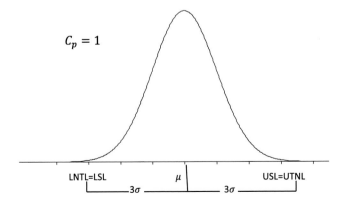

**Fig. 2.13** Perfect capable process

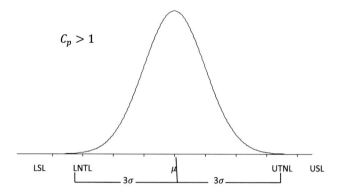

**Fig. 2.14** High capable process

of confidence in the producer provided that control charts are in regular use and maintain normal every time.

Note that the $C_{pk}$ index was introduced to take care of the $C_p$ index's limitations, which assumes that the mean of the distribution is equal to the normal size while estimating the proportion of defectives in the case of nominal-the-best (Chandra, 2016) situations. The percentage of defectives for the nominal-the-best type of characteristics depends upon the distances of the mean of the distribution from both the limits. Hence, for such a feature, an ideal index that indicates process capability and can be used to estimate the proportion of defectives should contain some information about both these distances. If one is constrained to select only one of these distances, then the value to choose would be the minimum of the two distances because the minimum represents the worst case in terms of the proportion of defectives.

***Example 2.4*** The chemical compound NaCl used in a medicine tablet used for treating a conventional flue has to be at least equal to 23 mg/cl. A random sample of 50 units selected confirmed an average of 32 mg/cl and a standard deviation of 2.2 mg/cl as the presence of NaCl. Discuss the capability of the process and decide whether the process is stable or not.

*Solution.* This is an example of characteristics with only one specification limit is given for capability analysis. So, according to the information provided: LSL = 23, $\mu = 32$, and $\sigma = 2.2$. The process capability is given by

$$
\begin{aligned}
C_{pl} &= \frac{\mu - \text{LSL}}{3\sigma} \\
&= \frac{32 - 23}{3 * 2.2} \\
&= 1.36
\end{aligned}
$$

Since the capability of the process is more than 1, it is possible to conclude that the measurements somewhat follow a normal distribution.

***Example 2.5*** The roughness of the ground surface of a component cannot exceed 0.02 units. A random sample of 25 parts was selected for inspection, yielded the mean roughness estimates as 0.01, and standard deviation as 0.003 units. Investigate the capability of the process, if the roughness measurements follow a normal distribution.

*Solution.* This is an example of characteristics with only one specification limit is given for capability analysis. So according to the information given: USL $= 0.02$, $\mu = 0.01$, and $\sigma = 0.003$. The process capability is given by

$$
\begin{aligned}
C_{pu} &= \frac{\text{USL} - \mu}{3\sigma} \\
&= \frac{0.02 - 0.01}{3 * 0.003} \\
&= 1.111
\end{aligned}
$$

Since the capability of the process is more than 1, it is possible to conclude that the measurements follow a normal distribution.

***Example 2.6*** A welding process associated with an automobile alignment has yielded the following information on coil diameter: Mean $= 3.3$ and standard deviation 0.35. Discuss the capability of the process, if the specification limits are set in the interval $(0.97, 4.86)$. Also, compute the performance index of the process.

*Solution.* As per the information are given, $\mu = 3.3$ and $\sigma = 0.35$. The process capability ratio is provided by

$$
\begin{aligned}
C_p &= \frac{\text{USL} - \text{LSL}}{6\sigma} \\
&= 1.85
\end{aligned}
$$

The performance index is given by $C_{pk} = \min(C_{pl}, C_{pu})$, where

$$
C_{pu} = \frac{\text{USL} - \mu}{3\sigma} = 0.743
$$

and

$$
C_{pl} = \frac{\mu - \text{LSL}}{3\sigma} = 1.109
$$

Hence, the performance index is given by $C_{pk} = \min(C_{pl}, C_{pu}) = 0.74$. Note that $C_p \geq C_{pk}$. This is true in most cases.

The process capability of center thickness measurement data is shown in Fig. 2.14. We can see that the overall mean and standard deviation of the data is 6.7985 and 0.079. Since the LSL = 6.6 and USLs = 7.0, we obtain the overall process capability ratios as

$$P_p = \frac{USL - LSL}{6\sigma}$$
$$= \frac{7.0 - 6.6}{6 * 0.079}$$
$$= 0.8438$$

$$P_{pk} = \min(P_{pl}, P_{pu})$$
$$= \min\left(\frac{\mu - LSL}{3\sigma_{LT}}, \frac{USL - \mu}{3\sigma_{LT}}\right)$$
$$= \min(0.8373, 0.85038)$$
$$= 0.8373$$

These values match the capability values obtained in Fig. 2.15. Since $C_{pk}$ and $P_{pk}$ are both less than unity, we conclude that the process is not capable.

***Example 2.7*** Consider two different normal processes A and B, as shown in Fig. 2.16, with the same variance 25. The lower and upper specifications limits are 5 and 25, respectively. Investigate the process capability indices and conclude.

The process capability indices for process $A$ are:

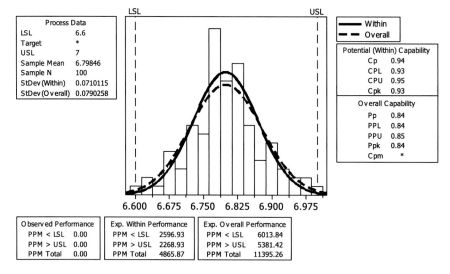

**Fig. 2.15** Process capability of center thickness

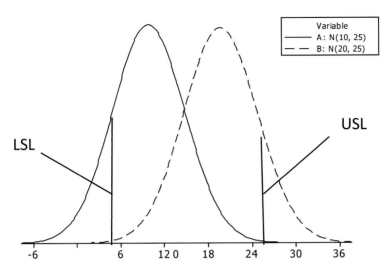

**Fig. 2.16** Two normal processes with the same variance

$$C_p = \frac{\text{USL} - \text{LSL}}{6\sigma}$$
$$= 0.667$$

The performance index is given by $C_{pk} = \min(C_{pl}, C_{pu})$, where

$$C_{pu} = \frac{\text{USL} - \mu}{3\sigma} = 1$$

and

$$C_{pl} = \frac{\mu - \text{LSL}}{3\sigma} = 0.333$$

Hence, the performance index is given by $C_{pk} = \min(C_{pl}, C_{pu}) = 0.333$.

Since the variance is the same for both $A$ and $B$, the capability indices for process $B$ will be the same as that of Process $A$. Note that $C_p \geq C_{pk}$. Similarly, if we consider two normal processes with equal mean and different variances, the capability values will be identical. Let us consider two normal processes with various means and deviations.

***Example 2.8*** There are two normal processes, $A$ and $B$, with different mean and variance, as shown in Fig. 2.17. The specification limits are 1.0 and 25.0, respectively. Investigate the capability indices.

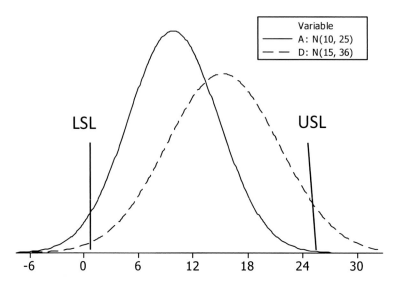

**Fig. 2.17**  Normal processes with a different mean and standard deviation

*Solution.* The PCR corresponds to Process *A* and *B* is respectively obtained as 0.8 and 0.67. The performance index corresponds to Process *A* is given by $C_{pk} = \min(C_{pl}, C_{pu}) = 0.6$. Similarly, the performance index corresponds to process *B* obtained as 0.57. In every case, $C_p$ is larger than $C_{pk}$ in both cases.

Capability rations are calculated for non-normal distributions as well. According to Luceno, a capability index for non-normal distribution is defined as

$$C_{pc} = \frac{\text{USL} - \text{LSL}}{6\sqrt{\frac{\pi}{2} E|X - T|}} \tag{2.1}$$

where the process target value $T = \frac{1}{2}(\text{LSL} + \text{USL})$. Similarly, the process capability ratios for a general family of distributions based on quantiles (Kotz & Lovelace, 1998; Rodriguz, 1992) are defined as

$$C_p(q) = \frac{\text{USL} - \text{LSL}}{x_{0.99865} - x_{0.00135}} \tag{2.2}$$

For normal distribution $x_{0.99865} = \mu + 3\sigma$ and $x_{0.00135} = \mu - 3\sigma$, and hence $C_p(q)$ reduces to $C_p$. The performance index corresponds to a non-normal process is

$$C_{pk}(q) = \min\left[\frac{X_{0.50} - \text{LSL}}{x_{0.99865} - x_{0.50}}, \frac{\text{USL} - X_{0.50}}{X_{0.50} - x_{0.00135}}\right] \tag{2.3}$$

Also, the process capability ratio corresponds to a standard process which takes care of the centering effect adequately is given by

$$C_{pm} = \frac{USL - LSL}{6\sqrt{\sigma^2 + (\mu - T)^2}} \tag{2.4}$$

One may refer to Montgomery (2003) and Chandra (2016) for a detailed discussion of process capability and technical perceptions.

**Example 2.9** A process capability study on a manufacturing process has yielded the following information: Mean = 3, standard deviation = 0.68, Target = 3.2. The specification limits are 2.08 and 4.56, respectively. Compute $C_{pm}$, if the process follows a normal distribution.

*Solution.* As per the notations, given $\mu = 3$, $\sigma = 0.68$, T = 3.2, LSL = 2.08, and USL = 4.56. Then

$$\begin{aligned} C_{pm} &= \frac{USL - LSL}{6\sqrt{\sigma^2 + (\mu - T)^2}} \\ &= \frac{4.56 - 2.08}{6\sqrt{0.68^2 + (3 - 3.2)^2}} \\ &= 0.88 \end{aligned}$$

To understand the process capability of a non-normal process, we consider the failure time of an electronic item. The data in Table 2.12 is the time to failure of 100 electronic chips with specification limits (1.8, 15.8):

According to MINITAB software, we have carried out a six-pack capability analysis, and the output is shown in Fig. 2.18. The histogram represents an exponential distribution. The process performance index, as per the study, is 0.23.

**Table 2.12**  Time to failure of electronic chips

| 3.36 | 3.82 | 3.39 | 52.29 | 9.13 | 0.60 | 1.65 | 20.22 | 0.05 | 2.19 |
|------|------|------|-------|------|------|------|-------|------|------|
| 28.80 | 4.59 | 9.93 | 7.49 | 7.38 | 2.72 | 5.57 | 6.85 | 12.17 | 10.27 |
| 0.82 | 11.88 | 11.16 | 3.65 | 12.08 | 13.57 | 5.15 | 4.52 | 29.99 | 0.88 |
| 4.20 | 8.65 | 0.05 | 6.66 | 0.59 | 1.55 | 2.76 | 15.03 | 32.67 | 36.94 |
| 2.75 | 14.95 | 4.01 | 2.76 | 10.69 | 1.79 | 10.51 | 41.75 | 2.48 | 3.97 |
| 4.62 | 9.38 | 5.29 | 32.11 | 10.95 | 16.09 | 3.11 | 5.15 | 5.37 | 8.18 |
| 3.10 | 13.65 | 2.53 | 8.07 | 24.60 | 25.22 | 18.16 | 0.42 | 2.22 | 14.29 |
| 7.34 | 3.30 | 21.63 | 15.19 | 5.09 | 8.71 | 1.50 | 3.72 | 8.23 | 0.83 |
| 7.36 | 2.13 | 4.68 | 5.14 | 3.76 | 1.66 | 7.84 | 1.79 | 5.69 | 0.60 |
| 1.47 | 0.62 | 15.15 | 14.51 | 3.26 | 9.90 | 1.12 | 3.92 | 5.19 | 2.15 |

**Table 2.13**  Quality Walk checklist form

| Waste Types | Details and observations specific to | | | Ideas to eliminate | Ideas to improve | Actions and Recommendations |
|---|---|---|---|---|---|---|
| | Lean | Green | Clean | | | |
| Overproduction | | | | | | |
| Inventory/Materials | | | | | | |
| Defects | | | | | | |
| Processing | | | | | | |
| Waiting | | | | | | |
| Motion | | | | | | |
| Transportation | | | | | | |
| People | | | | | | |
| Energy | | | | | | |

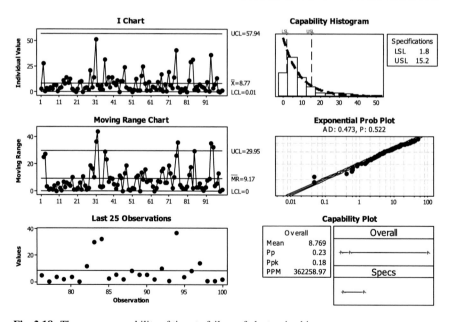

**Fig. 2.18**  The process capability of time to failure of electronic chips

## 2.4   Lean, Green, and Clean Quality Walk

A lot of quality perceptions have been discussed in this chapter. Process capability and process performance are the ingredients of the quality process, as they quantify the process statistically. The philosophy of LGC also recommends the quantification of all quality characteristics through measurements. Let us put all those into action now. Here, we discuss an outdoor activity through which we try to capture the quality essentials through waste identification and management. Whether it is a corporate level board meeting, ministerial-level appraisal meeting, university-level executive council meeting, research, and development meeting, infrastructure development meeting, campus development meeting, high power policy decision-making meeting, quality assessment meeting, etc., improving quality is the core of all those deliberations. Every time, the discussions are centered on quality, productivity, and safety improvement related matters. The type of meeting, level of expertise, frequency of meeting, number of people invited for the conference, is all pre-decided and can vary as per the urgency of the situation and requirements. Meetings are often held in well-furnished technology-enabled closed door locations routinely to discuss some of the common objectives like:

- To ascertain the progressive development of the programs and functions
- Enhancement of outputs and infrastructure
- Assessment of performance and efficiency of man, machine, and methods
- Stock-taking of stakeholder's welfare and well beings
- Evaluate institutions' future and growth.

Quality control, assurance, improvement, monitoring, and deployment are the focus of all meetings. It is time to have one of these meetings held while walking, where the rank and file will meet the service users and make all its effort to focus on quality improvement. In this process, an influential quality culture is built into the system. People involved with the exercise will have the courage to challenge the existing quality issues and suggest novel methods of containing the lapses.

*Quality Walk*, as the name suggests, is a holistic method of finding solutions to problems by walking around the workplace and communicating directly with the people with a purpose to detect, prevents, and mitigate the quality issues. The issues which are generally not addressed through board room meetings or exchange of communications can be quickly brought into action with Quality Walk. Imagine that board meets three times a month to discuss a company's problem. Instead of having all three meetings held in the meeting room, have one such meeting realized through this Quality Walk on an experimental basis. If it works, then increase the frequency of trail with an increased stretch goal with more participation from all quality concerned people. The target can be anything like productivity improvement, process capability assessment, cycle time reduction, inventory control, the efficiency of supply chain management, the welfare of the stakeholders, and so on.

In short, Quality Walk is an outdoor activity performed under shadows and in the open air. Here the members interact with each other in the group and the people around while walking through the premise or campus. The members will be given a *Quality Walk checklist*, where the checklist items are charted by quality professionals (the so-called think tank of the organization) based on the lean, green, and clean wastes. See Table 13 for the Quality Walk checklist form. The members are first trained to understand the various lean and green wastes and their impact on quality. While walking, they view, observe, perceive, discuss, and record everything in brief based on the subject of the meeting. The attention will be to record the quality issues happening in various processes associated with the topic of the meeting. For instance: identify existing safety hazards, the accuracy of standard operating procedures (SOPs) followed, the status of machinery and equipment conditions, enquire about the practiced standards, gain knowledge about the work status and build relationships with employees. If the members feel that there is a genuine quality problem that needs immediate attention, then solutions can also be sought instantly or recommended for instant approval and action.

There are many advantages of Quality Walk. It is an opportunity for senior members and staff to stand back from their day-to-day tasks to walk the floor of their workplace to identify wasteful activities. The walk will facilitate everyone's attention to safety and quality-related issues that hinder the organization's outcomes. By increasing communication and interaction, the problems can be resolved with immediate attention and recommended for implementation. The walk is generally carried out in an unstructured way without any protocol. The exercise will facilitate members' mobility, which is anyway good for health, increase confidence, and boost morale to address the problems better and find their solutions. Since the personnel hierarchy is absent (otherwise in board rooms this is maintained strictly) in this walk, members will be free for open discussions and ready for critical comments on the problems. This climate can enhance a platform for sharing ideas, acknowledging others' contributions, and exchanging pleasantries on a cordial note. There is an excellent chance that everybody in the group participates and contributes, which are again suitable for constructive criticism and conflict resolutions. Maybe a solution, which was never in the vicinity, becomes a reality if everybody accepts the problem and recommends a unified solution favorable to all the people.

If the meeting is held generally for two hours in-door, then let the Quality Walk also be conducted for a couple of hours by restricting the coverage. There is no need to cash in extra time of any members and their commitments. If the walk can be completed in a shorter time, that is also good. As per LGC philosophy, a Quality Walk can result in fast, sound, and waste-free results. The suggestions, recommendations, and improvements will be sustainable and flexible, having a high rate of acceptability among the stakeholders. Members can also get into brainstorming and problem-solving techniques, in case there are many problems to address and looking for solutions. The quality checklist may be appropriately modified to incorporate the necessary changes to accommodate this. The Quality Walk can be terminated conveniently either at the board room (need to walk back) or a place selected conveniently to discuss and then disperse. All the issues, such as queries, problems, logistics, and solutions, and so on raised by the members, will be compiled and analyzed further for implementation and deployment.

As such, there is no abstract reference for "Quality Walk" so far. Upon a thorough literature survey, we found that Butler et al. (2009) proposed a senior leader guide to improving cost and throughput through a 100-day workout method for quality improvement for healthcare professionals. According to Mozena et al. (1999), traditional improvement models devote the least amount of time to implementation, whereas effective models spend at least 25% of the time to implementation. General electric, which popularized the workout methodology, claims that the 100-day Workout allocates over 75% to implementation. Compared to other methods, effective deployment of a 100-day workout will require the most fabulous transformation among managers about the best way to manage the organization (Senge et al., 1999). Another vital reference to mention here is the toolkit developed to support health service professionals to focus on quality and safety provided by Health Service Executive (https://www.hse.ie/eng/: last accessed on 18.04.2020). The quality and safety walk-rounds offer a structured process to bring senior managers and front line staff together to have conversations with a purpose to prevent, detect, and mitigate patient/staff harm (Feitelberg, 2006; Morello et al., 2012; and Singer & Tucker, 2014). Safety walk-rounds have helped many organizations make a significant impact on their safety culture quality.

In the context of lean manufacturing, the *Gemba* or "workplace management" is referred to as the place where truth (or problem) is created. The location can be a construction site, sales floor, or where the service provider interacts directly with the customer. The objective is to bring managers to look for waste and opportunities for improvement by feeling the impact of the problem. The Gemba walk, much like management by walking around (MBWA) and was introduced by Taiichi Ohno, an executive at Toyota, to understand the value stream and its problems rather than review results or make superficial comments.

Let us understand the constitution of members of this Quality Walk. As discussed earlier, many often the list of members of any meeting is fixed (members are elected, selected, or invited) and they deliberate on a fixed agenda items. For a twenty-member Quality Walk, there should have a leader (CEO or MD or any person at the highest responsibility), eight to ten executive council members (Apex body members), HR managers, public relations officer, quality experts, a statistician, an office assistant, and few others as per situations demand. It is advised that all the members will contribute as per their professional capacity and point out at least one quality issue that needs some improvement. Since most of the members are preparing the quality checklist, the minutes can be arranged at the office level at a later stage.

Some intangible benefits of Quality Walk are:

- Improve the familiarity of processes, products, and services
- Increase commitment to quality and safety
- Increase a better understanding of the operation and supply chain
- Increase people engagement and participation
- Strengthen data-based information reporting for quality and safety

- Develop a culture of open communication and transparency
- Identify, acknowledge, and share good practice
- Support a proactive approach to minimizing risk
- Speed up the process of reporting and feedback on time
- Strengthen commitment and accountability for quality and safety
- Encourage good governance and best business practices.

According to Subir Chaudhuri (https://thinkers50.com/blog/walk-the-talk-of-qua lity/: last accessed on 20.04.2020), many problems with quality can be addressed by implementing a robust quality management program, but that is not enough. To change organizational processes and drive a substantial and sustainable increase in quality, you must also tap into people's power. Organizations that have successfully ignited people's power share one important characteristic. They have leaders who walk the talk of quality—at every moment, every encounter, and every level of the organization. When every executive, manager, and supervisor walks the talk of quality, they can set off an acculturation process that will sweep through the entire organization.

## 2.5   Exercises

2.1.   What are the features of a lean process? How does it differ from the usual process?

2.2   What are the symptoms of wastes? How to identify them?

2.3.   Distinguish various types of wastes encountered in the manufacturing process. Suggest methods for their elimination.

2.4.   Discuss various lean tools and their importance.

2.5.   What is green or sustainable products? What are the characteristics of a green product?

2.6.   Suggest the essential features of a green service.

2.7.   State all the quality dimensions of LGC and its significance.

2.8.   What is Six Sigma? How does it compare with Lean Six Sigma?

2.9.   What are the metrics of Lean Six Sigma?

2.10.   State all the characteristics of Lean Six Sigma.

2.11.   Discuss the methods of improving the clean attributes of a business.

2.12.   Define uncertainty and their sources.

2.13.   How does uncertainty influence a business activity?

2.14.   What is the statistical significance of the Histogram? How does it explain the variation in a process?

2.15.   The life (in hrs) of 60 electric bulbs is given below.

| 511 | 911 | 1177 | 1016 | 600 | 777 | 895 |
|-----|------|------|------|------|------|------|
| 749 | 1067 | 980 | 923 | 1314 | 1108 | 1137 |

(continued)

(continued)

| 906 | 1230 | 1090 | 1242 | 803 | 1131 | 918 |
|------|------|------|------|------|------|------|
| 1240 | 1057 | 980 | 997 | 763 | 759 | 1394 |
| 1111 | 1117 | 1143 | 808 | 948 | 857 | 962 |
| 922 | 817 | 1057 | 665 | 1171 | 936 | 1068 |
| 750 | 813 | 1139 | 1127 | 1163 | 934 | 515 |
| 907 | 1061 | 1198 | 1027 | 1081 | 991 | 1155 |
| 1199 | 806 | 950 | 1262 | | | |

Construct a histogram and interpret the data.

2.16. Discuss various process accuracy measures and their practical significance.

2.17. Fifty families were interviewed to know their numbers of siblings they have. The data is given as follows: Suggest a suitable measure of accuracy. What will be your conclusion about the population from which this sample is drawn?

| 3 | 2 | 2 | 4 | 1 | 1 | 2 | 3 | 4 | 1 |
|---|---|---|---|---|---|---|---|---|---|
| 2 | 0 | 1 | 2 | 1 | 0 | 4 | 2 | 1 | 0 |
| 0 | 1 | 3 | 0 | 3 | 2 | 2 | 3 | 0 | 3 |
| 2 | 5 | 0 | 1 | 2 | 1 | 4 | 3 | 0 | 5 |
| 2 | 0 | 1 | 1 | 2 | 6 | 1 | 2 | 1 | 5 |

2.18 The annual profit (in '000 Rs.) of 12 different diary branches of India are as follows:

1216, 1374, 1167, 1232, 1407, 1453, 1202, 1374, 1278, 1141, 1221, 1329. Calculate

(i)    Mean, median, and mode
(ii)   Mean deviation about mean, median, and mode
(iii)  Standard deviation
(iv)   Quartile deviation
(v)    Coefficient of variation.

2.19. The weekly earnings of a random sample of women workers in the year 2015 were as follows.

| Weekly earnings (in Rs '000) | Number of women |
|------------------------------|-----------------|
| Less than 15 | 1 |
| 15–20 | 4 |
| 26–25 | 28 |
| 25–30 | 42 |
| 30–35 | 33 |
| 35–40 | 18 |

(continued)

(continued)

| Weekly earnings (in Rs '000) | Number of women |
|---|---|
| 40–45 | 13 |
| 45–50 | 9 |
| 50 and above | 2 |

Calculate the appropriate measure of dispersion from the above data. Justify the selection of the measures. Discuss whether the wages are according to normal distribution.

2.20.  The following data shows the observations on nerve conductivity speeds. (Conductivity speeds in m/s)

| Healthy subjects (patients) | | | | | | |
|---|---|---|---|---|---|---|
| 52.50 | 53.81 | 53.68 | 54.47 | 54.65 | 52.45 | 54.43 |
| 54.06 | 52.85 | 54.12 | 54.17 | 55.09 | 53.91 | 52.95 |
| 54.41 | 54.14 | 55.12 | 53.35 | 54.40 | 53.49 | 52.52 |
| 54.39 | 55.14 | 54.64 | 53.05 | 54.31 | 55.90 | 52.23 |
| 54.90 | 55.64 | 54.48 | 52.89 | | | |
| **Nerve disorder subjects** | | | | | | |
| 50.68 | 47.49 | 51.47 | 48.47 | 52.50 | 48.55 | 45.96 |
| 50.40 | 45.07 | 48.21 | 50.06 | 50.63 | 44.99 | 47.22 |
| 48.71 | 49.64 | 47.09 | 48.73 | 45.08 | 45.73 | 44.86 |
| 50.18 | 52.65 | 48.50 | 47.93 | 47.25 | 53.98 | |

Obtain the summary statistics to compare, the conductivity speeds of the "patients with nerve disorder" with the "healthy patients." Draw histogram compare and comment.

2.21.  What is process capability? Suggest the importance of capability measures.
2.22.  Distinguish between long-term and short-term variation.
2.23.  Distinguish between capability and performance indices.
2.24.  Discuss various measures of process capability of a normal process.
2.25.  Discuss capability measures of a non-normal process.
2.26.  A manufacturer of a wide variety of seals, gaskets, and O-rings used in a number of industries, such as automotive, chemical processing, oil refining, medical, and aerospace.
    The O-rings are classified by two important characteristics: cross-sectional width and inside diameter. The particular O-rings is specified to have a cross-sectional width of 0.275 in. with an inside diameter of 4.725 in. The lower and upper specifications for the inside diameter are 4.692 and 4.758 in. The company has established strict quality control procedures for all its O-ring processes. The following table presents the inside diameter measurements of O-ring for 30 sub-groups of size 5.

| Sub-groups | $X_1$ | $X_2$ | $X_3$ | $X_4$ | $X_5$ |
| --- | --- | --- | --- | --- | --- |
| 1 | 4.7227 | 4.7274 | 4.7189 | 4.7201 | 4.7201 |
| 2 | 4.7204 | 4.7133 | 4.7111 | 4.7335 | 4.7279 |
| 3 | 4.7129 | 4.7219 | 4.7222 | 4.7384 | 4.7235 |
| 4 | 4.7275 | 4.7302 | 4.7430 | 4.7183 | 4.7221 |
| 5 | 4.7346 | 4.7314 | 4.7355 | 4.7237 | 4.7293 |
| 6 | 4.7228 | 4.7197 | 4.7383 | 4.7188 | 4.7114 |
| 7 | 4.7235 | 4.7289 | 4.7265 | 4.7290 | 4.7214 |
| 8 | 4.7253 | 4.7197 | 4.7313 | 4.7312 | 4.7197 |
| 9 | 4.7293 | 4.7449 | 4.7285 | 4.7369 | 4.7324 |
| 10 | 4.7268 | 4.7306 | 4.7381 | 4.7279 | 4.7259 |
| 11 | 4.7213 | 4.7340 | 4.7077 | 4.7353 | 4.7381 |
| 12 | 4.7224 | 4.7356 | 4.7304 | 4.7166 | 4.7213 |
| 13 | 4.7357 | 4.7241 | 4.7267 | 4.7288 | 4.7363 |
| 14 | 4.7229 | 4.7160 | 4.7274 | 4.7295 | 4.7221 |
| 15 | 4.7145 | 4.7231 | 4.7349 | 4.7288 | 4.7253 |
| 16 | 4.7311 | 4.7262 | 4.7278 | 4.7322 | 4.7304 |
| 17 | 4.7096 | 4.7229 | 4.7329 | 4.7201 | 4.7299 |
| 18 | 4.7231 | 4.7262 | 4.7296 | 4.7289 | 4.7167 |
| 19 | 4.7386 | 4.7366 | 4.7288 | 4.7321 | 4.7183 |
| 20 | 4.7251 | 4.7400 | 4.7191 | 4.7162 | 4.7288 |
| 21 | 4.7297 | 4.7125 | 4.7137 | 4.7262 | 4.7206 |
| 22 | 4.7223 | 4.7120 | 4.7294 | 4.7372 | 4.7200 |
| 23 | 4.7242 | 4.7282 | 4.7362 | 4.7401 | 4.7343 |
| 24 | 4.7233 | 4.7259 | 4.7243 | 4.7275 | 4.7360 |
| 25 | 4.7236 | 4.7243 | 4.7296 | 4.7225 | 4.7368 |
| 26 | 4.7319 | 4.7255 | 4.7251 | 4.7283 | 4.7251 |
| 27 | 4.7100 | 4.7234 | 4.7156 | 4.7233 | 4.7370 |
| 28 | 4.7275 | 4.7161 | 4.7281 | 4.7256 | 4.7263 |
| 29 | 4.7212 | 4.7198 | 4.7334 | 4.7225 | 4.7241 |
| 30 | 4.7263 | 4.7379 | 4.7124 | 4.7308 | 4.7200 |

(i) Check whether the characteristic under study is according to Normal.

(iii) Carry out a process capability study and interpret the results.

(jjj) Find the short-term Sigma level to ascertain the quality of the process.

# References

Allen D. G. (2006). Do organizational socialization tactics influence newcomer embeddedness and turnover? *Journal of Management, 32*(2), 237–256.

Butler, G., Caldwell, C., & Poston, N. (2009). *Lean Six Sigma for healthcare: A senior leader guide to improving cost and throughput.* Wisconsin: ASQ Quality Press.

Chandra, M. J. (2016). *Statistical quality control.* FL: CRC Press.

D'Amato, A. W., Bradford, J. B., Fraver, S., & B. J. (2009). Effects of thinning on drought vulnerability and climate response in north temperate forest ecosystems. *Ecological Applications, 23*(8), 1735–1742. 2013.

Feitelberg, S. (2006). Patient safety executive walkarounds. *The Permanente Journal, 10*(2), 29–36.

Garvin, D. A. (1987). Competing in the eight dimensions of Quality, *Harvard Business Review* (September–October).

Gryna, F. M., Chua, R. C. H., & Defeo, J. (2007). *Juran's quality planning & analysis for enterprise quality.* New Delhi: Tata McGraw-Hill.

Kotz, S., & Lovelace, C. R. (1998). *Process capability indices in theory and practice.* London: Arnold.

Kubiak, T. M., & Benbow, D. W. (2010). *The certified Six Sigma Black Belt Handbook* (2nd ed.). Dorling Kindersley (India) Pvt. Ltd.

Montgomery, D. C. (2003). *Introduction to statistical quality control.* Wiley-India.

Montgomery, D. C., & Woodall, W. H. (2008). An overview of Six Sigma. *International Statistical Review, 76*(3), 329–346.

Morello, R., Lowthian, J., Barker, A., McGinnes, R., Dunt, D., & Brand, C. (2012). Strategies for improving patient safety culture in hospitals: A systematic review. *British Medical Journal Quality and Safety*, on line published on 21st July 2012 as https://doi.org/10.1136/bmjqs-2011-000582.

Mozena, J., Emerick, C., & Black, S. (1999). *Stop managing costs* (pp. 37–38). Milwaukee: ASQ Quality Press.

Muir, A. (2006). *Lean Six Sigma way.* New York: McGraw Hill.

Muralidharan, K. (2015). *Six Sigma for organizational excellence: A statistical approach.* New Delhi: Springer Nature.

Muralidharan, K. (2018). Lean, green and clean: Moving towards sustainable life. *Communiqué, Indian Association for Productivity, Quality and Reliability, 40*(4), 3–4. October–December.

Muralidharan, K., & Syamsundar, A. (2012). *Statistical methods for quality, reliability and maintainability.* PHI Learning in PVT, India.

Oakland, J. S. (2005). *Statistical process control.* New Delhi: Elsevier.

Rampersad, H. K., & El-Homsi, A. (2008). *TPS-lean Six Sigma.* New Delhi: SARA Books, Private Limited.

Rodriguz, R. N. (1992). Recent developments in process capability analysis. *Journal of Quality Technology, 24.*

Senge, P., Kleiner, A., Roberts, C., Ross, R., Roth, G., & Smith, B. (1999). *The dance of change.* New York: Currency Doubleday.

Singer, S. J., & Tucker, A. L. (2014). The evolving literature on safety Walk Rounds: Emerging themes and practical messages. *British Medical, Journal Quality and Safety, 23*, 789–800.

Sobek, D. K. II, Liker, J. K., & Ward, A. C. (1998). Another look at How Toyota integrates product development. *Harvard Business Review* (July–August), 36–49.

Womack, J. P., Jones, D. T., & Roos, D. (1990). *The machine that changed the world: The story of lean production.* New York: Rawson Associates.

Woodall, W. H. (2006). Use of control charts in healthcare and public health surveillance (with discussion). *Journal of Quality Technology, 38*(2), 89–104.

# Chapter 3
# Lean, Green, Clean Sciences and Technologies

*Nature is based on harmony. So it says if we want to survive and become more like nature, then we have to understand that it's cooperation versus competition.*
—Bruce Lipton

## 3.1  Introduction

There are many reasons companies undertake radical environmental improvements: It is the right thing to do, as regulations and norms are getting tougher every day. Waste eliminated is money saved. Above all, greening any business activity is a matchless commercial opportunity. In market terms, the companies that are cleanest first will enjoy "a massive, once-in-a-century, competitive advantage." Those organizations that hesitate, preferring to follow the regulations rather than set the standards, may find themselves squeezed out by the leaders in greening (Hudson, 1991).

Wastes, in general, can be environmental, chemical, biological, and others. They are further classified into hazardous and non-hazardous. Hazardous wastes can be liquids, solids, gases, or sludge. They include discarded commercial products, such as cleaning fluids or pesticides, or the by-products of any manufacturing processes. Among the solid wastes, the e-waste (electronic waste) is posing many challenges for future generations. In all, Lean, Green, and Clean strategies reduce costs, risks, and safety hazards drastically. By being competitive, the management can improve environmental performance and bottom-line results. This author believes that, in the future, for all kinds of eco-labeling and certification programs of an organization, one may mandatorily go for the Lean, Green, and Clean compliances.

In Chap. 2, we have already seen how LGC procedures facilitate the identification of various types of wastes and their possible contribution in derailing the quality aspects from a process or product. There are many instances; people are using innovative methods and ideas for eliminating waste from our surroundings. Let us now discuss some scientific and technological advancement that happened over the last two decades and investigate how innovative thinking for quality has progressed and

K. Muralidharan, *Sustainable Development and Quality of Life*,
https://doi.org/10.1007/978-981-16-1835-2_3

to what extent it has affected everyone's life. Whatever may be the category of people; all new ventures in modern-day quality initiatives bring an opportunity for a better future-the future having less pain, sufferings, and anguish.

## 3.2  Green Energy

Being energy efficient is everybody's priority. Energy is an essential driver of economic growth and development. With lots of programs initiated worldwide, governments want to improve the availability and quality of energy so that people can live a healthy and peaceful life. Climate change concerns, coupled with high oil prices, are the main issues associated with energy consumption. It affects our ecology, economy, and society. Green energy can be extracted, generated, and consumed without any significant negative impact on the environment. Green energy can be defined in terms of creating power from a range of low-carbon renewable sources close to or at the point of use. The primary aim of adopting green energy is to achieve a shift from carbon-intensive energy sources principally based on coal or oil, either directly or indirectly, via the production of grid-based electricity.

The other related terms used for Green energy are sustainable energy and renewable energy. *Sustainable energy* is consumed at insignificant rates compared to its supply and with manageable collateral effects, especially environmental effects. Sustainable energy is an energy system that serves the needs of the present without compromising future generations' ability to meet their energy needs. While *renewable energy* is defined as energy that comes from natural resources. According to Kahle and Atay (2014), the energy naturally replenished on a human timescale should be available for the best human use. However, sustainable energy must not compromise the system in which it is adapted to the point of being unable to provide for future needs. However, not all renewable energy is sustainable. The energy-efficient methods, which are sustainable as well, are solar energy, wind energy, hydroelectricity, geothermal energy, bio-energy, etc. One can also generate renewable energy from recovered process waste energy, such as heat from refrigeration plants or air compressors. The low carbon alternative, such as natural gas and bio-diesel and nuclear, hydrogen, and fuel cells, are the other relevant sources of energy productions.

Due to increasing domestic demand for energy on the one hand, and the inadequacy of local supply on the other, India faces a formidable challenge in meeting its energy needs and in providing adequate energy of the desired quantity sustainably and at competitive prices (Planning Commission, 2006; Teddy, 2010). Energy security is, therefore, emerging as an essential concern in India. With increasing energy consumption, there is also growing concerned about energy-related emissions and their impact on the environment and climate change. India ranks among the top four emitters of greenhouse gas (GHG) emissions worldwide in aggregate terms after the USA, China, and Russia. According to some estimates, China will increase the world's $CO_2$ emissions by 39% by 2030, overtaking the USA's emissions in 2010!

Energy recovery can be made through solid and liquid wastes as well. For example, the waste materials collected through the pulp, paper, food sago/starch industries, distilleries, dairies, tanneries, slaughter-houses, poultry farms, and so on can be processed to generate renewable energies. The development of biogas up-gradation systems for converting biogas into natural gas fuel has resulted in the use of compressed natural gas (CNG). The promotion of solar energy is the other best way of harnessing green power. It is well known that a significant component of household energy in India is required for cooking. Fuelwood, kerosene, and lique-fied petroleum gas (LPG) are the significant sources that supply energy for food. However, there is a gradual shift in the use of fuelwood to LPG for cooking. LPG is a dense gas derived from petroleum refining and natural gas extraction and is stored in liquid form. Environmentally, the benefits CNG and LPG are sizeable reductions in nitrous oxides and particulate matter at the tailpipe. However, both LPG and CNG still originate from fossil fuels, so the well-to-wheel emissions of greenhouse gases are still high. Eyre et al. (2002) state that LPG does not offer lifecycle carbon benefits than diesel and conclude that LPG will have a somewhat marginal impact on either total road fuel energy demand or $CO_2$ emissions and does not provide a pathway to non-fossil fuel transport. Solar cooking offers a viable option for reducing energy use in the household. At the micro-level, it leads to financial savings for the consumer, and at the macro level, they help in the conservation of LPG and fuelwood (Teddy, 2010) (Fig. 3.1).

To sustain the business activities for future generations and to cope with the environmental concerns as mentioned above, organizations should promote quality concepts in every aspect of green energy activities. Some focused areas need special consideration are

- Ensure to get green quality certifications from an authorized agency

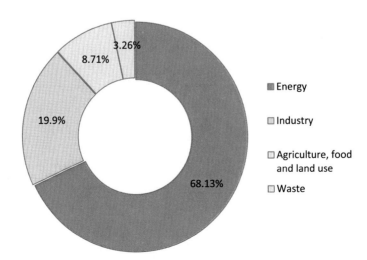

**Fig. 3.1** Sector-wise GHG emissions in India. *Source* GHG Platform

- Appraise GHG emissions' importance and its impact on the environment to all stakeholders (management, employees, customers, etc.) regularly
- Maintain health cards for all stakeholders for sustaining quality practices
- Maintain inventories of all items (repairable or non-repairable)
- Maintain dead-stock items, its reuse potential, and methods of its disposal
- Declare maintenance of raw materials and other items along with its frequency of use
- Maintain auditing of GHG emissions and carbon discharge of organizational activities
- Maintain the record of waste disposal and renewable source of energy
- Encourage e-business, e-logistics to save the environment
- Integrate Information technology in all day to day activities and assess the amount of carbon release in each activity
- Promote internet usage for all kinds of information based retrieval and dissemination of data for all in-house activities.

The influence of GHG on climate change is tremendous, and carbon dioxide plays a significant role as it is the most commonly produced. It is not practically possible for developing nations to reduce the emission of carbon dioxide as the countries need more industrialization. In this context, we have to invent technologies for reducing the amount of carbon dioxide in the atmosphere. Currently, the best way to save our future is to remove carbon dioxide through direct air capture, a process that involves pumping air through a system that removes carbon dioxide and either liquefies it and stores it or chemically turns it into a substance either inert or useful. Another method is carbon sequestration, which is the process of capturing and storing atmospheric carbon dioxide. There are mainly three types of sequestration:

1. *Ocean Sequestration:* Carbon stored in oceans through direct injection or fertilization.
2. *Geologic Sequestration:* Natural pore spaces in geologic formations serve as reservoirs for long-term carbon dioxide storage.
3. *Terrestrial Sequestration:* A large amount of carbon is stored in soils and vegetation, which sinks our natural carbon. Increasing carbon fixation through photosynthesis, slowing down or reducing decomposition of organic matter, and changing land-use practices can enhance carbon uptake in these natural sinks.

## 3.3  Green Biology

Green Biology is a subject-centered environment and plant sciences, which is an integral part of our ecosystem. Our ecosystem consists of a variety of species of plants and living organisms. Biodiversity boosts ecosystem productivity where each species, no matter how small, all have an essential role to play. Greater species diversity ensures natural sustainability for all life forms. Unfortunately, our Biodiversity is threatened by a variety of global changes resulting from human society's combined actions. The most direct threats are overharvesting and loss/disturbance of habitat resulting from the conversion of natural ecosystems to human use. However, other changes such as increased nutrient availability and elevated $CO_2$, with the resulting climate change, are long-term threats (Godfrey, 2004; Twidell & Weir, 2005; Hostetler, 2012).

There is no easy way to preserve our Biodiversity at the current level of diminishing returns. To some extent, loss of Biodiversity can be stopped by identifying hot spots of biodiversity, habitat and ecosystem restoration, utilization of managed forest, and setting aside various kinds of habitats for conservation reserves. There is a dire needs to propagate and educate the 3R principles, namely, reduce (maximum emphasis), reuse, and recycle, so that wastes, garbage, and pollution can be controlled to a manageable extent. The more we reduce our demand for new resources, the less habitat conversion will be necessary to get those resources or the energy to make the products we demand. The less waste goes into the landfill. While sensitizing the 3R policies, the 'recover' and 'resist' policies also to be practiced for the complete accountability and responsibility of what one does.

The biggest threat of today is global warming and natural calamities. Global warming is the continuous rise in the average temperature of the earth's atmosphere and oceans. Global warming is caused by an increased concentration of GHG in the atmosphere resulting from human activities such as deforestation and the burning of fossil fuels etc. These issues can only be addressed through the best management techniques of Lean, Green, and Clean concepts. To maintain a win–win situation for all, it is time to act positively and make our habitat a safer place for living. We believe that the principles underlined in LGC will help bring the habitat to its near normal by addressing the climate change issues mentioned above.

As mentioned before, Lean addresses wastes and its elimination. In line with the seven wastes identified in a Lean process, the seven Green wastes and their probable solutions are:

- *Energy*: Identify the use and source of energy in each activity, measure the quantity of energy used, minimize the use of energy, offset remaining energy use, the transition to self-harvested renewable energy.
- *Water*: Identify the use of water in the value stream and overall building, measure water, consumption and discharge in the whole building and value stream activities, measure the toxicity of water discharged in value stream activities, minimize the use of water, self-harvest rainwater, transition toward the continual reuse of water.
- *Materials*: Identify the input and output of stuff in your value stream, measure the recycled/recyclable and compostable content of each material input and output, classify materials as a technical nutrient, biological nutrient, or neither, assess the impact on the environment, phase out negative-impact substances, minimize materials usage, move toward fully released recycled/recyclable or transition to cent percent reuse.
- *Garbage*: Identify the creation of waste in your value stream, measure the makeup of waste in your value stream, measure the hazardous substances in your trash, minimize waste production, and move toward the total elimination of garbage.

- *Transportation*: Identify transportation within your value stream and overall building, measure the mode and distance of transportation, minimize traffic, offset remaining transit, and move toward the use of cent percent environmentally.
- *Emissions*: Identify the sources of emissions to produce your product or service, measure the type and amount of emissions used to create your product or service, identify the presence of emissions from the use of your product or service, measure the type and amount of emissions from the use of your product or service, minimize emissions, offset remaining emissions, move toward the total elimination of emissions.
- *Biodiversity*: Identify biodiversity waste, measure biodiversity destruction, minimize and eliminate biodiversity destruction by giving appropriate priority and concern.

Here are some simple things that will help reduce the environmental impact, and thereby reduce our adverse impact on Biodiversity (Hostetler, 2012). Many of them help in multiple ways, as well.

- Escalate the promotion of reducing, reuse, and recycle policies at all levels of organizational activities.
- Reduce the use of pesticides and fertilizers in lawn care. These runoffs of lawns flaw into adjacent lakes and streams with adverse effects for the plants and animals living there.
- Get involved with ecological restoration in your area by volunteering; one can help restore habitat for native species and eliminate invasive species.
- Aim for energy conservation in your home.
- Incorporate renewable energy and energy efficiency into homes. With some careful thought about the region, site, and needs, one can drastically reduce their energy consumption.
- Composting reduces the overall waste stream and thereby the need for landfill spaces and it provides natural slow-release fertilizer for flower or vegetable garden.
- Place limits on the amount of carbon that polluters are allowed to emit.
- Build a clean energy economy by investing in efficient energy technologies, industries, and approaches.
- Use environmentally friendly products for cleaning. This reduces chemical contamination of habitats both during manufacturing and when those chemicals go down the drain.
- Buy organic foods. This helps reduce inputs of fertilizers and pesticides into the environment, which in turn reduces negative impacts on nearby beneficial insects (for pollination and pest control) and adjacent aquatic Biodiversity.
- Buy sustainably harvested seafood. Many kinds of seafood, though delicious, are not harvested sustainably—either for the individual species itself or for those species that are unlucky enough to be trapped as "bycatch." Some trawlers destroy extensive seafloor habitat in catching fish; many shrimp farms destroy mangrove forests as crucial as nurseries for wild fish species.

The principle of reducing waste, reusing, and recycling resources and products is one of the flexible tools of Lean. Reducing means choosing to use things with care to minimize the amount of waste generated. Reusing involves the repeated use of items or parts of items that still have beneficial aspects. Recycling means the use of garbage itself as resources. Promoting 3Rs and educating people for Lean philosophy can have a tremendous impact on both waste identification and its minimization.

There is no doubt that the sustainability of all living beings depends on the climate system. Moreover, the study of the climate system is, to no small extent, the study of weather statistics; so, it is not surprising that quantitative or statistical reasoning, analysis, and modeling are pervasive in the climatological sciences. Quantitative analysis helps to quantify the effects of uncertainty, both in terms of observation and measurement and in terms of our understanding of the processes that govern climate variability. It also helps us to identify which of the many pieces of information derived from observations of the climate system are worthy of interpretation. Quantitative procedures are also integral to the vast majority of efforts that seek physically meaningful descriptions of observed climate variability. Thus, through the perfect use of Lean and Green principles and quantitative analysis, one can bring measurable changes in balancing the natural habitat and our ecosystem to better use and achievable target. At least, creating awareness for the adoption of these kinds of best practices can instill confidence in safeguarding our Biodiversity. This is the need of time.

## 3.4  Green Chemistry

Green chemistry, also called sustainable chemistry, is an area of chemistry and chemical engineering focused on the designing of products and processes that minimize the use and generation of hazardous substances. Green chemistry applies across the life cycle of a synthetic product, including its design, manufacture, use, and ultimate disposal. The concept is emerged from a variety of existing ideas and research efforts (such as atom economy and catalysis) in the period leading up to the 1990s, in the context of increasing attention to problems of chemical pollution and resource depletion (Linthorst, 2009; Woodhouse & Breyman, 2005).

In the United States, the Environmental Protection Agency (EPA) played a significant role in fostering green chemistry through its pollution prevention programs, funding, and professional coordination. At the same time in the United Kingdom, researchers at the University of York contributed to the establishment of the Green Chemistry Network within the Royal Society of Chemistry, and the launch of the journal Green Chemistry (Linthorst, 2009). According to the American Chemical Society (ACS), Green chemistry works with the following twelve essential principles:

- *Prevention.* Preventing waste is better than treating or cleaning up debris after it is created.

- *Atom economy.* Synthetic methods should maximize the incorporation of all materials used in the process into the final product.
- *Less hazardous chemical syntheses.* Synthetic methods should avoid using or generating substances toxic to humans and the environment.
- *Designing safer chemicals.* Chemical products should be designed to achieve their desired function while being as non-toxic as possible.
- *Safer solvents and auxiliaries.* Auxiliary substances should be avoided wherever possible and as non-hazardous as possible when they must be used.
- *Design for energy efficiency.* Energy requirements should be minimized, and processes should be conducted at ambient temperature and pressure whenever possible.
- *Use of renewable feedstock.* Whenever it is practical to do so, renewable feedstock or raw materials are preferable to nonrenewable ones.
- *Reduce derivatives.* Unnecessary generation of derivatives-such as the use of protecting groups-should be minimized or avoided if possible; such steps require additional reagents and may generate extra waste.
- *Catalysis.* Catalytic reagents that can be used in small quantities to repeat a reaction are superior to stoichiometric reagents (ones that are consumed in a response).
- *Design for degradation.* Chemical products should be designed so that they do not pollute the environment; when their function is complete, they should break down into non-harmful products.
- *Real-time analysis for pollution prevention.* Analytical methodologies need to be further developed to permit real-time, in-process monitoring, and control before hazardous substances form.
- *Inherently safer chemistry for accident prevention.* Whenever possible, the substances in a process, and the forms of those substances, should be chosen to minimize risks such as explosions, fires, and accidental releases.

A relatively similar concept is *Environmental Chemistry*, where it focuses on the effects of polluting chemicals on nature. Green chemistry focuses on the environmental impact of chemistry, including technological approaches to preventing pollution and reducing the consumption of nonrenewable resources (Clark et al., 2012; Sheldon et al., 2007). Vert et al. (2012) discuss the engineering concept of pollution prevention and zero waste at the laboratory and industrial scales. It encourages the use of economic and eco-compatible techniques that improve the yield and bring down the cost of disposal of residues at the end of a chemical process. Similarly, the *Lean and Chemicals Toolkit* prepared for the US Environmental Protection Agency by Ross and Associates Environmental Consulting, Ltd., in association with Industrial Economics, Inc. (https://www.epa.gov/sustainability/lean-che micals-toolkit: last accessed on 18.04.2020), provides practical strategies for using Lean manufacturing to reduce chemical wastes while improving the operational and environmental performance of manufacturing and industrial businesses.

According to the toolkit, the chemical management strategies that support Lean goals include:

- *Chemical Management Services*: Having chemicals delivered when you need them in the amounts that you need supports Lean goals while also reducing risks and wastes. Contracting with companies that provide chemical management services can remove even more waste.
- *Right-Sizing*: Limit unnecessary use of chemicals and increase process efficiency by making equipment and containers the right size for the task.
- *Applying Lean to hazardous Waste Management*: Use Lean to improve chemical and waste management "support" processes, such as waste-collection processes and compliance-reporting activities.
- *Point-of-Use Storage*: Adopt proper maintenance and control procedures to prevent regulatory compliance and worker health and safety issues associated with point-of-use storage of chemicals.
- *Visual Management*: Reinforce best practices for using and disposing of chemicals and hazardous wastes with visual controls, standard work, 5S, and Total Productive Maintenance.

The concepts like 5S and Total Productive Maintenance (TPM) etc. are very efficient for process orientation and control, thereby maximize the uptime of process equipment. Such an efficient process helps speed up the entire organizational processes by removing clutters in the process and leading to instant results (for detailed discussion, see Chap. 4). The objective of TPM is to allow the process to maintain its inherent capability. In the absence of adequate maintenance, equipment breakdowns increase both defects and variation around a nominal value (see Sect. 4.4 for a detailed description of the concept).

## 3.5   Green Supply Chain Management

Green supply chain management (GSCM) is a relatively new concept. The concept of GSCM is to integrate environmental thinking into supply chain management. GSCM aims to minimize or eliminate wastages including hazardous chemicals, emissions, energy, and solid waste along the supply chain such as product design, material resourcing, and selection, manufacturing process, delivery of the final product and end-of-life management of the product (Rao, 2006; Srivastava, 2007). The characteristics of GSCM are Green Procurement, Green Manufacturing, Green Distribution, and Green Logistics.

- *Green procurement*: Green procurement is defined as a set of supply-side practices utilized by an organization to effectively select suppliers based on their environmental competence, technical and eco-design capability, environmental performance, ability to develop environmentally friendly goods and knowledge to support focal company's environmental objectives (Paulraj, 2011). Also, it encourages the practice of eco-labeling of products, ensures supplier's environmental compliance certification, and conducts auditing for supplier's internal environmental management, etc. (Lee et al., 2012).

- *Green manufacturing*: Green manufacturing is a production process which converts inputs into output by reducing hazardous substances, increasing energy efficiency in lighting and heating, practicing 3Rs (reduce-reuse-recycle), minimizing waste, actively designing and redesigning green processes (Green et al., 2012; Holt & Ghobadian, 2009; Lee et al., 2012; Ninlawan et al., 2010; Zhu et al., 2005). According to the authors, green manufacturing requires manufacturers to design products that facilitate the reuse, recycle and recovery parts and material components; avoid or reduce the use of hazardous products within the production process, and minimize consumption of materials and energy.
- *Green Distribution*: Green distribution is all about a sustainable way of preparing consignment for its delivery. It consists of green packaging with the aims to downsize packaging; use green packaging materials; promote recycling and reuse programs; cooperate with the vendor to standardize packaging; encourage and adopt returnable packaging methods; minimize material uses and time to unpack; use recyclable pallet system; and save energy in warehouses (Holt & Ghobadian, 2009; Ninlawan et al., 2010).
- *Green logistics/Transportation*: Green logistics/transportation, is about delivering goods directly to the user site, using alternative fuel vehicles and grouping orders together, rather than in smaller batches, investing in vehicles that are designed to reduce environmental impacts, and planning vehicle routes, etc. (Holt & Ghobadian, 2009; Ninlawan et al., 2010). As stated by Laosirihongthong et al. (2013), Green logistics is about Reverse logistics (see Sect. 6.3 for details). It includes collecting used products and packaging from customers for recycling, returning packaging, and products to suppliers for reuse, and requiring suppliers to manage their packaging materials.

A supply chain is a network that consists of all stakeholders involved in the process directly or indirectly, in producing and delivering products or services to ultimate customers–both in upstream and downstream sides through physical distribution. SCM's primary focus is to provide the right product to the right customers at the correct cost, right time, high quality, proper form, and right quantity. Besides, the short-term strategic goal of SCM is to reduce cycle time and inventory and thus to increase productivity. In contrast, the long-term goal is to enhance profits through market share and customer satisfaction (Tan, 2002). The author of this book has introduced the concept of "Green statistics" as an alternative to supply chain management purely from a scientific perspective. The focus of Green statistics will be to have a measurement-based evaluation of all supply chains for yielding high throughput and reliable products (see Chap. 6 for more details).

The working model of SCM, in general, is the *Plan-Do-Check-Act* of Edward Deming (see Fig. 3.2). These steps facilitate managing the natural variability of the supply chain effectively and balance the quality of the company. Note that the planning phase takes care of the chain's procurement activity, which includes market research, supplier contracts, value, and quality of the suppliers. Thus, green SCM is a tool for promoting the tenets of Lean principles in its true spirit. One of the essential principles of SCM is to coordinate the raw materials and components flow efficiently

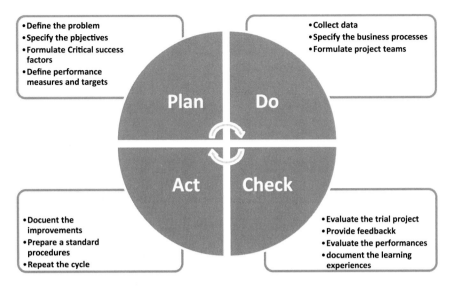

- Define the problem
- Specify the pbjectives
- Formulate Critical success factors
- Define performance measures and targets

- Collect data
- Specify the business processes
- Formulate project teams

**Plan** **Do**

**Act** **Check**

- Docuent the improvements
- Prepare a standard procedures
- Repeat the cycle

- Evaluate the trial project
- Provide feedbackk
- Evaluate the performances
- document the learning experiences

**Fig. 3.2** Deming's PDCA cycle

from various suppliers to manufacturing companies to convert raw materials into finished products and fulfill customers' value expectations. Suppliers' capabilities are directly linked to the firm's ability to produce a product with higher quality and lower costs while meeting the delivery promises. To achieve organizational sustainability, firms need to pay attention to supply-side practices (Chin et al., 2015).

According to Paulraj (2011), the environmental collaboration includes cooperating with suppliers to achieve ecological objectives and improve waste reduction initiatives, providing suppliers with design specification that include ecological requirements for purchased items, encouraging suppliers to develop new source reduction strategies, working with suppliers for cleaner production and helping suppliers to provide materials, equipment, parts, and services that support organizational goals.

The Lean Green and clean aspects of GSCM are the following:

- Prepare a Green mission and vision statement for all supply chain dynamics.
- Prepare a quality document for all supply materials.
- Strictly follow environmental regulations in its true spirit.
- Inculcate quality practice with an emotional touch.
- Think for the betterment of clients in all transactional processes.
- Be responsible for all climate-related causes and effects.
- 'Act of God' should not be a liability for the clients and customers.

The management approach closely related to SCM is the Lean Supply Chain Management (LSCM). It is all about reducing costs and lowering waste as much as possible. This methodology is vital for organizations involved with high volumes of purchase dealings, as debris and expenses can accumulate quickly. Thus, LSCM is a new way of thinking about supplier networks, where supplier partnerships and strategic alliances become a collaborative mechanism for sustaining competition and cooperation. A methodology that helps organizations to adapt to changing situations is called Agile Supply Chain Management (ASCM). Implementing ASCM allows organizations to react to unanticipated external economic changes, such as economic swings, changes in technology, or customer demand changes.

Integrating Lean and Green into SCM leads to the following working model for quality supply chain management (see Fig. 3.3. for LGSCM diagram):

- *Plan*—Plan all the aspects of sales and operations, including supplies, procurement, etc.
- *Identify*—Identify and eliminate all the non-value added activities in the supply process to assess the wastes.
- *Evaluate*—Evaluate the procurement pattern, assess the inventories for its natural control, and test the supply chain metrics, etc.
- *Implement*—Implement the solutions and prepare a viable quality improvement program for the ongoing supply chain.

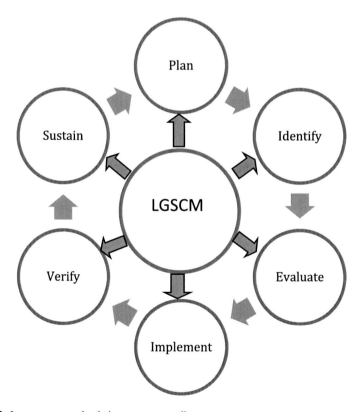

**Fig. 3.3**  Lean green supply chain management diagram

- *Verify*—Verify and ensure that the ongoing supply chain meets the customer requirements and specifications.
- *Sustain*—Sustain the new supply chain's development and be aware of the need for further chain.

Strategic management of quality in an organization will always be most effective when all the people in the organization understand the necessary tools for quality improvement. Towards this, elementary statistical concepts that form the basis of process control and the minimum knowledge of data analysis become desirable qualifications for all those executives and higher-ups responsible for creating a better business environment. There are many statistical tools applied to SCM, where the efficiency of the supply chain can be well assessed through a data-based measurement system. Hence, exploring the Green Statistics concept is warranted to understand better the tenets of LGC sciences. This is well addressed in Sect. 6.3 of Chap. 6.

One of the significant developments in SCM recently is the emergence of *Blockchain*, also referred to as Distributed Ledger Technology (DLT) concepts. As supply chains grow more complex, involve diverse stakeholders, and mainly rely on several external intermediaries, blockchain emerged as a strong contender for de-tangling all the data/documents/communication exchanges happening within the supply chain ecosystem. According to Paul Brody, *"At its most basic level, the core logic of blockchains means that no inventory can exist in the same place twice."* Through blockchains, companies gain a real-time digital ledger of transactions and movements for all participants in their supply chain network. Further, blockchains can track and manage resources at the ecosystem level; the payoff should be much higher accuracy and, from there, better forecasts and fewer inventories to maintain the same service level can be achieved (http://www.digitalistmag.com/tag/blockchain-and-supply-chain, last accessed on 10.05.2020).

A blockchain-based database can store relevant data from all your partners, giving your company a 360-view of the total volume of purchases, regardless of who managed the purchase activity. There will be no need for individual users to continually share operational data and someone else to crosscheck it (https://medium.com/@infopulseglobal_9037/, last accessed on 10.05.2020). A simple analogy for understanding blockchain technology is a Google Doc. When we create a document and share it with a group of people, it is distributed instead of copied or transferred. This creates a decentralized distribution chain that gives everyone access to the document at the same time. No one is locked out awaiting changes from another party, while all modifications to the doc are being recorded in real-time, making changes completely transparent (https://builtin.com/blockchain, last accessed on 10.05.2020).

Blockchain is the technology that enables the existence of *Cryptocurrency*, among other things. Bitcoin is the name of the best-known cryptocurrency, the one for which blockchain technology was invented. A cryptocurrency is a digital asset designed to work as a medium of exchange wherein individual coin ownership records are stored in a digital ledger or computerized database using strong cryptography to secure transaction record entries, to control the creation of additional digital coin records, and to verify the transfer of coin ownership (Greenberg, 2011). One of the significant ways the blockchain can improve supply chains' efficiency is its ability to collect and provide data. Companies can then have analysts or artificial intelligence evaluate the data to look for ways to improve efficiency. The blockchain can store an unlimited amount of data, from the route a product takes to how long it spends in each location (https://medium.com/@mainfinex/how-cryptocurrency-and-blockc hain-are-disrupting-supply-chains, last accessed on 20.05.2020).

## 3.6   Green Technologies

Green technology or sustainable technology is an environmentally friendly process used to conserves natural resources and the environment. While nature's resources and ecosystem services are declining, the human demand for such supplies is increasing. Green technology is to find ways to produce in such a manner that the earth's natural resources are not damaged or depleted. The main objectives of Green technology are sustainability by meeting the needs of the society without damaging the natural resources, recycle and reuse of materials, reducing waste, and pollution by changing patterns of production and consumption (Kalam & Pillai, 2013). The Green technology solution is facilitated by applying topics like environmental science, green chemistry, environmental monitoring, and electronic devices to monitor, model, and conserve the natural environment and resources, and curb the negative impacts of human involvement.

The use of renewable energy is highly encouraged for sustainable use, as they can be replenished quickly. Water, air purification, sewage treatment, and solid waste management are the best ways of controlling $CO_2$ emission and environmental pollution. Taesler (1991), encourages the E-gain forecasting for assessing the weather's impact on the buildings. The United States Patent 6098893 (Stefan Berglund) states that by adjusting the heat based on the weather forecast data, the system eliminates excessive use of heat, thus reducing the energy consumption and greenhouse gas emissions.

According to Kalam and Pillai (2013), the growth of high technology will increase exponentially due to the current status of technology advancement and the progress made to develop the infrastructure. Therefore, sustainable development should address society; the environment and economy in one go for better livelihood and habitat. The authors also recommend the following Green technology solutions:

- *Green technologies for Agriculture*: Agricultural technologies should be sound enough to provide maximum yield and minimum environmental problems. Hence assistive technologies such as design farm vehicles and farm equipment can be run on cent percent bio-fuels or electrically operated to take care of environmental consideration. Development of intelligent farm machines like harvester recording the yield per unit area while harvesting, smart mechanization weeding out only weeds and not the plants will help the farmers with reduced dependence on laborers and increased management of the yield.

- *Climate Engineering*: Climate engineering is all about the reduction of $CO_2$ in air-carbon sequestration. The heat from the earth is trapped in the atmosphere due to high levels of $CO_2$ prohibiting it from releasing into space ("Greenhouse effect"). Trees are the best means of reducing the impact of $CO_2$. Trees remove (sequester) $CO_2$ from the atmosphere during photosynthesis and return oxygen to the atmosphere. Moreover, planting trees in the fields protects the crops against wind and sandstorms and leaves shed by the trees to fertilize the soil and improve the microclimate that becomes the most critical factor for a better harvest. It is recommended to promote geoengineering projects like cool roof projects, ocean iron fertilization projects, solar radiation management projects for capturing $CO_2$, and its storage in geological formations.
- *Power and fuel from plastic wastes*: Waste could be solid, liquid, plastic, or metal. Attempts are being made to generate fuel from plastic waste through Plasma Pyrolysis Technology (PPT). The process is based on the random depolymerization of waste plastics in the presence of a catalyst into liquid fuel. A commercial plant is established at Butibori, Nagpur, to produce 5MT fuel per day. Plastic tar can also be used for road construction. The bitumen and gravel mix used for laying roads could be combined with flakes or granules made from domestic chemical wastes like carrying bags, teacups, and a variety of domestic plastics. The advantages are excellent skid resistance and texture values, reasonably substantial with good even surface, no potholes, rutting, raveling or edge flaw, for a more extended period, and higher binding strength of the mix. Salem, an industrial town, is the first city to lay a plastic-tar road in the country.
- *Power through solid waste*: Municipal solid waste can be very well converted to generate energy and power. A company based at Hyderabad (M/s. Selco International Limited) is the first to establish a plant for producing electricity from municipal solid waste and garbage. The alternate sources of clean energy are solar, ocean (tidal, wave, current and thermal), wind, etc.
- *Liquid waste management*: Benefits of liquid waste management through wastewater treatment include lakes & water bodies that would be clean and pristine, treated sewage water can be used for irrigation, recycled wastewater reduces freshwater demand, environment and sanitation of surrounding improves significantly, biological sludge could be used as mature. The significant presence of dyes in effluents from the dyeing and bleaching units causes water contamination. This is a primary environmental concern. A possible solution is through enhancing the existing effluent plant through reverse osmosis or switching over to cleaner production technologies such as the Advanced Oxidation Process.
- *Power through Methane gas*: Production of Synthetic Natural Gas (SNG), also called methane, is a promising alternative to petroleum due to its potential as a transportation fuel. Methane is produced in a specially built reactor by fusing $CO_2$ and hydrogen. $CO_2$ is obtained from the exhaust, from fossil fuel or chemical fuel like ethanol. Water is dissolved into its elemental components, hydrogen, and oxygen, through the electrolysis method. Hydrogen is blended with $CO_2$, and the mixture is passed through a particular methane reactor. The catalyst-driven reaction produces methane that is chemically equivalent to natural gas. Methane

gas is an ideal hydrocarbon fuel with minimal environmental impact as it consumes nearly fifty-seven tones of $CO_2$ to produce one million cubic feet of methane.

- *Artificial Intelligence*: Artificial intelligence, also called machine intelligence, is intelligence demonstrated by machines, in contrast to the natural intelligence displayed by humans and other animals. According to the Chief Environment Officer of Microsoft, Lucas Joppa, Artificial intelligence converts information into insightful knowledge and helps fix environment-related problems across the world. Artificial intelligence is all about data management, which is vast, uncontrolled, and intelligible, and then what you make out of it so that it becomes information that translates into knowledge. AI integrates three core subject areas: viz. Statistics, Mathematics, and Information Science. It generates models from inductive (empirical) learning systems to deductive (analytical) methods by suitably selecting algorithms for analytical reasoning without human intervention. Hence the name artificial intelligence.

- *Machine Learning*: Machine learning, or learning using empirical data, is one of the fast-developing fields in computer science. They are aimed at solving real-world problems by extracting features for building models, and hence the application ranges from interactive knowledge acquisition to automated theory formation (Alpaydin, 2014; Waldinger, 1977). Machine learning methods employ the right statistical modeling and inferences for feature selection (Japkowicz & Shah, 2011). Statistical tests of a hypothesis are used for assessing the validity of assumptions and their analyses of goodness-of-fit of the underlying data. Machine learning also helps to construct valid models even if the data extracted are not random and do not follow normal distributions. The use of AI and Machine language can boost our efforts to map and catalog the region's unique, resource-rich ecosystem. It can help farmers protect their crops and avoid losses through smart decision making (The Speaking Tree, September 18, 2018). It is not too far, where these cloud-based applications entirely control our Biodiversity.

- *Cloud Computing*: Cloud Computing is a new way of looking at technology and services. It makes data and applications available through the internet. According to Ried et al. (2010), Cloud Computing (CC) is a "standardized IT capability (network, servers, storage, software or infrastructure, services) delivered via Internet technologies in a pay-per-use, self-service way." The essential characteristics of CC are on-demand service, broad network access, and resource pooling. Some standard features include virtualization, service orientation, distribution and scaling, advanced security, and low-cost software, etc. The cloud service model *provides software as a Service* (SaaS), Cloud *Infrastructure as a Service* (IaaS), and *Platform as a Service* (PaaS) Venters and Whitley (2012), Yang and Tate (2012), Marston et al. (2011). The advantages of CC are improved performance, lower computer and software costs, instant software updates; document format compatibility; unlimited storage capacity; increased data reliability; universal access to software and programs, device independence, increased data security, and safety, and so on. Thus, CC technology is perfectly matching with the objective participation of LGC activities. The disadvantages of the uninterrupted requirement of internet and power supply, data loss, can decrease the system's speed because of many data on queue for storage and so on. The other computing technology variants are distributed computing, cluster computing, grid computing, and utility computing.

*Green nanotechnology* is considered to be a new upcoming area, where the technology can manipulate materials at the scale of the nanometer, which is one-billionth of a meter. Some scientists believe that mastery of this subject will transform how everything in the world is manufactured. Green nanotechnology is the application of green chemistry and green engineering principles to this field. To a more significant extent, the issues with space, time, infrastructure, workforce, and management will be resolved with the advent of these futuristic engineering principles and technologies. They can address many of the significant concerns of future generations associated with the climate and environment. The penetration of these new techniques can take the human brain to a super level of thinking to acquire an unimaginable level of goals that nobody thought earlier. Many old problems have already been revisited, and brighter solutions were found through these techniques. Scientists worldwide are now able to offer the best scientific reasoning for every modern-day problem without ambiguity. We expect more such opportunities in the future.

## 3.7  Exercises

3.1.  What is Green energy? Suggest methods of conserving Green energy.
3.2.  Distinguish between CNG and LPG.
3.3.  What are the components of GHG?
3.4.  What is sequestration? Discuss the type and impact of sequestration.
3.5.  What is Green Biology? State their relevance in maintaining the balance of the ecosystem.
3.6.  What are the Lean management tools applied in Green biology?
3.7.  State various Green wastes and their mitigation methods.
3.8.  What is Green chemistry? State all the principle steps involved in Green chemistry.
3.9.  What is the Lean management tools applied in Green chemistry?
3.10.  Discuss various characteristics of Green Supply Chain Management.
3.11.  Distinguish between Green Procurement and Green Logistics.
3.12.  Distinguish between Green Manufacturing Green Distribution.
3.13.  Discuss how the PDCA improves the Green Supply Chain.
3.14.  Discuss Lean Green and clean aspects of GSCM. How LGC principles improve the supply chain.
3.15.  What is Blockchain? How does it help to improve the supply chain?
3.16.  How Green technologies improve Agriculture and Industry?
3.17.  What is Climate Engineering?
3.18.  Distinguish between Artificial Intelligence and Machine Language. How do they help to enhance a transaction process?

3.19.   What are the uses of Cloud Computing? How do they help resolve the storage space?

3.20.   What is Green nanotechnology?

# References

Alpaydin, E. (2014). *Introduction to machine learning* (3rd ed.). The MIT Press.

Chin, T. A., Tat, H. H., & Sulaiman, Z. (2015). *Green supply chain management, environmental collaboration and sustainability performance.* In *12th Global Conference on Sustainable Manufacturing, Procedia, CIRP* (Vol. 26, pp. 695–699).

Clark, J. H., Luque, R., & Matharu, A. S. (2012). Green chemistry, biofuels, and biorefinery. *Annual Review of Chemical and Biomolecular Engineering., 3,* 183–207. https://doi.org/10.1146/annurev-chembioeng-062011-081014.PMID22468603.

Eyre, A. W., Keightley, P. D., Smith, N. G. C., & Gaffney, D. (2002). Quantifying the slightly deleterious mutation model of molecular evolution. *Molecular Biology and Evolution, 19*(12), 2142–2149. https://doi.org/10.1093/oxfordjournals.molbev.a004039.

Godfrey, B. (2004). *Renewable energy* (2nd ed.) Oxford University Press (ISBN: 0-19-926178-4).

Green, K. W., Jr., Zelbst, P. J., Meacham, J., & Bhadauria, V. S. (2012). Green supply chain management practices: Impact on performance. *Supply Chain Management: An International Journal, 17*(3), 290–305.

Greenberg, A. (2011). *Crypto currency.* Forbes.

Holt, D., & Ghobadian, A. (2009). An empirical study of green supply chain management practices among UK manufacturers. *Journal of Manufacturing Technology Management, 20*(7), 933–956.

Hostetler, M. (2012). *The green leap: A primer for conserving biodiversity in subdivision development.* University of California Press.

Hudson, D. A. (1991). *Business and the environment: Lean, clean, and green.* Video Publishing House, Inc.

Japkowicz, N., and M. Shah (2011). *Evaluating learning algorithms: A classification perspective.* Cambridge University Press.

Kahle, L. R., & Atay, E. G. (2014). *Communicating sustainability for the green economy.* M.E. Sharpe.

Kalam, A. P. J., & Pillai, S. (2013). *Thoughts for change-we can do it.* Pentagon Press.

Laosirihongthong, T., Adebanjo, D., & Tan, K. C. (2013). Green supply chain management practices and performance. *Industrial Management & Data Systems, 113*(8), 1088–1109.

Lee, S. M., Kim, S. T., & Choi, D. (2012). Green supply chain management and organizational performance. *Industrial Management & Data Systems, 112*(8), 1148–1180.

Linthorst, J. A. (2009). An overview: Origins and development of green chemistry. *Foundations of Chemistry, 12,* 55. https://doi.org/10.1007/s10698-009-9079.

Marston, S., Li, Z., Bandyopadhyay, S., Zhang, J., & Ghalsasi, A. (2011). Cloud computing—The business perspective. *Decision Support System, 51,* 176–189.

Ninlawan, C., Seksan, P., Tossapol, K., & Pilada, W. (2010). The implementation of green supply chain management practices in the electronics industry. In *Proceedings of the International Multi-Conference of Engineers and Computer Scientists,* March 17–19: Hong Kong.

Paulraj, A. (2011). Understanding the relationships between internal resources and capabilities, sustainable supply management, and organizational sustainability. *Journal of Supply Chain Management,* 2011; 47(1): 19–37.

Planning Commission (2006). Annual Report 2006–07, Government of India.

Principles of Green Chemistry—American Chemical Society. *American Chemical Society* (last accessed on 26.08.2018).

Rao, P. (2006). Greening of suppliers/in-bound logistics—In the South East Asian context. Greening the supply chain (pp. 189–204).

Ried, S., Kisker, H., & Matzke, P. (2010). *The evolution of cloud computing markets.* Technical Report, Forrester Research.

Schaeffer, J. (2007). *Real goods solar living sourcebook: The complete guide to renewable energy technologies and sustainable living* (30th-anniversary edition).

Srivastava, S. K. (2007). Green supply-chain management: A state-of-the-art literature review. *International Journal of Management Reviews, 9*(1), 53–80.

Sheldon, R. A., Arends, I. W. C. E., & Hanefeld, U. (2007). *Green chemistry and catalysis.* https://doi.org/10.1002/9783527611003.

Taesler, O. (1991). The bioclimate in temperate and northern cities. *International Journal of Biometeorology, 35*(3), 161–168.

Tan, K. C. (2002). Supply chain management: Practices, concerns, and performance issues. *Journal of Supply Chain Management, 38*(1), 42–53.

Teddy (2010). *TERI Energy Data Directory and Yearbook*, New Delhi.

Twidell, J., & Weir, T. (2005). *Renewable energy resources.* Routledge.

Venters, W., & Whitley, E. A. (2012). A critical review of cloud computing: Researching desires and realities, *Journal of Information Technology, 27*(3).

Vert, M., Doi, Y., Hellwich, K. H., Hess, M., Hodge, P., Kubisa, P., Rinaudo, M., & Schué, F. (2012). Terminology for bio-related polymers and applications (IUPAC Recommendations 2012). *Pure and Applied Chemistry, 84*(2), 377–410. https://doi.org/10.1351/PAC-REC-10-12-04.

Waldinger, R. (1977). Achieving several goals simultaneously. In E. W. Elcock & D. Michie (Eds.), *Machine intelligence* 8. New York: Halstead and Wiley.

Woodhouse, E. J., & Breyman, S. (2005). Green chemistry as a social movement? *Science, Technology and Human Values, 30*(2), 199–222. https://doi.org/10.1177/0162243904271726.

Yang, H., & Tate, M. (2012). A Descriptive literature review and classification of cloud computing research. *Communications of the Association for Information Systems, 31* (2012).

Zhu, Q., Sarkis, J., & Geng, Y. (2005). Green supply chain management in China: Pressures, practices, and performance. *International Journal of Operations & Production Management, 25*(5), 449–468.

www.epa.gov/lean.

http://www.ross-assoc.com.

www.qualitygurus.com.

https://www.epa.gov/sustainability/lean-chemicals-toolkit.

http://www.data.gov.in.

http://www.imd.gov.in.

https://www.ipcc.ch.

https://www.kaggle.com.

# Chapter 4
# Lean, Green, and Clean Quality Improvement Models

*All improvement takes place project by project … and in no other way.*
—Dr. Joseph M. Juran

## 4.1 Introduction

Optimizing the product features, functions, resources, cost, etc., is the minimum objective of any management projects. Management models are, therefore, crucial for improving the quality and financial part of any organization. The best way to improve the financial part is by setting proper business goals, sustainable business plans, and cost reduction methods. However, if the business goals are not set as per the societal need, then sustaining the business becomes difficult. Note that, lean is not merely about cutting waste or reducing fat, but also about emphasizing the need for creating value. Hence to improve value creation and quality improvement, one needs to practice total quality management, lean thinking, Six Sigma quality, the Theory of Constraints, etc., like efficient methods routinely.

The totality of features and characteristics of a product or service that bear on its ability to satisfy stated or implied needs is what *quality* means. However, the customer's perception of quality is that the product should have a proper design with a better look, feel, and function apart from its purpose of existence and longevity. In the earlier period, the inspection was used to assure quality. It was one of the manufacturing operators' duties to inspect the part and make a decision to accept or reject it. The modern quality concepts are entirely customer-driven, and hence, we have customer-focused products and services.

Simply put, the Father of our Nation, Mahatma Gandhi, in his own words describes the importance of the customer in this way: "*A customer is the most important visitor on our premises; he is not dependent on us. We are dependent on him. He is not an interruption in our work. He is the purpose of it. He is not an outsider in our business.*

K. Muralidharan, *Sustainable Development and Quality of Life*,
https://doi.org/10.1007/978-981-16-1835-2_4

*He is part of it. We are not doing him a favor by serving him. He is doing us a favor by allowing us to do so".*

Among many quality improvement theories available, the most important and widely used quality approach is W. Edwards Deming's quality philosophy introduced in the early 1920s. Most of his arguments are based on the logic that most product defects resulted from management shortcomings rather than careless workers. That inspection after the fact was inferior to designing processes that would produce better quality. According to Deming, the top management must be responsible for defining, disseminating, and supporting the organization's quality system. For doing the organizational jobs efficiently, he introduced the concept of *Plan-Do-Check-Act* (PDCA) as a quality cycle for routine practice for managers and engineers (see also Fig. 3.2). Deming is famous for articulating the best business management ideas captured in his fourteen points and seven deadly diseases (Deming, 1982, 1986). He revised his fourteen points in 1990 from the original version published in 1986. The revised fourteen versions of his quality philosophy are detailed below:

1.  Create and publish to all employees a statement of the company's aims and purposes or other organizations. The management must consistently demonstrate their commitment to this statement.
2.  Adopt the new philosophy. We are in a new economic age, created by Japan. We can no longer live with commonly accepted American management styles or with widely accepted levels of delays, mistakes, and defective products.
3.  Understand the purpose of inspection, for improvement of processes and reduction of cost. Eliminate the need for investigation on a mass basis by building quality into the product in the first place.
4.  End the practice of awarding business based on price tag alone. Instead, minimize total cost.
5.  Improve continually and forever, the system of production and service.
6.  Institute training on the job.
7.  Teach and institute leadership. The aim of supervision should be to help people, machines, and gadgets to do a better job. Control of management requires an overhaul, as well as monitoring of production workers.
8.  Drive out fear. Create trust. Create a climate for innovation.
9.  Optimize toward the company's aims and purposes, the efforts of teams, groups, and staff areas.
10. Eliminate slogans, exhortations, and targets for the workforce, asking for zero defects and new productivity levels.
11. a. Eliminate numerical quotas for production. Instead, learn and institute methods for improvement.
    b. Eliminate management by objectives. Instead, lean the capabilities of the system and how to improve them.
12. Remove barriers that rob the hourly worker of his right to pride in artistry.
13. Encourage education and self-improvement for everyone.
14. Take action in the company to work to accomplish the transformation. The transformation is everybody's job.

The above fourteen points of the management theory have the potential to transform the traditional boss-management style into lead management. Among the "Seven Deadly Diseases," two critical points still attract a lot of relevance nowadays. They are:

1. Emphasis on short-term profits, as it can defeat constancy of purpose and long-term growth and secondly
2. Mobility of top management—Deming observes that it is difficult for workers to remain committed to policy or achieve long-term improvement when they know that their manager's tenure is likely to be short.

Deming observes that companies employing annual ratings of their workers from their superiors are tempted to work for themselves, not for the company. Rewards and punishments do not motivate people to do their best; they do precisely the opposite, and pride in artistry is lost. Managers, in effect, are managers of defects, and the organization becomes the loser. The job of management is inseparable from the welfare of the company. If individuals are to work jointly toward a common end, they require time to build trust and synergy. Teamwork-along with team effectiveness-develops over time. According to Deming, for companies that provide health insurance, medical costs represent the most significant single expenditure. For example, medical expenses have a multiplier effect for any company because they are embedded in the prices of every other firm's products and services with which it does business.

According to Armand Feigenbaum (Total Quality Control, 1961): "quality is a customer determination based on the customer's experience with the product or service, measured against his or her requirements—stated or unstated, conscious or merely sensed, technically operational or entirely subjective and always representing a moving target in a competitive market." In short, the three principles of quality are: (1) quality leadership, (2) modern quality technology, and (3) organizational commitment. A quality flow diagram is shown in Fig. 4.1.

In quality management circles, Philip Crosby is known as an unbeaten quality deployment champion. Crosby promoted the concept of "zero defects" and "Cost of Quality." He emphasizes the delivery of defect-free products and services that conform to specifications and defined quality as conformance to requirements. Crosby's response to the quality crisis was the principle of *"doing it right the first time."* Another stalwart of quality, Dr. Joseph M. Juran propounded the universal approach to managing for quality through his "quality trilogy" philosophy. He is one of the critical people in Japan's quality transformation after World War II. The underlying concept of the quality trilogy is that managing for quality consists of three basic quality-oriented processes: quality planning, quality control, and quality improvement.

The man who brought quality engineering (QE) and robust design in the modern era is the famous statistician Dr. Genichi Taguchi in the late 1940s. Primarily, QE deals with activities performed to reduce variability in product/process function. The methodologies and philosophies developed by Taguchi for optimizing product and process functions are based on engineering concerns, not purely scientific or

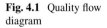

**Fig. 4.1** Quality flow diagram

statistical ones. They facilitate the engineer in his/her quest to answer the question "how" to maintain function and not necessarily the "why" role varies. The objective is to assist the engineer in the synthesis of the phenomenon rather than its analysis so that robust product/process functions are achieved. These methodologies enable the engineer to efficiently and rapidly obtain the technological capability required to keep an organization profitable and competitive in today's market.

Thus, for initiating any kind of quality improvement activities, the management support and favorable quality culture are vital. A robust quality culture can be built with a positive approach, even if organizational capabilities are not well-defined. In such situations, the management should prepare a quality manual or a dashboard that should be monitored and evaluated progressively. This dashboard can be an activity chart or checklists and should regularly remind the people of their duties and responsibilities. The organization can also frame a standard operating procedure (SOP) amenable to modifications and changes as per future requirements based on this quality checklist. The quality checklist should include:

- *Quality vision*: A quality vision is a general statement of what will become and how good the organization intends to be at the activity.
- *Quality strategy*: A quality strategy is usually expressed as part of the overall business's strategic plan. It includes the vision, information on what is essential to our customers, and how our customers view our quality.
- *Quality goals*: Quality goals are specific, measurable, and related to the quality of the goods and services provided by the organization. They may be set on several different bases like the needs and satisfaction of customers, cost of poor quality, etc.

- *Quality costs*: The costs associated with the creation of the overall quality plan, the inspection plan, the data systems, and quality assurance.
- *Quality council*: This refers to the steering committee needed to perform the functions required of upper managers jointly. Sometimes called part and parcel of the management.
- *Set priorities*: Quality improvement comes only through a project by project. So the most critical project must receive priority. Upper managers usually set these priorities in quality, just as they set priorities in other areas to maintain the progress in quality.
- *Resources*: The success of quality teams depends on the resources to do the work. The most critical resources include training, time for meetings, data collection, and its analysis, facilitator support, etc.
- *Maintaining the progress*: Establish cross-functional teams: The essential quality improvements almost always require cross-functional teams. Planning the quality of new or replacement goods and services also require cross-functional teams. These teams need to follow formal, mandatory, and structured processes to maintain progress.
- *Quality culture and attitude*: The information on the internal culture can help upper managers identify the strengths and weaknesses that they are facing. Such data can also help identify quality improvement opportunities that address employee needs and satisfaction. The leadership style you and the other members of the upper management team lead your organization will have a substantial impact. Quality culture and attitude can be changed through their actions and participation in different management activities.
- *Motivation*: Motivation can be brought into the system through recognition, promotion, and financial rewards. Recognition is a robust non-monetary incentive to adopt changes. Recognition must be in a form that is professionally valuable for the individual being recognized. Anybody who contributes to the quality should be financially rewarded also. If needed, a higher promotion in essential cases should be offered to the individual concerned.

Quality in a product or process is achieved through quality assurance and quality control. The former is considered to be the preventive method of delivering quality, and the latter is the detection of those negative characteristics which affect the quality of the product. Generally, both the methods are incorporated in an organization aimed at satisfying the customer. This is the objective of total quality management (TQM) as it involves the consideration of processes in all the major areas like marketing, design, procurement, operations, distributions, and invoicing. A sustainable quality improvement program also facilitates increased value creation, service enhancement, promotes innovative ideas, and trust among the stakeholders (see Chap. 8 for a detailed discussion on quality control and deployment tools).

Let us understand the concept of quality through some technical perspective.

## 4.2   Uncertainty and Probability Models

As discussed in Chap. 2, uncertainty can be associated with man, machine, methods, materials, management, measurement, and above all, environmental and climate (mother earth) factors. Uncertainty leads to variation, and variation leads to poor quality. So accounting for change is vital for any quality improvement. A measure of uncertainties can be well explained through the concept of *probability*. The chance of occurrence of an event associated with a random experiment, trial, or an investigation is called probability. Any action, whether drawing an item from a consignment, measurement of a product's dimension to ascertain quality, the launching of a new product in the market, deciding an outcome good or bad, rating a service, tossing a coin, tossing a dice, selection of an employee constitutes an experiment in the probability theory. Since each trial's outcome is unpredictable every time, we call such a *random experiment*. An experiment can be based on the whole items, say *population,* or a *sample* of subjects. However, sample experiment is preferred over the population study, because of its economic viability, sustainability, and sufficiency (refer to an excellent statistical book for understanding the reason behind it).

The set of all possible outcomes of an experiment is defined as the *sample space,* usually denoted by $S$. For instance, the set {success, failure} defines the sample space of a launching experiment of a satellite. Similarly, the set {1, 2, 3, 4, 5, 6} define the sample space of a dice (or die) tossing experiment. An *event* in probability theory constitutes one or more possible outcomes of an experiment. Thus, an event is a subset of the sample space, $S$. Suppose the event $A$ is equal to a number less than 5 in a single toss of a dice. Then the probability of an event $A$ denoted as $P(A)$ is simply the ratio of the number of favorable cases (here it is 1, 2, 3, 4) to the total number of possibilities. Therefore, $P(A) = 4/6$.

One of the aims of statistical science is to describe and predict real-world situations, which are generally subject to uncertainties, as described above. Probability models (or probability distributions) are often used to describe uncertainties about a situation. The probability models are constructed based on probability laws. And these laws are perceived through random experiments of cases. We are often interested in the behavior of specific quantities that take different values in different outcomes of the experiments. These quantities are called *random variables.* A random variable $X$ will be called *discrete* if the range of $X$ is countable. A random variable $X$ is said to be *continuous* if there exists a function $f_X(.)$ such that $F_X(x) = \int_{-\infty}^{x} f_X(u)\mathrm{d}u$ for every real number $x$.

A *probability distribution function* is a mathematical formula that relates the features' values with their probability of occurrence in the population. The collection of these probabilities is called a probability distribution. There are many discrete and continuous distributions explored in statistics to model various situations in life. Some of the most frequently used discrete probability distributions are the binomial distribution, Poisson distribution, geometric distribution, hypergeometric distribution, rectangular distribution, etc. And some of the continuous probability models

**Table 4.1** Summary of some frequently used probability distributions

| Distribution | Probability function | Application contexts |
|---|---|---|
| Binomial | $f(X = r) = \binom{n}{r} p^r q^{n-r}$, $r =$ <br> $0, 1, 2, \ldots, n$ <br> $n =$ number of trials <br> $r =$ Number of occurrences <br> $p = P(\text{success})$, $q = P(\text{failure})$, $p + q = 1$ | A Bernoulli trail (any experiment results in two outcomes: Success or Failure) is repeated for n times with constant probability of success in each independent trial |
| Poisson | $f(X = r) = \frac{e^{-\lambda}\lambda^r}{r!}$, $r =$ <br> $0, 1, 2, \ldots$ <br> $\lambda = np$ <br> $n =$ number of trials <br> $r =$ Number of occurrences <br> $p = P(\text{success})$ | A situation where the number of trials is enormous, but the probability of an event is tiny in each experiment. The experiment is performed similarly to the binomial distribution |
| Uniform or rectangular | $f(x) = \frac{1}{b-a}$, $a \leq x \leq b$ <br> $a$ and $b$ are some constants on the real line | A situation where probability is assigned uniformly throughout an interval. Provide theoretical support to other distributions |
| Exponential | $f(x) = \frac{1}{\theta}e^{-\frac{x}{\theta}}$, $x \geq 0$ <br> $\theta > 0 =$ Scale parameter | Applied in life testing and reliability studies where items/objects fail and constantly grow as time elapses |
| Weibull | $f(x) = \frac{\beta}{\theta}x^{\beta-1}e^{-\frac{x^\beta}{\theta}}$, $x \geq 0$ <br> $\theta > 0 =$ Scale parameter <br> $\beta > 0 =$ Shape parameter | They are potentially applied in life testing and reliability studies where items/objects deterioration and growth can be assessed as time elapses |

are exponential distribution, normal distribution, gamma distribution, Weibull distribution, triangular distribution, uniform distribution, etc. See Table 4.1 for the probability function of few distributions along with their application areas. A detailed discussion on developing a statistical distribution (model) is presented in the next section.

## 4.2.1 Probability Model Development

Suppose an unbiased coin is tossed once. Then the sample space contains two outcomes, viz. $S = \{\text{Head, Tail}\}$ or for simplicity, let us write $S = \{H, T\}$. That is, the probability of getting a head or tail is 0.5 each. Suppose the coin is tossed four times. Then the sample space contains $2^4 = 16$ outcomes as $S = \{HHHH, HHHT, HHTH, HHTT, HTHH, HTHT, HTTH, HTTT, THHH, THHT, THTH, THTT, TTHH, TTHT, TTTH, TTTT\}$. Let us compute some probabilities:

(i)  $P(\text{all heads}) = P(HHHH) = 1/16 = 0.0625$

(ii)  $P(\text{one head}) = P(HTTT \text{ or } THTT \text{ or } TTHT \text{ or } TTTH) = 4/16 = 0.25$

(iii)  $P(\text{Three heads}) = P(HHHT, THHH, HTHH, HTHH) = 4/16 = 0.25$

(iv)  $P(\text{at most three heads}) = P(\text{there is no head or one hear}$
$$\text{or two head or three head})$$
$$= 1 - P(\text{all four heads}) = 1 - P(HHHH)$$
$$= 15/16 = 0.9375$$

(v)  $P(\text{At east two heads}) = P(\text{two heads or three heads or four heads})$
$$= P(HHHH, HHHT, HHTH, HHTT, HTHH,$$
$$HTHT, HTTH, THHH, THHT, THTH, TTHH)$$
$$= 11/16 = 0.6875$$

and so on. Writing sample space and finding probabilities in this manner are not very easy if the coin is tossed a large number of times. To simplify this, we associate a random variable with the experiment of tossing a coin. Let $X$ is the random variable denoted as "*the number of heads*" in the experiment. Then $X$ takes values 0, 1, 2, 3, and 4, with the following outcomes and probabilities:

Table 4.2 is easy to interpret, and the computation of probabilities is much more comfortable in this case. The calculation of probabilities was also easy in this case because the outcomes are available without any difficulties here. Now, imagine the coin is tossed ten times. Then listing the results, as shown above, is not easy as there are about $2^{10} = 1024$ outcomes. To simplify the situation, we use the probability model expressed in some mathematical form to describe the experimental result.

Consider the coin-tossing experiment as described above and to generalize the situation. We know that a coin has two faces (or outcomes), where the probability of occurrence of each face is 1/2. As defined earlier, let $X$ be the random variable

**Table 4.2**  Distribution of four tosses of an unbiased coin

| $X$ | Outcomes | Probability, $P(X = x)$ | $F(x) = P(X \leq x)$ |
|---|---|---|---|
| 0 | *TTTT* | 0.0625 | 0.0625 |
| 1 | *HTTT, THTT, TTHT, TTTH* | 0.2500 | 0.3125 |
| 2 | *HHTT, HTHT, HTTH, THHT, THTH, TTHH* | 0.3750 | 0.6875 |
| 3 | *HHHT, HHTH, HTHH, THHH* | 0.2500 | 0.9375 |
| 4 | *HHHH* | 0.0625 | 1.0000 |

denoting the "number of heads." Assume that this coin is tossed "$n$" times independently, with the probability of success being $P$(number of heads). Call this probability as saying, $p$, such that $P(\text{Failure}) = q = 1 - p$. If $p$ is constant throughout the trial (or toss), then the random variable $X$ follows a binomial distribution with probability density (or mass) function as

$$f(X = r) = \binom{n}{r} p^r (1 - p)^{n-r}$$

$$= \frac{n!}{r!(n - r)!} p^r (1 - p)^{n-r},$$

$$r = 0, 1, 2, \ldots, n, 0 < p < 1, p + q = 1 \tag{4.1}$$

As mentioned in Table 4.1, a binomial distribution is suitable for modeling situations involving any two outcomes: success or failure; true or false; male or female; defective or non-defective; presence or absence; and so on. Any experiment results in two outcomes are called the Bernoulli trail. Hence, a binomial distribution is an extension of Bernoulli distribution repeated for $n$ times with a constant probability of success in each independent trial. The graph of the distribution for various values of $n$ and $p$ is shown in Fig. 4.2.

The mean and variance of binomial disruptions are, respectively, $np$ and $np(1 - p)$. The cumulative probability function is computed as

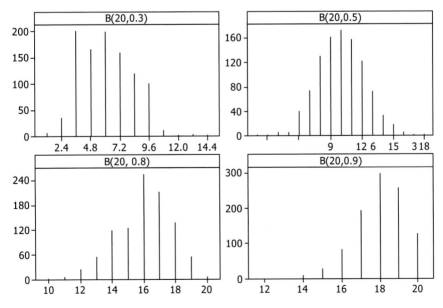

**Fig. 4.2** Graph of binomial distribution

$$P(X \le j) = \sum_{x=0}^{j} \frac{n!}{x!(n-x)!} p^x (1-p)^{n-x} \tag{4.2}$$

Appendix Table 1 provides the values of (4.2) for various combinations of $n$ and $p$. Now it is easy to answer the questions raised in the coin-tossing experiment discussed above. As per the binomial distribution language, we have $n = 4$ and $p = 0.5$. Then

(i) $P(\text{all heads}) = P(X = 4)$

$$= \frac{4!}{4!(4-4)!}(0.5)^4(1-0.5)^{4-4}$$

$$= 0.0625$$

(ii) $P(\text{one head}) = P(X = 1)$

$$= \frac{4!}{1!(4-1)!}(0.5)^1(1-0.5)^{4-1}$$

$$= 0.25$$

(iii) $P(\text{three heads}) = P(X = 3)$

$$= \frac{4!}{3!(4-3)!}(0.5)^3(1-0.5)^{4-3}$$

$$= 0.25$$

(iv) $P(\text{at most three heads}) = P(X \le 3)$

$$= \sum_{x=1}^{3} \frac{n!}{x!(n-x)!} p^x (1-p)^{n-x}$$

$$= 1 - P(X = 4)$$

$$= 1 - \frac{4!}{4!(4-4)!}(0.5)^4(0.5)^{4-4}$$

$$= 0.9375$$

(v) $P(\text{at least two heads}) = P(X \ge 2)$

$$= \sum_{x=2}^{4} \frac{n!}{x!(n-x)!} p^x (1-p)^{n-x}$$

$$= \frac{4!}{2!(n-2)!}(0.5)^2(0.5)^{4-2} + \frac{4!}{3!(4-3)!}(0.5)^3(0.5)^{4-3}$$

$$+ \frac{4!}{4!(4-4)!}(0.5)^4(0.5)^{4-4}$$

$$= 0.6875$$

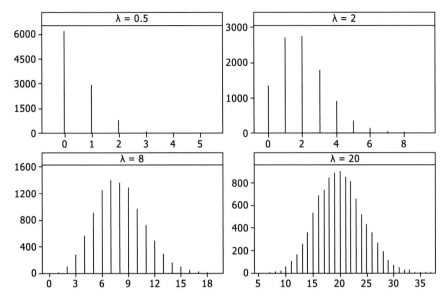

**Fig. 4.3**  Graph of Poisson distribution

Note that all probability estimates match with the calculation done earlier. The advantage here is that we do not need to specify the actual outcomes as done early. This way of finding estimates for probability can be made even simple by referring to the values provided in Appendix Table 1.

Now, suppose the experimental trail $n$ is very large (i.e., $n \to \infty$) in the case of binomial distribution, and the probability of occurrence is minimal (i.e., $p \to 0$), then the binomial distribution can be approximated to a Poisson distribution with mean $\lambda = np$. A Poisson distribution has the probability mass function:

$$f(X = r) = \frac{e^{-\lambda}\lambda^r}{r!}, \quad r = 0, 1, 2, \ldots, \infty \tag{4.3}$$

The mean and variance of this distribution are the same and is equal to $\lambda$. The cumulative probability function is computed as

$$P(X \leq j) = \sum_{x=0}^{j} \frac{e^{-\lambda}\lambda^r}{r!} \tag{4.4}$$

For the various value of $\lambda$, the values correspond to (4.4) are presented in Appendix Table 2. The graph of probability mass function for different values of $\lambda$ is shown in Fig. 4.3. If $\lambda < 1$, the graph of the probability function decreases steadily, whereas if If $\lambda > 1$, the chart increases steadily to the value at the mode.

Let us consider the same coin-tossing experiment discussed previously. We have $n = 4$ and $p = 0.5$. Although $n$ is not very large and $p$ is not very small, we still attempt to find the probabilities asked in the question using a Poisson distribution approximation with a parameter, $\lambda = np = 2$. Hence,

(i) $P(\text{all heads}) = P(X = 4)$

$$= \frac{e^{-2}2^4}{4!}$$

$$= 0.0902$$

(ii) $P(\text{one head}) = P(X = 1)$

$$= \frac{e^{-2}2^1}{1!}$$

$$= 0.2706$$

(iii) $P(\text{Three heads}) = P(X = 3)$

$$= \frac{e^{-2}2^3}{3!}$$

$$= 0.1804$$

(iv) $P(\text{at most three heads}) = P(X \leq 3)$

$$= \sum_{x=1}^{3} \frac{n!}{x!(n-x)!} p^x (1-p)^{n-x}$$

$$= 1 - P(X = 4)$$

$$= 1 - \frac{e^{-2}2^4}{4!}$$

$$= 0.908$$

(v) $P(\text{at least two heads}) = P(X \geq 2)$

$$= \sum_{x=2}^{4} \frac{e^{-2}2^x}{x!}$$

$$= \frac{e^{-2}2^2}{2!} + \frac{e^{-2}2^3}{3!} + \frac{e^{-2}2^4}{4!}$$

$$= 0.5413$$

The values are slightly different because the Poisson assumption does not hold here correctly. At least the amount of $n$ should be huge to keep the assumption irrespective of the value of $\lambda$. However, the application of Poisson distribution is plenty and is used for modeling rare events whose probability of occurrence is minimal. Some examples of random variables that usually obey Poisson law are:

- Number of errors per 1000 invoices
- Number of surface defects in an iron casting
- Number of faults of insulation in a specified length of cable
- Number of visual defects in a bolt of cloth
- Number of absenteeism in a specified time
- Number of accidental death happens in a city per day
- Number of breakdowns of a computer per month
- PPM of toxicant found in water or air emission from a manufacturing plant, etc.

**Example 4.1** An acceptance sampling plan is carried out to assess the quality of the manufactured product of iron casting. On average, 5% of defective castings are found to be in every batch of items selected for inspection. What is the probability that a packet chosen at random contains, (i) no defective, (ii) 2 or more defective, (iii) less than five defective?

*Solution.* According to the information given $\lambda = 0.05$. Then

(i)  $P(\text{at least two heads}) = P(X \geq 2)$

$$= \sum_{x=2}^{4} \frac{e^{-2}2^x}{x!}$$

$$= \frac{e^{-2}2^2}{2!} + \frac{e^{-2}2^3}{3!} + \frac{e^{-2}2^4}{4!}$$

$$= 0.5413$$

(ii)  $P(2 \text{ or more defective}) = P(X \geq 2)$

$$= 1 - [P(X = 0) + P(X = 1)]$$

$$= 1 - \left[ \frac{e^{-0.05}0.05^0}{0!} + \frac{e^{-0.05}0.05^1}{1!} \right]$$

$$= 1 - [0.2865 + 0.3581]$$

$$= 0.0012$$

(iii)  $P(\text{less than 5 defective}) = P(X < 5)$

$$= P(X = 0) + P(X = 1) + P(X = 2)$$

$$+ P(X = 3) + P(X = 4)$$

$$= 0.9999$$

## 4.2.2 Normal Distribution

The use of *normal distribution* in a scientific study is immense. This distribution is used in both theory and applied sciences. In Chap. 2, we have initiated the study of normal distribution descriptively. Policy decision-makers always look for data being normal and address their quality issues based on the normal distribution. Normal distribution was first discovered by De-Moivre in 1733 and was also known to Laplace in 1774. Later it was derived by Karl Friedrich Gauss in 1809 and used it for the study of errors in astronomy, and hence, the distribution is often called Gaussian distribution. The normal distribution is often characterized by two parameters called mean $(\mu)$ and variance $(\sigma^2)$, and hence, it is universally denoted as $N(\mu, \sigma^2)$. Mathematically, it is expressed as:

$$f(x) = \frac{1}{\sqrt{2\pi}\sigma} e^{-\frac{1}{2\sigma^2}(x-\mu)^2}, \quad -\infty < x < \infty, \ -\infty < \mu < \infty, \ \sigma > 0 \qquad (4.5)$$

Here the total area bounded by the curve (4.5) is one. The value of $\pi = 2.14159$. Hence, the area under the curve between two ordinates $X = a$ and $X = b$, where $a < b$ represents the probability that $X$ lies between $a$ and $b$. This is denoted by $P(a < X < b)$. Since $X$'s value can be anything between $-\infty$ and $\infty$, finding the probability of any such variable will require numerical integration over time. This is simplified by using the standardized distribution of $X$ by using the transformation, say, $= \frac{x-\mu}{\sigma}$. In such a case, we say that $Z$ is a standard normal variate or $Z$ is *normally distributed with mean zero and variance one* and is denoted as $N(0, 1)$. As an illustration, suppose $X \sim N(50, 25)$, then the standardized variable $Z \sim N(0, 1)$. The graph corresponds to $X$, and $Z$ is shown in Fig. 4.4.

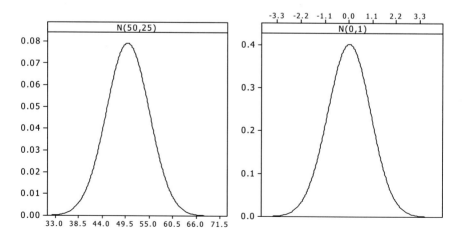

**Fig. 4.4** Graph of normal distributions $X$ and $Z$

The standardization simplifies the computation of probabilities to a great extent. The standardized values are computed and presented in Appendix Table 3. This table values can be used for finding the area between any two ordinates by using the symmetry of the curve about $z = 0$. This is explained through a few examples below.

***Example 4.2*** Suppose the monthly expenditure of a student staying in a hostel is normally distributed with average expenses as Rs 5500 and a standard deviation equal to Rs 500. What percentage of the time the student spent (i) less than Rs 5000, (ii) more than Rs 7000, (iii) between Rs 4500 to Rs 6000. Suppose the student is very cautious about saving some money for unexpected expenditures every month, for which he limits his expenses to seventy percent of the money he receives. What was the maximum he spends under this condition?

*Solution*: Given $X \sim N(5500, 25000)$.

$$\text{To find } P(X < 5000) = P\left(\frac{X - 5500}{500} < \frac{5000 - 5500}{500}\right)$$

(i)
$$= P(Z < -1)$$
$$= 0.1587$$

That is roughly 16% of the time the student spends less than Rs 5000 monthly. The graphical presentation of the area is presented in Fig. 4.5.

(ii) To find $P(X > 7000) = P\left(\frac{X - 5500}{500} > \frac{7000 - 5500}{500}\right)$
$$= P(Z > 3)$$
$$= 0.0013$$

That is hardly zero expenses go beyond Rs 7000. The diagrammatic presentation is shown in Fig. 4.6.

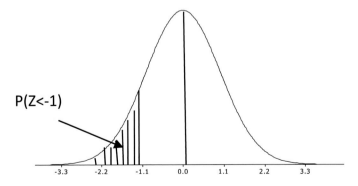

Fig. 4.5  The area correspond to $P(Z < -1)$

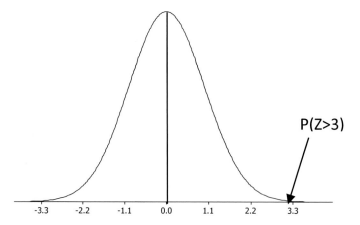

**Fig. 4.6** The area correspond to $P(Z > 3)$

(iii) To find $P(4500 < X < 60000) = P\left(\dfrac{4500 - 5500}{500} < \dfrac{X - 5500}{500} < \dfrac{6000 - 5500}{500}\right)$

$$= P(-2 < Z < 1)$$
$$= 0.3641$$

That is roughly 36% time; the expense goes between Rs 4500 and Rs 6000. The diagrammatic presentation of the area covered is shown in Fig. 4.7.

(iv) Let the maximum expenses be $x_M$, So according to the given facts, we have been given that $P(X < x_M) = 0.70$, which is equivalent to $P\left(\dfrac{X-5500}{500} < \dfrac{x_M-5500}{500}\right) = 0.70$, which is equal to $P(Z < Z_1) = 0.70$, where

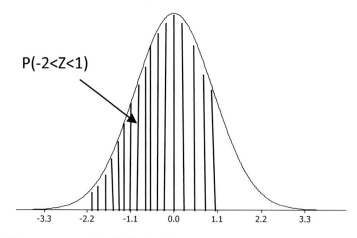

**Fig. 4.7** The area correspond to $P(-2 < Z < 1)$

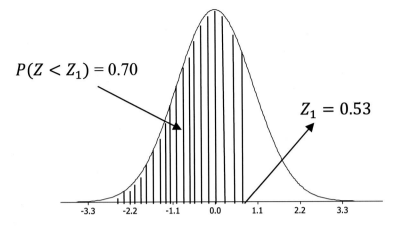

**Fig. 4.8** The area correspond to $P(Z1) = 0.70$

$Z_1 = \frac{x_M - 5500}{500}$. Figure 4.8 presents the situation. Using the standard normal table, we can find the value of $Z_1$ as 0.53. Solving, we get $x_M = 5765$. So if the maximum expense is kept at Rs 5900, the student can save enough money for his future requirements.

Note that normal distribution is the basis of all significant sample inference in statistical studies. The normal distribution is the foundation of parametric inferences, confidence interval, regression analysis, and control chart construction. Another important reason for using normal distribution is due to its *area property*, which is stated as follows: If the population is normally distributed, then approximately (i) 68.26% of the entire values lie within $\mu \pm \sigma$ limits, (ii) 95.44% of all the values lie within $\mu \pm 2\sigma$ boundaries, and (iii) 99.73% of all the values lie within $\mu \pm 3\sigma$ limits. To some extent, this property justifies the notion that normal distribution can be used as a qualified distribution. This is described in Fig. 4.9.

It is believed that any process can run as per normal distribution, provided there is a favorable climate of executing an organizational activity. However, because of the complexity of the process and products and the environment we live in, things will not work as per our wishes and choices every time. So normality may remain as only an assumption, rather than a reality. Despite all these challenges, normal destruction remains a quality distribution because of its natural evolution. At various places in this book, this is illustrated. To begin with, we consider an example to demonstrate the area property of normal distribution.

***Example 4.3*** The quality control department of an oven manufacture is monitoring the amount of radiation (in mW/cm$^2$) emitted when the doors of the ovens are closed. Observations of the radiation emitted through closed doors of 80 randomly selected ovens were as follows: Discuss the area property of the data.

| 0.2489 | 0.2497 | 0.2296 | 0.2428 | 0.2974 | 0.2088 | 0.2296 | 0.2139 | 0.2413 | 0.2831 |
| 0.2538 | 0.2830 | 0.2907 | 0.2944 | 0.2380 | 0.2499 | 0.2257 | 0.2750 | 0.2772 | 0.2175 |
| 0.1707 | 0.2248 | 0.1376 | 0.2680 | 0.3216 | 0.2899 | 0.2333 | 0.3074 | 0.2809 | 0.2693 |
| 0.1976 | 0.2832 | 0.2584 | 0.1939 | 0.2019 | 0.2151 | 0.2714 | 0.2168 | 0.1582 | 0.2257 |
| 0.3699 | 0.2811 | 0.2448 | 0.2604 | 0.2437 | 0.2825 | 0.2546 | 0.2839 | 0.2581 | 0.2062 |
| 0.2908 | 0.3096 | 0.2267 | 0.1923 | 0.1654 | 0.2694 | 0.2634 | 0.3213 | 0.2519 | 0.2880 |
| 0.2217 | 0.3047 | 0.2651 | 0.2619 | 0.1968 | 0.2691 | 0.1407 | 0.2048 | 0.1400 | 0.2183 |
| 0.1842 | 0.2548 | 0.1453 | 0.1562 | 0.2369 | 0.3304 | 0.2341 | 0.3091 | 0.2654 | 0.3037 |

*Solution.* The graphical summary of the data is shown in Fig. 4.10. The graph panel includes a histogram with a normal plot, box plot, and confidence interval plots (a detailed discussion on these plots and their use are presented in Chap. 8). The descriptive statistics summary is shown on the right-side panel. The mean and standard deviation of the data are 0.246 and 0.048, respectively. Similarly, the $1\sigma$, $2\sigma$, and $3\sigma$ limits are, respectively, obtained as $(0.198, 0.294)$, $(0.150, 0.342)$, and $(-0.21072, 0.390)$. These intervals include 75%, 94%, and 100% observations, respectively, as

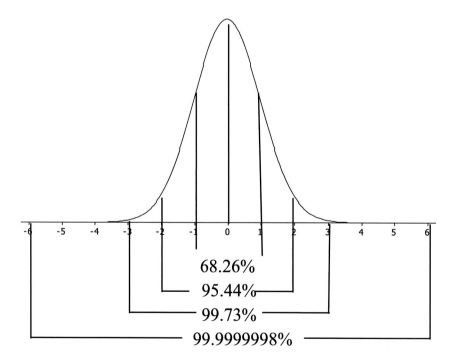

**Fig. 4.9**  Area under normal curve

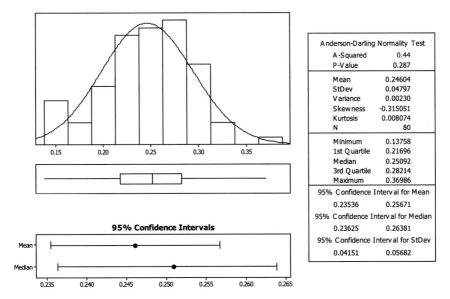

| Anderson-Darling Normality Test | |
|---|---|
| A-Squared | 0.44 |
| P-Value | 0.287 |
| Mean | 0.24604 |
| StDev | 0.04797 |
| Variance | 0.00230 |
| Skewness | -0.315051 |
| Kurtosis | 0.008074 |
| N | 80 |
| Minimum | 0.13758 |
| 1st Quartile | 0.21696 |
| Median | 0.25092 |
| 3rd Quartile | 0.28214 |
| Maximum | 0.36986 |
| 95% Confidence Interval for Mean | |
| 0.23536 | 0.25671 |
| 95% Confidence Interval for Median | |
| 0.23625 | 0.26381 |
| 95% Confidence Interval for StDev | |
| 0.04151 | 0.05682 |

**Fig. 4.10** Graphical summary of the amount of radiation

the given data follows a normal distribution (see $p$-value $= 0.287$). Therefore, we can conclude that the data is consistent with controlled variation and hence has quality.

### 4.2.3 Central Limit Theorem Explained

Like area property, there are several other important properties for normal distribution. For instance, most of the discrete distributions tend to normal as $n$ (sample size) increases ($n \to \infty$). Many sampling distributions ($x^2, t, F$) can be obtained from a normal distribution with appropriate transformations (Readers are advised to consult some good books on statistical methods to understand these theoretical concepts). One of the essential motivations behind the extensive use and application of normal distribution lies in the most celebrated theorem in the statistical study called the *Central Limit Theorem* (CLT). The theory is stated as follows:

*Statement.* If $x_1, x_2, \ldots, x_n$ are $n$ random variables from any population which is independent and have the same distribution with mean $\mu$ and standard deviation $\sigma$, then as $n \to \infty$, the sampling distribution of the mean ($\bar{x}$) is approximately a normal distribution with a mean of $\mu$ and a standard deviation of $\sigma/\sqrt{n}$ (standard error of the mean). Technically, the CLT is written as:

$$\frac{\bar{x} - \mu}{\sigma/\sqrt{n}} \to N(o, 1) \tag{4.6}$$

**Table 4.3**  Binomial samples

| Sample numbers | $x_1$ | $x_2$ | $x_3$ | $x_4$ | $x_5$ | Mean |
|---|---|---|---|---|---|---|
| 1 | 7 | 5 | 8 | 9 | 8 | 7.4 |
| 2 | 8 | 8 | 8 | 6 | 9 | 7.8 |
| 3 | 8 | 7 | 10 | 8 | 10 | 8.6 |
| ... | ... | ... | ... | ... | ... | ... |
| ... | ... | ... | ... | ... | ... | ... |
| ... | ... | ... | ... | ... | ... | ... |
| 300 | 8 | 6 | 9 | 10 | 7 | 8 |

Thus, the CLT states that under rather general conditions, sums and means of random measurements drawn from any population tend to possess, approximately, a bell-shaped distribution in repeated sampling. If the sample size is large, then the convergence to normal is fast. The concept of a large and small sample is a tricky statement. However, in statistical studies, a sample size of more than 30 is generally considered significant, and anything less than 30 is a small sample. Since many of the estimators used to make inferences about the characteristics of a population are sums or means of sample measurements, we can expect the estimator to be approximately normally distributed in repeated sampling, when $n$ is sufficiently large. For normally distributed random variables, the means of the sample will be normal again, which can be proved theoretically. Let us illustrate the working principle of CLT thorough a couple of non-normal samples.

The first illustration is based on the binomial distribution. Here we generate 300 random samples of size five from a binomial distribution with parameter $n = 10$ and $p = 0.7$. The random numbers can be generated through any statistical software. See Table 4.3 for the summary of samples and the mean of each of the five samples in the last column. Note that the mean of the population (here binomial distribution) is $np = \mu = 7$, variance $= npq = \sigma^2 = 2.1$, and standard deviation is 1.4491.

Figure 4.11 is the plot of the histogram of the sample statistic $\bar{x}$. Note that the histogram is a normal distribution with a mean of 7.467 ($\cong \mu$) and a standard deviation of 0.6101 ($\cong \sigma/\sqrt{5}$). The error in approximation can be nullified if the sample size is very large.

The second example is based on a continuous skewed distribution. We generate 300 samples, each of size five from an exponential distribution with a mean 10 ($= \mu$). The standard deviation of this distribution is 10 ($= \sigma$). A portion of the data is shown in Table 4.4. The plot of the histogram of the sample statistic $\bar{x}$ is shown in Fig. 4.12. Note that the histogram shows a normal distribution with a mean of 10.37 ($\cong \mu$) and a standard deviation of 4.738 ($\cong \sigma/\sqrt{5}$). The error in approximation can be nullified if the sample size is very large.

From a practical context, one can visualize that, if a process runs according to a normal distribution, roughly one out of one hundred units will not meet the specifications (see $3\sigma$ limits and area property discussed earlier). This is as good as saying that the process is $3\sigma$-capable from a quality control perspective. The items which do not

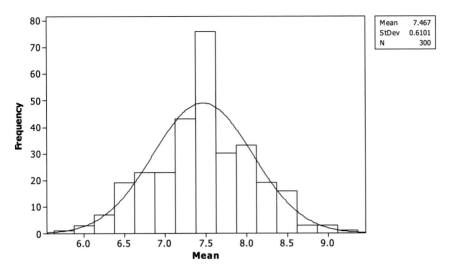

**Fig. 4.11**  Plot of sample statistics from binomial distribution

**Table 4.4**  Exponential samples

| Sample number | $x_1$ | $x_2$ | $x_3$ | $x_4$ | $x_5$ | Mean |
|---|---|---|---|---|---|---|
| 1 | 1.0131 | 5.8021 | 21.0653 | 2.3401 | 2.1462 | 6.4734 |
| 2 | 8.5427 | 9.2516 | 3.8966 | 11.6292 | 33.6977 | 13.4036 |
| 3 | 8.5933 | 12.1314 | 17.0683 | 31.2138 | 43.5807 | 22.5175 |
| 4 | 1.1507 | 15.8512 | 6.0318 | 9.8768 | 3.8799 | 7.3581 |
| 5 | 1.0245 | 17.5683 | 16.4670 | 26.8966 | 18.2392 | 16.0391 |
| 6 | 10.6979 | 0.5381 | 41.2135 | 39.4346 | 4.2977 | 19.2364 |
| 7 | 3.9989 | 6.7427 | 1.2275 | 11.4231 | 22.7300 | 9.2244 |
| 8 | 5.2517 | 4.0743 | 1.7940 | 14.5044 | 1.9700 | 5.5189 |
| 9 | 7.3321 | 3.0222 | 12.8174 | 12.1748 | 17.4551 | 10.5603 |
| 10 | 16.1328 | 6.3708 | 7.9816 | 8.8678 | 6.4430 | 9.1592 |
| … | … | … | … | … | … | … |
| … | … | … | … | … | … | … |
| 300 | 1.3989 | 8.4798 | 32.3122 | 14.1604 | 0.7094 | 11.4122 |

meet specifications are called defectives, affecting the quality of the product. More
the defective, less the quality is. So normal distribution represents an ideal situation
for any transaction, service, manufacturing, production, or any such kind of process
activities. In this situation, the managers or process owners will be able to realize
a reasonably good quality in their process. This is why normal distribution became
the manager's special, and they base all their decision to improve the organizational
activities. Compromise on this level of quality will be not advisable and, of course,

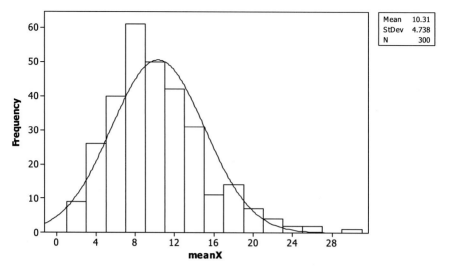

**Fig. 4.12**  Plot of sample statistics from an exponential distribution

a matter of concern for both producers and customers. With the advent of scientific innovations and technological progress, the customer's perception of quality has also changed enormously. They want everything in perfection. This paves the way for quality excellence in every aspect of the organization (Muralidharan, 2015).

Quality conscious products, value-based services, and customer-focused products are the norm of the day. Hence, the demand for lean, green, and clean products is on the rise. Lean manufacturing, green supply chains, Six Sigma quality programs, and many such innovative programs are now on-demand and are available in every producer's menu of the offering. For advanced quality planning, knowledge of statistical tools is equally important than the management tools for suggesting a planning and control plan. Keeping these points on priority, in various sections below, we discuss several quality improvement tools of LGC process management.

## 4.3   Total Quality Management

Total quality management (TQM) or continuous process improvement is defined as a management approach to an organization centered on quality, based on the participation of all its members and aiming at long-term success through customer satisfaction and benefits to the members of the organization and society. The core objective is that the customer's requirements must be first identified, defined, and classified. The philosophy entails that the procedures and systems are established to monitor, control, and improve the organization's activities. It does not matter whether; those input variables are directly or indirectly involved in the production of uniform products and the delivery of consistent services.

Joseph Jablonski, the author of *Implementing TQM,* identified three characteristics necessary for TQM to succeed in an organization: participative management, continuous process improvement, and team utilization. Participative management refers to the intimate involvement of all company members in the management process, thus deemphasizing traditional top-down management methods. In other words, managers set policies and make critical decisions only with the subordinates' input and guidance, which will have to implement and adhere to the directives. This technique improves upper management's grasp of operations and, more importantly, is an essential motivator for workers who begin to feel as if they have control and ownership of the process in which they participate. The working philosophy of TQM is the *Plan-Do-Study-Act* (PDSA) or *Plan-Do-Check-Act* (PDCA). TQM can also be best described with the 5Ps as *People–Process–Product–Planning–Performance* (see also Fig. 4.13).

To bring synergy between processes, it is necessary to have project planning to establish overall communication. Project planning becomes the tracking system to ensure that all the elements, tools, processes, and interactions are brought together as a whole system for doing the work that is necessary for the workplace. An effective project planning requires skills in the following areas (Kerzner, 1995):

- Information processing
- Communication
- Resource negotiations
- Securing commitments
- Incremental and modular planning
- Assuring measurable milestones
- Facilitating top management involvement.

**Fig. 4.13** TQM process diagram

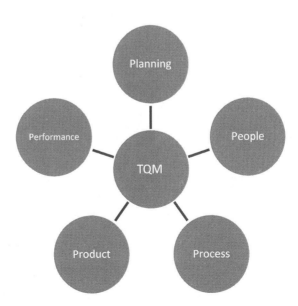

Continuous process improvement, the second characteristic of TQM, entails recognizing small, incremental gains toward the goal of total quality. Significant benefits are accomplished by minor sustainable improvements over the long term. This concept necessitates a long-term approach by managers and the willingness to invest in the present for benefits that manifest themselves in the future (Muralidharan & Syamsundar, 2012). A result of continuous improvement is that workers and management develop an appreciation for, and confidence in, TQM over time. TQM also supports cultural interventions of product variety and service delivery along with the empowerment of people associated with a quality process (Jabnoun, 2002). The central cultural values that underlie empowerment, customer satisfaction, and continuous improvement are innovation and challenge (Zeitz et al., 1997), Openness (Steyn, 1999), humbleness (Gupta, 1996), respect for people (Seiling, 1999), integrity (Goetsh & Davis, 2000), trust (Axline, 1991), empathy (Parasuraman et al., 1985), and cooperation (Oakland, 1997). These are precisely the nuances of lean, green, and clean philosophies.

TQM is an organization-wide commitment to getting things right. TQM is founded upon quality control, inspection, and statistical process control. Since TQM is a process-oriented approach to quality control, it is also called total quality control (TQC). Both are instrumental in increasing communication and culture in the organization. It is always desirable for a company to have developed some control and quality assurance systems for overall performance by focusing on quality. Commitment is the foundation of an effective TQM program. Without proper obligation, even the best plan may not work. In some companies, top management fails to understand the level of commitment required for an effective TQM program. The responsibilities of leadership can be summarized as follows:

- Search for challenging opportunities to change, grow, innovate, and improve
- Experiment, take the risk, and learn from the mistakes
- Have a great vision for the future
- Make others participate in the vision plan by convincing them
- Trust others by sharing all relevant information and delegate authority wherever possible
- Plan for small successes that can promote constant progress and bring about commitment
- Recognize individual contributions to the success of every project
- Celebrate team accomplishments and success regularly.

One of the best strategies to improve the process is through problem-solving. Stamaitis (1996) provides the eight disciplined approach to problem-solving as

- Use a team approach
- Describe the problem
- Start and check interim actions
- Define and check root causes
- Check corrective actions
- Start permanent corrective action

- Stop future problems
- Congratulate your team.

The project personals can use various charts and diagrams to identify and classify the causes affecting improvement efforts. For monitoring an ongoing process, there are many quality control and improvement tools available in Management and Statistics literature. Some of them are histogram, Pareto chart, cause-and-effect diagram, process flow diagram, scatter diagram, control charts, failure mode effect analysis (FMEA), etc. (see Chap. 8 for a detailed discussion on these aspects). Pareto chart organizes data to show which items or issues have the most significant impact on the process, system, or service. It is a pictorial representation of the data stratified according to the largest to the lowest number of items according to the frequency of occurrences in each group. This works on the principle of the Pareto principle and is also called 80:20 rules. That is, 80% of the problems or effect(s) are due to 20% of the causes. This chart's objective is to develop a plan to work on issues that will give us the most significant return for process improvement efforts. Once a Pareto chart has been prepared, the office worker and team can start developing plans and actions for improving the process, beginning with the most frequently occurring issue or problem. As that problem is improved, the team can go to the next most frequently recurring problem and work on that, and so on.

Another broad approach to improvement is the *Theory of Constraints*. A constraint is the weakest link in a system (process) and should be the focus of improvement. Constraints are mostly policy (procedures, past practices, etc.) matters but may also be physical (machines, people, and other resources). Theory of Constraints (TOC) identifies the constraints and analyzes it to ensure that all parts are aligned and adjusted to support the maximum effectiveness of the constraint (Goldratt, 1977). For a Six Sigma project, the most efficient way of formulating the goal statement is done through TOC. The marketing professionals also use TOC to develop marketing solutions once the constraint for improvement is identified. Below we discuss similar improvement tools, which can encourage organizational leaders to promote the LGC concepts toward a better environment and living.

## 4.4  Total Productive Maintenance

Total Productive Maintenance (TPM) is the way of achieving the goal of TQM (Nakajima, 1988). It is a system of maintaining and improving the integrity of production and quality systems through the machines, equipment, processes, and employees that add business value to the organization. The TPM program's goal is to increase production markedly and to drive all waste to zero: zero accidents, zero defects, zero breakdowns, zero absenteeism, etc. Maintainability is one of the critical factors which lead to the lowering of maintenance costs. Maintainability is a design criterion. It defines the ease with which systems can be maintained economically and efficiently when they fail or are about to fail. This, in turn, decides the life-cycle cost of the systems, affecting the industry's ability, in which the systems are employed, to remain competitive in today's market. Therefore, designers must take maintainability into account while designing the systems.

There are various methods of maintenance. *Preventive maintenance* (is the work carried out on the system on time or age-based cycle before a system failure occurs) and *corrective maintenance* (is the work carried out on a scheme to restore its functionality on failure) are the two essential activities carried out to assess the deterioration of the machines and diagnose the equipment condition to prevent all types of variations and wastes. There is another type of maintenance called *condition-based maintenance*, where the maintenance work is carried out on the system based on need after monitoring and assessment of system condition. This can be found in an inspection work wherein the system is physically checked to assess its situation or based on the information generated by various condition monitoring indicators mounted on it (Muralidharan & Syamsundar, 2012).

If TQM attempts to increase the quality of goods, services, and accompanying customer satisfaction by raising awareness of quality concerns across the organization (Wienclaw, 2008), then TPM focuses on the equipment used to produce the products by preventing equipment breakdown, improving the quality of the equipment and by standardizing the equipment. Promotion of TPM enhances the volume of production, employee morale, and job satisfaction (Prabhuswamy et al., 2013). It is a fact that all systems deteriorate over some time and fail to perform their intended functions due to the wear and tear that they undergo. Once a system fails to perform its intended function, whether partially or fully, as per its design specification, it requires to be maintained. Hence, TPM's main objectives are to increase the overall equipment effectiveness (OEE) of plant equipment. Where OEE is a combined measurement that shows the impact of equipment availability, equipment performance, and quality of output. The metric is calculated by multiplying *Availability * Performance * Quality*. For instance, if the availability is 98%, performance is 95%, and quality is 96%, then OEE is $0.98 \times 0.95 \times 0.96 = 89.37\%$.

The other service-intensive quality improvement programs are

- *Total Service Management*: High-performance service management aims to optimize the service-intensive supply chains, which are usually more complex than the typical finished-goods supply chain. This tool helps management to optimize inventory, avert overstocking, or stock out situations and also improves the right quantity at the right time through integration of Kanban (continuous improvement) and Poka-Yoke (mistake-proofing) with service management. The benefits include: high service costs can be reduced, inventory levels of service parts can be reduced, customer service or parts/service quality can be optimized, increase service revenue, and above all, improve customer satisfaction levels.
- *Total Human Resource Management*: This is a people-focused management system that aims to increase customer satisfaction at a continually lower cost. Human resource management is the strategic and coherent approach to managing an organization's most valued assets; the people working there who individually and collectively contribute to the achievement of the business's objectives.

- *Total Flow Management*: The idea of total flow management is to achieve excellence with continuous improvement in lean supply chains. It creates flows between production, internal logistics, creating external logistics, and finally to supply chain flow design (Coimbra, 2009).

Rampersad and El-Homsi (2008) combine the power of Lean Six Sigma within an approach that stresses the importance and need of developing an organizational culture and philosophy that combines the goal of and aspirations of the individual with those of the company and call that approach as Total Performance Scorecard (TPS). It links the driving force and motivation of individuals within the organization with the organizational vision, mission, and critical success factors to achieve a competitive advantage. According to the authors, TPS is an all-encompassing management approach that leverages corporate associates' motivation and capabilities and applies them to the realization of the highest corporate objectives.

According to Armand Feigenbaum, an organization should set their priorities as per the customer's call. For this, he advocates that organization should be excellence-driven rather than the defect drove. According to this version, quality is an expected characteristic than a desired one. Therefore, quality management should escalate to processes along with products and services to realize this. This is the concept of total quality control (TQC). It is a system for overall improvement, as it covers the entire company-wide operating work structure. Further, TQC guarantees operating and financial benefits, if processes are intended to perform its life cycle thoroughly. It also promotes the idea of practicing quality at the source.

## 4.5 Kaizen

Here we discuss another beautiful philosophy, which is considered to be an improvement tool covering the essentials of total quality management (TQM) and total quality control (TQC). Kaizen or continuous improvement is the foundation of Japanese management. The concept was introduced by Masaaki Imai of Japan and has received worldwide fame and acceptability in improving ongoing products and processes. Kaizen is a combination of two Japanese words *kai* (change) and *Zen* (good). The Kaizen philosophy assumes that our way of life–be it our working life, our social life, or our home life deserves to be continuously improved. It involves everyone, including both managers and workers, for ongoing improvement. The concept has helped Japanese companies generate a process-oriented way of thinking and develop strategies that assure continuous improvement involving people at all levels of the organizational hierarchy. The message of the Kaizen strategy is that not a day should go by without some kind of improvement being made somewhere in the company (Muralidharan & Syamsundar, 2012).

Kaizen is an umbrella concept covering those "uniquely Japanese" practices like customer orientation, QC circles, suggestion system, automation, discipline in the workplace, TPM, just-in-time, zero defects, small-group activities, cooperative labor-management relations, and new product development. According to Kaizen philosophy, management has two major components to work on continuously: maintenance and improvement. Maintenance refers to activities directed toward maintaining current technological, managerial, and operating standards; improvement refers to those directed toward improving current standards. Under its maintenance functions, management performs its assigned tasks so that everybody in the company can follow the established *standard operating procedure* (SOP). This means that management must first create policies, rules, directives, and procedures for all primary operations and then see to it that everybody follows the procedures. If people can follow the standard, management must introduce discipline to sustain that. If people are unable to follow the norm, management must either provide training or review or revise the rule to support it.

It has been observed that the organizational structure generally involves top or senior management, middle management, supervisors, and workers category of peoples. Kaizen applies to all. The role of senior management is critical for the organization's overall success (Deming, 1986; Juran, 1989). Deming (1986) argues that management is responsible for more than 90% of the quality problems. People in the other hierarchy should maintain the improvement and sustain the benefit of it. Further,

*Top management*

- Be determined to introduce Kaizen as a corporate strategy
- Provide support and direction for Kaizen by allocating resources
- Establish a policy for Kaizen and cross-functional goals
- Realize Kaizen goals through policy deployment and audits
- Build systems, procedures, and structures conducive to Kaizen.

*Middle management*

- Deploy and implement Kaizen goals as directed by top management
- Use Kaizen in functional capabilities
- Establish, maintain, and upgrade standards
- Make employees Kaizen conscious through intensive training programs
- Help employees develop skills and tools for problem-solving.

*Supervisors*

- Use Kaizen in functional roles
- Formulate plans for Kaizen and guide workers
- Improve communication with workers and sustain high morale
- Support small group activities and the individual suggestion system
- Introduce discipline in the workplace
- Provide Kaizen suggestions.

*Workers*

- Engage in Kaizen through the suggestion system and small-group activities

- Practice discipline in the workshop
- Engage in continuous self-development to become better problem-solvers
- Enhance skills and job performance expertise with cross-education.

Lee et al. (2000), believe that Kaizen can be beneficial in understanding manufacturing settings across the industry. Various metrics employed to determine the outcomes of Kaizen blitz are workflow improvement, improvement in ideas, increased quality levels, safe work environment, non-value-added time reduction, optimum floor space, flexibility in processes, etc. Kaizen can be best achieved through the 5S + Safety approach (Muralidharan, 2015), which is considered to be one of the flexible waste elimination tools described below.

## 4.6  5S + Safety and Security Techniques

5S describes how to organize a workspace for efficiency and effectiveness by identifying and storing the items used, maintaining the area and articles, and sustaining the new order ("cleanliness begins at home"!). It is a planned, systematic follow-up of activities that become an essential feature of any TQM process and helps to emphasize the organizational commitment to quality improvement continuously. The 5S's stands for five Japanese words: Seiri (structurize or sort), Seiton (systemize or straighten), Seiso (sanitize or scrub), Seiketsu (Standardize), and Shitsuke (self-discipline or sustain). These practices are useful not only in improving their physical environment but also in their thinking processes.

- *Seiri* is about separating the things which are necessary for the job from those that are not and keeping the number of the former as low as possible and at a convenient location.
- *Seiton* means to put things in order. Items must be kept in order so that they are ready for use when needed. This can be achieved by deciding where things belong, analyze the status quo, putting things back where they belong, etc.
- *Seiso* means to keep the workplace clean. Everybody in the organization is responsible for this.
- *Seiketsu* means continually and repeatedly maintaining neatness and cleanliness in the organization.
- *Shitsuke* means instilling the ability to do things the way they are supposed to be done. That is sustaining the previous four steps with improvement every time.

The 5S practices' success depends not only on the management but also on the people who are part of the entire quality process. It is the commitment and the driving force, enabling a good practice of neatness, cleanliness, and standardization in an organization (Muralidharan, 2015).

Perhaps the significance of the 5S approach is its simplicity. The benefits are apparent, as all the tools are easy to understand and apply. *Safety* and *security* are also paramount to realize the advantages of 5S. Working under a safe and secure environmental condition is every worker's right. Providing the necessary precautions arising from fire, water, chemical, and air-related hazards is management's commitment. This will help workers to communicate with the organization properly. Thus 5S + safety and security are the most crucial impetus of LGC philosophy for finding solutions and corrective action for successful process execution. The impact will be so huge if everyone in the organization considers this as a routine matter.

## 4.7  3Ms Technique

The three Ms considered being a part of Kaizen activities to reduce waste and to put a break on frequent occurrences of waste and inconsistencies in the process. The three interconnected waste components are *Muda* (waste), *Muri* (overburden), and *Mura* (unevenness). All are not good for a process and affect the quality of planning, control, and improvement.

*Muda* is non-value-added, wasteful activities that cause long cycle time and excess inventory of a production process. There are seven kinds of Muda, namely overproduction, waiting, transportation or conveyance, over processing, defects, inventory, and motion of people, materials, and vehicles. Muda is the result of Mura. *Mura* (unevenness) in production will result in excess inventory and overburden. When different transactional projects are handled simultaneously, then the consequences of changing priorities of transactions while they were in progress and having to correct defects in transactions by reworking them can cause unevenness in the process. These behaviors create enough problems in the system that specialized expediters may be responsible for hand carrying emergency orders through the system. Even though these particular orders are handled quickly, it is done at the expense of every other transaction in the system.

*Muri* is nothing but overburdening the people, machinery, or processes that result in quality or safety problems. Failing to accommodate variation in customer demands can result in swings of over and underproduction. The emphasis for production in times of massive need can be made at the expense of overtime, canceled training, or system upgrades. These factors can result in stress for the workers. The result can be impatience with customers, canceled transactions, or an increase in errors and Muda. Through the systematic implementation of 5S, it is possible to minimize Muda, Mura, and Muri's effect. While these techniques work quite well, they should best be applied to the areas where you have identified the problems during the project's measurement and analysis phases.

## 4.8  Poka-Yoke

Poka-Yoke is a Japanese concept that means "mistake-proofing." Its objective is to eliminate product defects by preventing, correcting, or drawing attention to human errors. The philosophy allows building quality into the process and simplifies the identification of root causes of a particular problem. According to Shigeo Shingo, *"Defects arise because errors are made; the two have a cause and effect relationship.... Yet errors will not turn into defects if feedback and action take place at the error stage."*

Poka-Yoke recognizes that as long as humans are involved in process activities, there will be errors. If a proper analysis is done, you can implement Poka-Yoke to prevent failure. Poka-Yoke systems create a process in which a worker cannot generate an error. Parts and procedures are designed so that desired results are inevitable. Some of the Poka-Yoke systems types of mistakes and their *safeguards* are:

- *Forgetfulness*: Forgetting a step or a part (Safeguards: use checklist, visual standard operating procedure, etc., to recall)
- *Errors due to misunderstanding*: Not very familiar with the required operation (Safeguards: continuous training, visual SOP, etc.)
- *Errors in identification*: Problems in identification or clearness of required steps or parts (Safeguards: training, visual training, standardization, etc.)
- *Errors due to lack of experience*: New employees (Safeguards: skill building and training, work standardization, etc.)
- *Errors due to lack of standards*: No clear way to perform the task or job (Safeguards: standard operations, visual instructions, etc.)
- *Errors due to machine readability*: Machine out of specifications (Safeguards: TPM, critical parts list, maintain equipment, history list, etc.).

The main principles of Poka-Yoke are:

- Build quality into the process
- All inadvertent errors and defects can be eliminated
- Stop doing it wrong and start doing it right now
- Do not think of excuses; think about doing it right
- An 80% chance of success is good enough—focus on implementation
- Attack defects and errors as a team
- Ten heads are better than one—brainstorm ideas, discuss, and think
- Seek out the root cause, use the 5 Whys.

Thus, Poka-Yoke activities enable process control by eliminating the source of variation and helping reduce the occurrence of rare events. Apart from the above tools, many other simple techniques can help to reduce waste in a process. Even the *pull* and *push* system also helps the management to produce the inventories in an equilibrium manner and a zero-waste condition. Another related technique is Baka-Yoke, a term used for manufacturing techniques for preventing mistakes by designing the manufacturing process, equipment, and tools so that an operation literally cannot be performed incorrectly. This technique also provides a warning signal of some sort for incorrect performance. Interestingly, all waste reduction methods have now developed into lean thinking of manufacturing (Muralidharan, 2015).

## 4.9  Setup Reduction

Setup reduction or single-minute exchange of dies (SMED) is one of the many lean production methods for reducing waste in the manufacturing process. It provides a fast and efficient way of converting a manufacturing process from running the current product to running the next product, thereby completing the work in less than a single digit minute (Shingo, 1985). It advocates teamwork, effective visual factory control, performance measurement, 5S, and continuous improvement (Kaizen) in every lean production activities.

According to Shingo (1985), SMED consists of three conceptual stages: separating internal and external setup, converting internal setup to external setup, and streamlining all aspects of the setup operations. Here the *interior setup* is that part of the setup which must be done while the machine is shut down, for example, removing or attaching dies. The *external setup* is that part of the setup, which can be done while the device is still running, for example, preparing a die to be used for the next run (Rubrich & Watson, 2004). Thus, SMED reduces the non-value-added time by streamlining and standardizing the operations for exchange tools, using simple techniques and secure applications (Alves & Tenera, 2009). Michels (2007) concludes that SMED can be a useful tool to provide improved changeover methods, resulting in reductions in overall time and labor and improving the organization's ability to enhance customer satisfaction through better utilization of plant resources.

## 4.10  Just-in-Time Approach

JIT is a production strategy that strives to improve a business's return on investment by reducing in-process inventory and associated carrying costs. It is a Japanese management philosophy that has been applied since the early 1970s in many Japanese manufacturing organizations. It was first developed and perfected within the Toyota manufacturing plants by Taiichi Ohno to meet consumer demands with minimum delays. Taiichi Ohno is frequently referred to as the father of JIT. The other perceptions of JIT are short cycle manufacturing (SCM) (Heard, 1987), IBM's continuous-flow manufacturing (CFM) (Barkman, 1989), and demand flow manufacturing (DFM) (Roebuck, 2011).

JIT defines wastes in a more general form, and continuous improvement is the core of minimizing all types of wastes and elimination of defects from a manufacturing process. JIT promotes the elimination of wastes from overproduction, waiting time, transportation, excess inventory, process, motion, and mistakes. Apart from this, other elements of JIT are preventive maintenance, good housekeeping, setup time reduction, balanced flow management, Kanban, skill diversification of workers, flexible changeover approaches, organization and discipline of materials and products, etc.

## 4.11 Kanban

Kanban is one method through which JIT is achieved. This is a system developed by Taiichi Ohno at Toyota to improve and maintain a high production level. Kanban is a visual signal for overproduction. It eliminates wastes in a controlled way. A Kanban is a system that signals the need to replenish stock or materials or to produce more of an item. This philosophy also helps the manufacturers to reduce waste in their process. This principle works on a "pull" system as opposed to the traditional "push" system. Kanban systems need not be elaborate, sophisticated, or even computerized to be effective. A method is best controlled when material and information flow smoothly and rationally into and out of the process. If process inputs arrive before they are needed, unnecessary confusion, inventory, and costs generally occur. If process outputs are not synchronized with downstream operations, the result is often delays, disappointed customers, and associated costs. A Kanban system may be used to simplify and improve resupply procedures.

When a business process has perfect one-piece flow, each business process will generate a "pull" signal to the upstream business process to supply one article for processing. This is extremely difficult to achieve, so a small batch of material is usually "pulled" instead of a single piece. This unit, called a *Kanban*, is not the same as the size of the pitch unit. When the intermediate supply of an article decreases to a predetermined minimum amount (safety stock), a Kanban signal is sent upstream to generate a batch of medium material. The downstream process can keep using the safety stock until the next batch arrives. The calculation of these parameters depends on some economic parameters, the size of the batch, stock-out-costs, and replenishment cycle time, etc. Variation in inventory delivery cycle time, discounts for large batch forward buying, and changes in consumption rate complicate the calculation. When the inventory of these intermediate articles increases, the ability to economically make quick design changes get degraded. The effect of variation in customer demand can be smoothed out in manufacturing processes by allowing changes in inventory levels. Using an adequately designed deterministic inventory model, one can find the optimum cycle time, which reduces the overall variation in the process.

## 4.12 Six Sigma Concepts

A Six Sigma initiative is a customer-focused problem-solving approach with reactive and proactive improvements of a process leading to sustainable business practices. Sustainable business practices include innovation, development, competition, environmental compliance, customer satisfaction, and growth of the organization (Muralidharan, 2015). The main goal of Six Sigma is to identify, isolate, and eliminate variation or defects. Six Sigma is a business process tool to achieve customer satisfaction by reducing variations. To make a product defect-free, it is essential to

**Table 4.5** Area under the normal curve and the corresponding PPM

| Range | Area covered in % | The area outside in % | PPM |
|---|---|---|---|
| $(\mu - \sigma)$ to $(\mu + \sigma)$ | 68.26 | 31.74 | 317,400 |
| $(\mu - 2\sigma)$ to $(\mu + 2\sigma)$ | 95.44 | 4.56 | 45,600 |
| $(\mu - 3\sigma)$ to $(\mu + 3\sigma)$ | 99.73 | 0.27 | 2700 |
| $(\mu - 4\sigma)$ to $(\mu + 4\sigma)$ | 99.99366 | 0.00634 | 63.4 |
| $(\mu - 5\sigma)$ to $(\mu + 5\sigma)$ | 99.9999426 | 0.0000574 | 0.574 |
| $(\mu - 6\sigma)$ to $(\mu + 6\sigma)$ | 99.9999998 | 0.0000002 | 0.002 |

minimize the variation of a process. Variation can be technically controlled using a normal (distribution) process.

In Sect. 4.2 above and previous chapters, we laid many exciting discussions on the suitability of normal distribution as an ideal distribution for any process. We have also seen that the $3\sigma$ limits provide a situation of controlled variation in a process, as $\sigma$ is a measure of variation or spread or dispersion. So to have an ideal quality, the probability of rejection should be as low as possible. Table 4.5 presents the area under the normal curve and the corresponding units in parts per million (PPM). If we translate the probability of rejection to the number of units produced, then $3\sigma$ limits offer one defective out of 100 units produced, equivalent to 10,000 defectives out of a million ($10^6$) units produced. This is a huge number, so accommodating such a large number of defective units could lead to a considerable loss. These days, generating millions of units is very common as improved technology, sophisticated machines, durable infrastructure, and people's expertise are available in abundance. In such an environment, allowing 10,000 defectives is not ideal and advisable at all. So by extensive reduction of variation, if we can reduce the number of defective items to a shallow level, then it is guaranteed to improve the quality of the process further. This is where the Six Sigma concepts and philosophy developed.

Six Sigma as a quality improvement tool advocates measuring the variability of a process, which may then be controlled by continuous improvement. The application of Six Sigma begins with translating a practical problem into a statistical one. An optimal solution is found using appropriate analytical tools and then implemented as a possible solution to a real-life situation (Harry & Schroeder, 2000). Figure 4.14 explains this.

If a process is described as within "Six Sigma," the term quantitatively means that it produces fewer than 3.4 *defects per million opportunities* (DPMO). That represents an error rate of 0.0003%; conversely, a defect-free rate of 99.9997% useful items can be realized. It may be noted that the sigma measure compares the performance

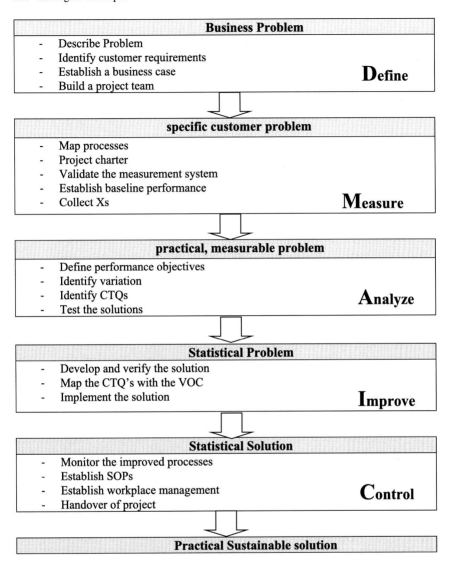

**Fig. 4.14**  Six Sigma problem-solving approaches

of the company to customer requirements (defined as a target), and the requirements vary with the type of industry or business. The sigma value of a process describes the quality level of that process. The computation of the sigma level as a measure of process capability and performance will be attempted later in this chapter.

Among all perceptions of the Six Sigma concept, the widely accepted definition of a Six Sigma process is a process that produces 3.4 defective parts per million opportunities. This is based on the fact that a process that is normally distributed will have 3.4 parts per million beyond a point that is 4.5 standard deviations above or

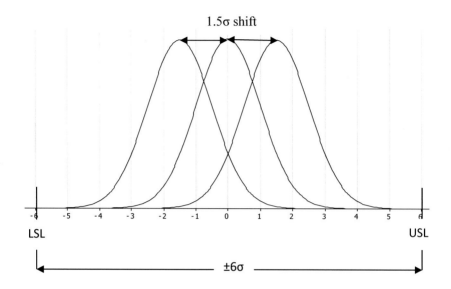

**Fig. 4.15**  Six Sigma process with 1.5σ shift

below the mean. So the 3.4 DPMO, also called *parts per million* (PPM) of a Six Sigma process, corresponds to 4.5 sigma, namely Six Sigma minus the 1.5σ shift introduced to account for long-term variation (see Fig. 4.15 for a Six Sigma process with 1. 5σ shift). This allows for the fact that special causes may result in deterioration in process performance over time and is designed to prevent underestimating the defect levels likely to be encountered in real-life operation. The possible causes that could result in process shift are:

- Changing equipment
- Changing operators and shifts
- Machine life (viz. wear and tear, etc.)
- Climate conditions (viz. humidity, temperature changes, etc.)
- Process calibration
- Equipment breakdown or repairs
- Location and setup changes.

The purpose of Six Sigma is to generate organizational performance improvement to a high level. It is up to the organization to determine, based on customer expectations, the appropriate sigma level of a process. The purpose of the sigma level is to determine whether a process is improving, deteriorating, stagnant, or non-competitive with others in the same business. The value of the sigma level is ascertained through the DPMO or PPM. Table 4.6 presents the sigma level with a 1.5 Sigma shift corresponds to different DPMOs. For a successful implementation, a Six Sigma approach requires companies to consider changes in methodologies across the enterprise, introducing new linkages and communications.

**Table 4.6**  Sigma level versus DPMO (includes a 1.5 Sigma shift)

| Sigma level | DPMO | Sigma level | DPMO | Sigma level | DPMO |
|---|---|---|---|---|---|
| 0.00 | 933,193 | 4.05 | 5386 | 5.05 | 193 |
| 0.50 | 841,345 | 4.10 | 4661 | 5.10 | 159 |
| 0.75 | 773,373 | 4.15 | 4024 | 5.15 | 131 |
| 1.00 | 691,462 | 4.20 | 3467 | 5.20 | 108 |
| 1.25 | 401,294 | 4.25 | 2980 | 5.25 | 89 |
| 1.50 | 500,000 | 4.30 | 2555 | 5.30 | 72 |
| 1.75 | 401,294 | 4.35 | 2186 | 5.35 | 59 |
| 2.00 | 308,537 | 4.40 | 1866 | 5.40 | 48 |
| 2.25 | 226,627 | 4.45 | 1589 | 5.45 | 39 |
| 2.50 | 158,655 | 4.50 | 1350 | 5.50 | 32 |
| 2.75 | 105,650 | 4.55 | 1144 | 5.55 | 26 |
| 3.00 | 66,807 | 4.60 | 968 | 5.60 | 21 |
| 3.25 | 40,059 | 4.65 | 816 | 5.65 | 17 |
| 3.50 | 22,750 | 4.70 | 687 | 5.70 | 13 |
| 3.60 | 17,865 | 4.75 | 577 | 5.75 | 11 |
| 3.70 | 13,904 | 4.80 | 483 | 5.80 | 9 |
| 3.75 | 12,225 | 4.85 | 404 | 5.85 | 7 |
| 3.80 | 10,724 | 4.90 | 337 | 5.90 | 5 |
| 3.90 | 8198 | 4.95 | 280 | 5.95 | 4 |
| 4.00 | 6210 | 5.00 | 233 | 6.00 | 3.4 |

Let us understand the computation of the sigma level of the business process. In the first case, we consider an example based on a continuous measurement.

***Example 4*** A process that produces titanium forgings for automobile turbocharger wheel is examined for its dimensional accuracy. A sample of 70 forgings was inspected and measured for its inner diameter of the wheel (see Table 4.7). It was decided that the computation of the current Sigma level of the process will determine

**Table 4.7**  Data on wheel dimension

| | | | | | | | | | |
|---|---|---|---|---|---|---|---|---|---|
| 85.37 | 73.70 | 61.67 | 66.93 | 68.87 | 50.20 | 82.55 | 66.57 | 67.46 | 89.46 |
| 72.05 | 72.46 | 62.44 | 88.19 | 84.96 | 80.41 | 83.63 | 78.81 | 76.20 | 62.14 |
| 97.45 | 82.36 | 71.59 | 77.84 | 61.58 | 74.04 | 71.67 | 90.77 | 56.22 | 74.85 |
| 73.81 | 68.02 | 70.83 | 74.18 | 77.60 | 82.48 | 66.66 | 72.64 | 75.81 | 56.03 |
| 80.57 | 80.78 | 68.04 | 78.99 | 67.01 | 59.14 | 87.19 | 74.94 | 77.80 | 86.01 |
| 85.00 | 99.27 | 76.21 | 72.49 | 64.42 | 69.66 | 78.06 | 77.11 | 72.09 | 70.06 |
| 86.42 | 79.35 | 80.48 | 85.27 | 70.50 | 53.45 | 73.50 | 78.08 | 72.14 | 76.80 |

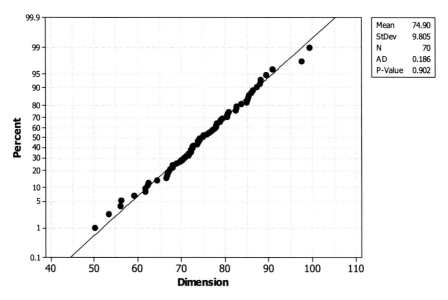

**Fig. 4.16**  Normal probability plot of wheel dimension

the extent of wheel alignment necessary for the assembly of forgings. The lower and upper diameter levels are, respectively, given as 52 and 97.

We first investigate whether the data follows a normal distribution or not. Figure 4.16 shows that the data follows normal probability law (see $p$-value $= 0.902$), with mean dimension ($\bar{x}$) as 74.9 and standard deviation ($\sigma$) as 9.805. Now we compute the total proportion of non-conformance of the size as per the specification limit as $P(Z < Z_{\text{LSL}}) + P(Z > Z_{\text{USL}})$, where

$$P(Z < Z_{\text{LSL}}) = P\left(Z < \frac{\text{LSL} - \bar{x}}{\sigma}\right)$$
$$= P\left(Z < \frac{52 - 74.9}{9.805}\right)$$
$$= P(Z < -2.33554)$$
$$= 0.00964$$

Similarly,

$$P(Z > Z_{\text{USL}} = P\left(Z > \frac{\text{USL} - \bar{x}}{\sigma}\right)$$
$$= P\left(Z < \frac{97 - 74.9}{9.805}\right)$$
$$= P(Z < 2.2539)$$
$$= 0.01222$$

Therefore,

$$P(\text{non - conformance}) = P(Z < Z_{\text{LSL}}) + P(Z > Z_{\text{USL}}),$$
$$= 0.02186 \quad (\text{according to Appendix table 3})$$

Since, $P(\text{non-conformance}) + P(\text{conformance}) = 1$, we get $P(\text{conformance})$ as 0.97814. Referring to the standard normal distribution table again, we find that this probability corresponds to the ordinate 2.02, which is the long-term sigma level ($Z_{\text{LT}}$) of the process. Hence, the short-term sigma level ($Z_{\text{ST}}$) is obtained as $Z_{\text{LT}} + 1.5 = 3.52$.

The long-term process capability of the process is given by $P_{pk} = \min(P_{pl}, P_{pu})$, where

$$P_{\text{pl}} = \frac{\mu - \text{LSL}}{3\sigma}$$
$$= \frac{74.9 - 52}{3*9.805} = 0.7785$$

and

$$P_{\text{pu}} = \frac{\text{USL} - \mu}{3\sigma}$$
$$= \frac{97 - 74.9}{3*9.805} = 0.7531.$$

Therefore, the overall process capability is $P_{pk} = 0.7531$. Figure 4.17 presents the process capability of the wheel dimension as per MINITAB software. The parts per million (ppm) total corresponds to the overall performance is 21855.57, which according to the DPMO Table 4.6 (also see Appendix Table IV), corresponds to the 3.53 level of Sigma.

The Six Sigma concept, as described above, is a data-driven technique for achieving the desired quality. It is a customer-driven process improvement system that puts our customers and their requirement first. It focuses on eliminating defects and reducing variations. It combines proven statistical and analytical tools and techniques into an integrated system. It is accelerated through a rigorous management commitment and devotion. This, coupled with sustainable business practices and environmental concern, is why Six Sigma quality improvement programs are gaining importance across the worldwide industrial organizations (Muralidharan, 2015).

As discussed above, the Six Sigma philosophy works under a structured problem-solving approach. It actively coordinates with the 5Ms (Man–Machine–Methods–Materials–Measurement) process, as they are all subject to random variations and human errors. Therefore, the problems are generally concerned with eliminating variability, defects, and waste in a product or process, all of which undermine customer satisfaction. The working philosophy of Six Sigma philosophy is generally called Define–Measure–Analyze–Improve–Control (DMAIC) method, where

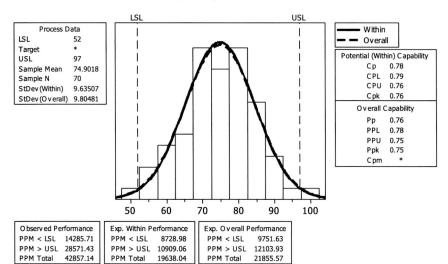

**Fig. 4.17** Process capability of wheel diameter

- *Define*—Define the problem and project goals that need to be addressed. The tools specific to the define phase are project charter, Gantt chart/timeline preparation, process mapping, flow chart, Pareto chart and control charts, QFD/house of quality, cost of quality, trend analysis, suggestions/complaints, surveys/interviews/focus group. The initial sigma level estimation is computed here to understand the current state of the process. The project charter generally contains information about the problem statement, the people involved in the project, process details, and the timeline of the project.
- *Measure*—Measure the problem and process from which it was produced. The team depends more on statistical tools to facilitate the sample size determination, data collection, and preparation of the *Supplier–Input–Process–Output–Customer* (SIPOC) diagram, and so on. The processed data will be subject to all kinds of descriptive statistics, and charts and plots will be prepared to see the quality of the data. The other tools are normal probability plots, capability analysis, Pareto charts, and to assess the accuracy and precision of data measurement system analysis (MSA) and Gage studies are carried out. Process map/flow chart, project charter, Gantt chart/timeline preparation are primarily prepared for regular monitoring of project progress and assessments.
- *Analyze*—Analyze data and process to identify the root causes of defects and opportunities. The tools used for identifying root causes are the cause-and-effect (C&E) diagram, failure mode effect analysis (FMEA), and high-level process map. The use of histogram, Pareto charts, scatter diagram, and correlation study are common for identifying the relationship between variables and their effects. Testing the hypothesis, confidence intervals, control charts, and time series charts

are the standard tools for confirmatory analysis and monitoring variation in the process. For predicting the future behavior of output variables, regression analysis, analysis of variance (ANOVA), design of experiments (DOE), response surface methods (RSM) are performed. Multivariate techniques facilitate the grouping and factorizing the variables. Analyze phase also uses many engineering concepts and reliability analysis for correctly setting the process parameters and their specifications.

- *Improve*—Improve the process by finding solutions to fix, reduce, and prevent future problems. Many advanced level statistical methods including testing of hypothesis, confidence intervals, ANOVA, multivariate analysis of variance (MANOVA), multi-vari chart, general linear models, logistic regression, probit analysis, DOE, factor analysis, FMEA, etc., are used as confirmatory tools for future process improvement. A high-level process map and flowchart are drawn again to ascertain the real process improvement. The simulation technique and trial and error are used to create a working model for implementation and validation. The existing standard operating procedure (SOPs) will be modified and revised accordingly.

- *Control*—Implement, control, and sustain the improvement's solutions to keep the process on the new course. Many tools discussed above are revised here again. Notably, the control charts and process capability analysis are the standard tools used in this phase. Along with the monitoring and control, these techniques focus the attention on output's efficiency of the process. Poka-Yoke or mistake-proofing, standardization for work instructions, process dashboards, capability studies, etc., are frequently carried out for quality sustenance. The yield calculation and final sigma level estimation are done again to assess the performance of the process and cost of poor quality. This phase also focuses on the improved SOPs for documentation, final reporting, and presentation.

The DMAIC philosophy applies to both process improvement and product improvement or even design redesign efforts. The critical elements in a DMAIC project are team discipline, structured use of metrics and tools, and execution of a well-designed project plan that has clear goals and objectives. It is expected that the goal statement should be SMART:

- *Specific*—Goals should be precise and to the point. Generic and philosophical goals can confuse and mislead the project people.
- *Measurable*—Unless the team has measurable goals, it will not know whether it is making progress or succeed.
- *Achievable*—Goals should be under possible targets. One can have an aggressive campaign for meeting the attainable target.
- *Relevant*—The goal must be linked explicitly to the strategic objectives of the enterprise.
- *Timely*—The goals must make sense in the time frame in which the team must work.

Muir (2006) advocates GRPI (goals, roles and responsibilities, processes and procedures, and interpersonal relationships) model for managing a successful team over the length of the project. Here also, the goals should be SMART as described above and should be included in the project charter. Roles and responsibilities are essential for the overall impact on the target. With specified tasks, the team can communicate efficiently and impartial way to its completion. Six Sigma projects' success depends on all these aspects (see also Kubiak & Benbow, 2010). For a detailed discussion on leadership roles and responsibilities, see Sect. 9.3.

### 4.12.1  Six Sigma Project Essentials

A Six Sigma project is executed through a team of experts and highly responsible personnel of organizations. Such a plan is initiated after various stages of brainstorming, training, consultations, and briefing with the stakeholders of the organizations. The success of any Six Sigma project depends on its ability to deliver world-class products and services with a high-performance index and capability. Therefore, the expertise and experience matter a lot to the project's success. The system adopts different belts training to facilitate this. The value of Six Sigma is well-explained in Fig. 4.18.

A ten-member (not fixed) Six Sigma project includes a project leader (also called champion or sponsor)), a Six Sigma Master Black Belt (MBB), two Black Belts (BB), and remaining Green Belts (GB) or executives. The leading roles associated with a specific belt of a Six Sigma projects can be summarized as follows:

**Fig. 4.18**  Value of Six Sigma

- *Champion*: The role of Champions is to ensure that the right projects are being identified and worked on, that teams are making good progress, and that the resources required for successful project completion are in place.
- *Master Black Belt (MBB)*. An individual well-trained in statistical methods and have the management expertise to lead a team. They have established a record of carrying out BB projects and BB training. They have the authority to certify GBs and BBs within an organization. They often write and develop training modules and are deeply involved in project definition and selection, and work closely with business leaders called champions.
- *Black Belts (BB)*. They are the backbone of the success of every Six Sigma projects. They are trained statisticians and have the record of executing high yield Six Sigma projects in their organization. They are generally the full-time employee and have the experience to train and mentor GBs.
- *Green Belt (GB)*. A Six Sigma role associated with an individual who retains his or her regular position within the firm but is trained in the tools, methods, and skills necessary to conduct Six Sigma improvement projects either individually or as part of larger teams.

The success of any Six Sigma project starts with the identification of a suitable project. There can be many prospective projects that need the simultaneous attention of the officials. It is ubiquitous to run multiple projects in large organizations, either simultaneously or in isolation. A Six Sigma project selection involves lots of exercises involving company think tank and consultants. They are identified based on many criteria and features. See Table 4.8 for the compilation of various projects weighted and ranked on various criteria.

The weights are scores measured on a scale of 1–10, where 1 is the least important and 10 being highly important features. In the same way, each project is measured on three different scales: 1 for projects of less importance, 3 for projects of moderate importance, and 9 for a project of high importance with respect to various criteria and features. The total score for each project is calculated as the sum of products of project score and the corresponding weight of the criteria. For instance, the total score corresponds to Project-3 which is $8 * 3 + 7 * 3 + \cdots + 9 * 3 = 357$. Similarly, scores for other projects are computed. Finally, the projects are ranked as per the total score. It is now easy to identify which project should be given priority for execution. Projects with low scores may be either deferred or discarded.

A critical component of the Six Sigma project is the project charter. A *project charter* is a document stating the purposes of the project. A project charter is as good as a job sheet, which includes the details of the problem, problem statement, the objectives required to realize each activity's timeline, and the aspects of the customer and process in brief. A typical project charter is presented in Table 4.9.

There are many items in the project charter. One of the issues that need special care is the problem definition and the statement of the project. A problem can be identified through formal or informal discussions with the think tank of the organization. Through brainstorming sessions and root cause analysis also, this can be realized.

**Table 4.8**  Project identification process

| Criteria and features | Weight | Projects and their score | | | | | |
|---|---|---|---|---|---|---|---|
| | | Project-1 | Project-2 | Project-3 | Project-4 | Project-5 | Project-6 |
| Process improvement and innovation | 8 | 9 | 1 | 3 | 3 | 9 | 9 |
| Quality improvements | 7 | 9 | 9 | 3 | 9 | 3 | 9 |
| Customer satisfaction | 9 | 3 | 3 | 9 | 1 | 3 | 3 |
| Cost reduction | 7 | 1 | 1 | 3 | 3 | 1 | 3 |
| Productivity enhancement | 8 | 3 | 3 | 3 | 3 | 1 | 1 |
| Design and features | 9 | 3 | 3 | 9 | 9 | 9 | 3 |
| Sustainable development | 10 | 9 | 9 | 3 | 9 | 9 | 3 |
| Capacity building | 6 | 3 | 3 | 1 | 1 | 3 | 9 |
| Reliability and maintenance | 7 | 1 | 1 | 3 | 1 | 1 | 1 |
| Employee growth and satisfaction | 7 | 3 | 3 | 3 | 1 | 3 | 1 |
| Alignment with vision and mission | 9 | 3 | 9 | 3 | 9 | 9 | 3 |
| **Total score** | | **383** | **373** | **357** | **413** | **433** | **343** |
| **Rating of projects** | | **3** | **4** | **5** | **2** | **1** | **6** |

Regardless of the technique used, one should involve 5 W (*who–what–when–where–why*) technique along with the *how's* the problem to identify a prospective case for the project. Table 4.10 summarizes the questions which are frequently asked to identify a plan for improvement.

Another critical component of the Six Sigma project is its balanced scorecard. The project balanced scorecard (BSC) is a top-down management instrument that is used for making the LSS project's ambition operational at all levels. So project BSC is a balanced and holistic project charter. The elements of project BSC, according to Rampersad and El-Homsi (2008), are shown in Table 4.11.

A snapshot of the Six Sigma process is depicted through a SIPOC diagram. A *SIPOC* diagram (sometimes called high-level process map) is an extended version of the IPO (input-process–output) diagram where the supplier–customer interface

**Table 4.9** A project charter for optimization of the cycle time of pump generation process

| Project charter | | | |
|---|---|---|---|
| *General project informations* | | | |
| Project name | To improve the time and accuracy of pump KPM generation process | | |
| Project manager | Dr. XYZ | Start date: 05.05.2019 | End date: 10.10.2019 |
| Organizational unit | Pump generation plant | | |
| *Project essentials* | | | |
| Business case | Improvement of the cycle time of project engineering activities to improve customer satisfaction as notified by the inspection team | | |
| Problem statement | The compiled information for 2018 indicates the need for improvement in the timely release of the bill of material and reducing error. This includes project cycle time, wrong selection of parts, typological errors, etc. | | |
| Goal statement | To identify the present bottlenecks in the KPM generation process and develop a standard design procedure by clearly defining the activities, with specific time frames, to ensure a smooth and a system dependent process for ensuring timely completion of bare pump designing and with 98% accuracy | | |
| Project KPOVs | 1. Pump generation at batch plant<br>2. Number batches processed | | |
| Assumptions | The size of the batch should be uniform throughout the Year | | |
| Constraints | Nil | | |
| Project scope (in scope/out scope) | In scope: API 3700 pump model KPM generation process after order confirmation<br>Out scope: other processes included in the lead time of the pump | | |
| Expected benefits | Cycle time reduction and productivity improvement | | |
| | **Key milestone/timelines** | **Start date** | **Completion date** |
| Project schedule | Review and scope | 05.05.2019 | 16.05.2019 |
| | Plan and kickoff | 17.05.2019 | 20.05.2019 |
| | Define phase | 21.05.2019 | 28.05.2019 |
| | Measure phase | 29.05.2019 | 15.06.2019 |
| | Analyze phase | 16.06.2019 | 19.06.2019 |
| | Improve phase | 21.07.2019 | 30.07.2019 |
| | Control phase | 01.09.2019 | 10.10.2019 |
| | Summary and closure | 11.10.2019 | 10.10.2019 |
| Project team | Mr. Ullas (MBB), Mr. Surya (BB), Dr. Varki (BB), Mr. Saroj (GB), Mrs. Vaidehi (GB), Mr. M. Patel (GB), Mrs. Rupal, Mr. Ajay, Mr. R. Ganatra | | |

**Table 4.10** Essential questions to ask in defining the problem

| | |
|---|---|
| Who | – Who is causing the problem?<br>– Who says this is a problem?<br>– Who is impacted by this problem?<br>– Who is associated with this problem? |
| What | – What are the symptoms of the problem?<br>– What are the impacts of the problem?<br>– What are the side effects of the problem?<br>– What will happen if this problem is not solved? |
| Where | – Where does this problem occur?<br>– Where does this problem have an impact?<br>– Where was this problem identified? |
| When | – When does this problem occur?<br>– When did this problem first start occurring?<br>– When was it first signaled for variation?<br>– When was the last time the problem occurred? |
| Why | – Why is this problem occurring?<br>– Why did the process allow it to happen?<br>– Why did the process owner take a note of it? |
| How | – How the problem got located?<br>– How should the process or system work?<br>– How are people currently handling the problem?<br>– How many people are involved with the problem? |

is brought in to entire improvement activities of the organization. Such involvement often results in joint economic planning, joint technological planning, and cooperation during contract execution. It is generally used in the early phase of the define phase to help specify the project scope. A typical SIPOC diagram is shown in Fig. 4.19.

According to Pande et al. (2003), the benefits of SIPOC are:

- It displays a cross-functional set of activities in a single and straightforward diagram.
- It uses a framework applicable to processes of all sizes.
- It helps to maintain a "big picture perspective" to which additional detail can be added.
- A well-defined SIPOC can synchronize other critical sub-processes to achieve the stretch goal in a very efficient way.

LSS modifies the Six Sigma-DMAIC approach by emphasizing speed. However, a SIPOC diagram is not mandatory for LSS. Instead, the process flow is established through value stream mapping activities (see Sect. 10.2 for details). LSS also focuses on streamlining a process by identifying and removing non-value-added steps. A leaned production process eliminates waste, and hence, the targeted metrics include zero wait time, zero inventory, scheduling using customer pull, cutting batch sizes to improve flow, line balancing, and reducing overall process time. The tools specific to LSS are workplace management, Kaizen, one-piece flow, visual management,

**Table 4.11** Project balanced score card

| Project vision | – The project vision provides a desired and possible future of the project and the route needed to reach the vision<br>– It indicates what is essential for the project's success, envisions the future after executing the plan, the long-term project intentions, and which process changes lay ahead |
| --- | --- |
| Project mission | – The project mission consists of a project's identity and indicates its reason for existing<br>– An expertly charted mission creates a sense of unity in the behavior of project members, strengthens their like-mindedness, and improves both communication and the atmosphere within the project team |
| Project values | – Project values determine how one must act to realize the project vision<br>– They function as the guiding principles that support team member's behavior during the execution of the LSS projects |
| Project critical success factors | – The critical factors are derived from project ambition<br>– A project critical success factor is one that is of overriding importance to the project's success<br>– The critical success factors form the link between project, vision, mission, and core values and the remaining project BSC elements |
| Project objectives | – Project objectives are measurable results that must be achieved<br>– They are derived directly from the project critical success factors and form realistic milestones<br>– The objectives form part of a cause and effect chain, resulting in the final project objective |
| Project performance measures | – Project performance measures are the standards by which the progress of the LSS project objectives is measured<br>– They provide team members with timely signals of project guidance, based on the measurement of process changes and the comparison of the measured results to the norm |

automation, and 5S. However, LSS also deploys DMAIC from a diluted format to enable a company-wide Six Sigma program without any technical expertise. This is the reason for promoting lean projects instead of a rigorous Six Sigma project. Ultimately, LSS's goal is to produce quality products that meet customer requirements as efficiently and effectively. The unique metrics associated with both LSS and Six Sigma are presented in Table 4.12.

LSS combines many tools, as given in the table above. The selection and usage of appropriate metrics in the context of LSS would enable organizations to identify and target the proper problems during LSS implementation projects, evaluate potential improvement initiatives and select action plans, establish baseline data for process improvement, communicate the improvement results and continuously monitor the deployment of LSS initiatives. Comparing different projects according to defects levels or sigma level is also typical in LSS initiatives. As mentioned above, the sigma level or sigma rating helps us to evaluate the performance of a process. The sigma

| Suppliers | | Input | Process | | Output | | Customers |
|---|---|---|---|---|---|---|---|
| 1 | Partners Clients | Trading Account and Demat form | | Start point Collection of Trading & Demat Application | 1 | Activated Demat account within defined TAT | 1 | End client: Account activation within given TAT, processing of form with minimum queries |
| | | Supporting documents | | Operational activities | 2 | Certificate of the transaction ID | 2 | Partner: Account activation within TAT, processing of form with minimum queries, the update of every status |
| | | | 1 | Application and document verification | | | 3 | Individual clients having the right to transact |
| | | | 2 | Inward entry | | | | |
| | | | 3 | In-person verification | | | | |
| | | | 4 | Receipt at HO | | | | |
| | | | 5 | Second level verification | | | | |
| | | | 6 | Activation of Demat Account | | | | |
| | | | 7 | Activation of Trading account | | | | |
| | | | 8 | Activation/Rejection of application | | | | |

**Fig. 4.19** A high-level Six Sigma process map. *Source* Muralidharan (2015)

**Table 4.12** Lean versus Six Sigma metrics

| Lean metrics | Six Sigma metrics |
|---|---|
| – Process flow | – DPMO |
| – Customer value | – Goal value |
| – Waste (all forms) | – Initial Sigma level |
| – VA-NVA activities | – Throughput yield |
| – First-time quality | – Rolled throughput yield |
| – Take time | – Cost of poor quality |
| – Lead time | – Process capability indices |
| – Cycle time | – Process performance indices |
| – Change over time | – Final Sigma level |

level is computed for quality characteristics that are not measurable by converting the total defectives into a million opportunities. *Defectives* are the total number of units containing different types of defects, and the d*efect* is any characteristic of the product that fails to meet customer requirements.

It is better to draw a Gantt chart or timeline chart for presenting the ongoing activities of a *Lean Six Sigma* (LSS) project according to time. They are used in process/project planning and control to display planned work and finished work with time. Gantt charts (see Fig. 4.20) were considered extremely revolutionary when first introduced by James (2003) in recognition of Henry Gantt's contributions. Modern Gantt charts also show the dependency (i.e., precedence network) relationships between activities and can be used to show current schedule status. A maintenance schedule of an industrial process can also be depicted in the form of Gantt charts for tracking the project activities at any phase. They are sometimes used

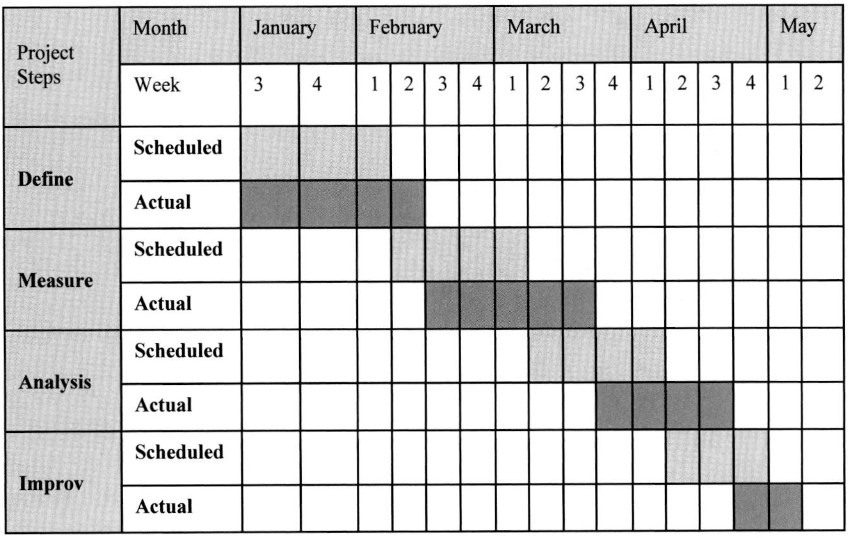

| Project Steps | Month | January | | February | | | | March | | | | April | | | | May | |
|---|---|---|---|---|---|---|---|---|---|---|---|---|---|---|---|---|---|
| | Week | 3 | 4 | 1 | 2 | 3 | 4 | 1 | 2 | 3 | 4 | 1 | 2 | 3 | 4 | 1 | 2 |
| **Define** | Scheduled | | | | | | | | | | | | | | | | |
| | Actual | | | | | | | | | | | | | | | | |
| **Measure** | Scheduled | | | | | | | | | | | | | | | | |
| | Actual | | | | | | | | | | | | | | | | |
| **Analysis** | Scheduled | | | | | | | | | | | | | | | | |
| | Actual | | | | | | | | | | | | | | | | |
| **Improv** | Scheduled | | | | | | | | | | | | | | | | |
| | Actual | | | | | | | | | | | | | | | | |

**Fig. 4.20** Gantt chart

to convey project information involving changes that are easier to draw or photograph than to explain in words.

Calculation of sigma level through DPMO involves the concept of defects, unit, and opportunity. *The unit* is any product, part, assembly, process, or service for which quality characteristics are desired. *Opportunity* (or chance) is a value-added feature of a unit that should meet the specifications proposed by the customer. The total opportunity (TOP) is given by number of units checked * number of opportunities of failure and

$$DPU = \frac{\text{No. of Defects}}{\text{No. of units checked}}$$

and

$$DPMO = \frac{DPU}{\text{No. of opportunity per unit}} * 10^6$$

The sigma level corresponds to each DPMO is given in Appendix Table 4. One can also use the standard normal distribution table given in Appendix Table 3 and get the long-term sigma level as

$$\sigma_{LT} = \Phi^{-1}\left(1 - \frac{DPMO}{10^6}\right)$$

where $\Phi^{-1}(.)$ is the inverse standard normal distribution. The short-term sigma level is then obtained by adding $1.5\,\sigma$ to the long-term sigma level. That is, if $\sigma_{LT}$ and $\sigma_{ST}$ are the long-term and short-term sigma levels, respectively, then $\sigma_{ST} = \sigma_{LT} + 1.5\sigma$.

As an illustration, consider 250 units are selected from a production process, and each one is tested for 20 possible opportunities. Suppose 75 defects are observed in the process, then the DPU is obtained $DPU = \frac{75}{250} = 0.3$. Hence, the defects per million opportunities are calculated as $DPMO = \frac{0.3}{20} * 10^6 = 15000$. That is, the process fails to meet specifications for 15,000 opportunities out of 1,000,000 (1 million) opportunities. This DPMO is corresponding to a short-term sigma level of $3.67\sigma$ as per Appendix Table 4.

For the service industry, Six Sigma philosophy helps impart flawless services and skills. It facilitates people excellence as well as technical excellence in terms of creativity, collaboration, communication, dedication, and, above all, increases the accountability of what one does in an organization (Henderson & Evans, 2000; Pande et al., 2003; Muralidharan, 2008a, b, 2010). Thus, the ultimate aim of Six Sigma philosophy is to involve all the people in the organization in the improvement of all internal and external processes. As a result, it brings a good quality culture amenable to changes, encourages originality and creativity, improved standards, and allow space for individual growth.

The proven benefits of LSS project or Six Sigma project management are as diverse as (Muralidharan, 2015):

- Reduce costs, defects, and cycle time and wastes
- Generates sustained success through reduced variation
- Sets a performance goal for everyone
- Enhances value to customers
- Accelerates the rate of improvement
- Encourage the development of the new process, product, and service
- Improve market share and growth
- Executes strategic change
- Encourages cultural change
- Promotes learning and cross-pollination across the organization.

In India, the number of companies and institutions that have integrated Six Sigma projects in their corporate activities are increasing financial profits and business enhancement tremendously. They include manufacturing, automobile, and service industries. The projects are generally addressed for cycle time reduction, throughput enhancement, quality of production, breakthrough improvements, service quality improvements, etc. Even many IT and IT-enabled service industries are also carrying out Six Sigma projects for massively improving their processes. See the discussion on some case studies presented in Sect. 9.4, where we connect the importance of LGC concepts in improving the quality of a process.

## 4.13  Capability Maturity Model Integration

The capability maturity model integration (CMMI) is a collection of characteristics of effective processes that guide for improving an organization's processes and ability to manage the development, acquisition, and maintenance of products or product components. The CMMI framework is the structure that organizes the ingredients used in generating models, training materials, and appraisal methods. The CMMI Product Suite is the full collection of models, training materials, and appraisal methods created from the CMMI framework. This concept was first introduced and popularized by the Software Engineering Institute (SEI) of Carnegie Mellon University, Pittsburg (http://www.sei.cmu.edu/cmmi). The CMMI Framework components are organized into groupings, called constellations, which facilitate the construction of approved models.

People use CMMI in process improvement activities as a

- Collection of best practices
- Improve delivery of performance, cost, and schedule
- Framework for organizing and prioritizing activities
- Support for the coordination of multi-disciplined activities that might be required to build a product successfully
- Means to emphasize the alignment of the process improvement objectives with organizational business objectives.

A process can only be evaluated within the context of its maturity level. The maturity level will determine how easily a process will adopt a new technology solution. Misjudging the process maturity can impact on the benefits gained from any investment. As the name suggests, there are various levels of maturity for process improvement. *Scales* are used in CMMI to describe an evolutionary path for an organization that wants to improve the processes it uses to develop and maintain its products and services. The five primary levels as per the maturity level focus on:

1. Initiating and understanding the process
2. Basic level project management
3. Process standardization
4. Quality management
5. Continuous process improvement.

The emphasis of the last two levels is on quantitative and measurement-based quality improvement. To improve interfaces between government and industry, one may use the *People CMM* to enhance the capability of organizations' workforces through improved management and human capital. The People CMM defines capability as the level of knowledge, skills, and process abilities available within each workforce competency of the organization to build its products or deliver its services. The People CMM is used as a guide in planning and implementing improved human capital management practices, improving organizational standards, and as a standard for analyzing and appraising an organization's implemented workforce practices (http://seir.sei.cmu.edu/seir).

People CMM helps organizations do the following:

• Develop the workforce required to execute a business strategy.
• Characterize the maturity of workforce practices.
• Set priorities for improving workforce capability.
• Integrate improvements in processes and the workforce.
• Become an employer of choice.

Thus, for improving organizational culture, one needs to explore all kinds of quality improvement programs as per viability and regulatory requirements. It is better to prepare standard operating procedures (SOPs) for each significant activity to facilitate the organization's functioning. An SOP can be anything like the operating guides, standard job practices, work instructions, job aids, level-three ISO 9001 documentation, and so on. An SOP can unfold the detailed description of how to carry out a task along with the person responsible and the type and nature of work. The SOP should be a living document, as it guides the people in the office to do the job without any problem and risk. If something changes in the system, the office worker should ensure that the SOP is updated. Once something changes in the process and a new, desirable level of quality is achieved, the office worker should help update all documents relating to that process appropriately and destroy the old records.

To ensure the improvement, the management can form a self-directed team. A self-directed team is a team of people who have been authorized by management to work on a processor to improve a process using all the resources available to them and to manage their own time and energy, as a self-operating unit, for the good of the organization. Self-directed teams tend to share leadership equally among the group members; thus, rather than one person being the leader, they are all responsible for the output of the team. These teams tend to be more critical in a company that has evolved into a lean manufacturing or lean office organization. The team members have shared goals and take responsibility for communication with the rest of the organization.

It is recommended that a cost–benefit analysis (CBA) is carried out frequently to judge the quality level of the organization. CBA is a financial tool that reports how quality levels are sustained in the office and an organization. This analysis uses those costs related to products or services like prevention costs, appraisal cost, and internal (or external) failure costs. Through this analysis, one can know the value of (poor) quality and the amount of waste over some time.

Thus, the quality improvement programs discussed above can create an organizational climate within which people are encouraged to take the initiative to be innovative and creative to develop new and better ways of doing things. It also includes the empowerment of people through delegation and other measures so that they could act and correct their own mistakes. Each program's results can facilitate establishing policies and procedures to define performance standards and acceptable ways of doing work, guided by quality, productivity, and environmental concerns.

## 4.14 Exercises

4.1. State all the fourteen principles of management as given by Dr. E. Deming.
4.2. Discuss the relevance of Philip Crosby's quality philosophy
4.3. What is the relevance of Quality Engineering philosophy of Dr. G. Taguchi?
4.4. Critically examine the quality trilogy of Juran's quality philosophy.
4.5. What are the components of quality checklists?
4.6. Define probability of an event. What are the features of a probability model?
4.7. Customers are used to evaluate preliminary product designs. In the past 95% of highly successful products received good reviews, 60% of moderately successful products received good reviews, and 10% of poor products received good reviews. In addition, 40% of products have been highly successful, 35% have been moderately successful, and 25% have been poor products.

   (i)   What is the probability that a product attains good review?
   (ii)  If a new design attains a good review, what is the probability that it will be a highly successful product?

4.8. Define a binomial distribution. State its applications.
4.9. Construct a probability distribution for each of the following situations:

(i)   A fair coin is tossed 5 times, and number of tails is recorded.

(ii)  Two fair dice are rolled and the difference of the numbers on both dice is recorded.

(iii) From the process of manufacturing one by one, five items are taken randomly and each item is inspected whether it is defective or non-defective. The number of defective items is recorded.

4.10.   The random variable X has the probability distribution as follows

$$P(X = x) = \begin{cases} cx, & x = 1, 2, 3, 4, 5 \\ c(10 - x), & x = 6, 7, 8, 9 \end{cases}$$

Find the value of $c$. Find $F(x)$. Find mean and variance of the distribution.

4.11.   Define a Poisson distribution. State its applications.

4.12.   Suppose the number of women comes to civil hospital for deliveries at night time is a Poisson random variable with mean 3.4 per day. What is the probability that at a random day 6 or more women come to the hospital for delivery? How many days in a year the hospital can expect no woman come for delivery?

4.13    A university found that 20% of its students drop out without completing the introductory statistics course. Assume that 20 students have registered for the course this quarter.

4.13.   (i)    What is the probability that two or fewer will dropout?

(ii)   What is the probability that exactly four will dropout?

(iii)  What is the probability that more than three will dropout?

(iv)   What is the expected number of withdrawals?

(v)    Use Poisson approximation to find (i)–(iv) and compare the results.

4.14.   $X \sim N$ (50, 100), Determine the following probabilities

(i)    $P(X < 40)$

(ii)   $P(40 < X < 65)$

(iii)  $P(X > 55)$

(iv)   $P(38 \leq X < 62)$

4.16.   Let $X \sim N(25, 100)$ find the values $X$ corresponding to the following probabilities:

(i)    $P(X > x) = 0.3251$

(ii)   $P(X > x) = 0.9382$

(iii)  $P(X < x) = 0.2859$

(iv)   $P(X < x) = 0.7340$

4.17.   In Big Mart store, the daily sales is assumed to be normally distributed random variable. The probability that the daily sales will be less than 38.6 thousand Rs and the probability that the daily sales will be more than 97.4

thousand Rs are 0.025. Find the average sales and standard deviation sales of the store.

4.18. In a rainy season, weekly rainfall in a city is assumed to be a normally random variable with the average rainfall 3 in. and standard deviation 0.8 in.

    (i)    What is the probability that on a randomly chosen week, the rainfall is 1 in. or less?

    (ii)   What is the probability that on a randomly chosen week, the rainfall is 5 in. or more?

4.19. Distinguish between total quality management and total quality control.

4.20. Discuss various characteristics of total preventive maintenance.

4.21. What is Kaizen? How does it help to improve a workplace?

4.22. State all the components of 5S and critically examine its used for improving a service process.

4.23. Describe the uses of 3 M-wastes.

4.24. Describe Poka-Yoke systems and their Safeguards.

4.25. Distinguish between setup reduction and Just-in-time approach.

4.26. Kanban is one method through which JIT is achieved. Justify the statement.

4.27. Describe the importance of Six Sigma concept in reducing the variation. How does the concept facilitates quality improvement?

4.28. Distinguish between Sigma as a measure and Sigma as level of performance of a process.

4.29. The efficiency measurement (measured in percentage) of 50 workers during a workout is given below:

| 94 | 83 | 78 | 76 | 88 | 86 | 93 | 80 | 91 | 82 |
|----|----|----|----|----|----|----|----|----|----|
| 89 | 97 | 92 | 84 | 92 | 80 | 85 | 83 | 98 | 103 |
| 87 | 88 | 88 | 81 | 95 | 86 | 99 | 81 | 87 | 90 |
| 84 | 97 | 80 | 75 | 93 | 95 | 82 | 82 | 89 | 72 |
| 85 | 83 | 75 | 72 | 83 | 98 | 77 | 87 | 71 | 80 |

    (i)    Perform a normality check on the data.

    (ii)   Calculate the Sigma level of the process, if lower and upper efficiency levels are, respectively, given as 76 and 93, respectively.

    (iii)  Compute various process capability measures.

4.30. The number of defective items produced by a manufacturing process is found to be according to Poisson distribution with mean 6. What is the probability that a randomly selected item is (i) non-defective, (ii) two defectives? Also obtain the yield of the process.

4.31. Discuss various approaches of Six Sigma in creating value for an organization.

4.32. What are various components of project charter?

4.33. What is capability maturity model integration? Discuss various types of the integration.

# References

Alves, A. S., & Tenera, A. (2009). Improving SMED in the automotive industry: A case study. In *Production and Operations Management Society 20th Annual Conference* Orlando, Florida, US.

Axline, L. L. (1991, July). TQM: A look in the mirror. *Management Review.*

Barkman, W. E. (1989). *In-process quality control for manufacturing.* Boca Raton, FL: CRC Press.

Coimbra, E. A. (2009). *Total flow management: Achieving excellence with Kaizen and lean supply chains.* Hardcover, Kaizen Institute (Editor).

Deming, W. E. (1982). *Quality, productivity, and competitive position.* Massachusetts Institute of Technology, Center for Advanced Engineering Study, 1982—Business & Economics.

Deming, W. E. (1986). *Out of the crisis.* Cambridge, MA: MIT Center for Advanced Engineering Study.

Feigambaum, A. V. (1961). *Total quality control.* New York: McGraw-Hill.

Goetsh, D. L., & Davis, S. B. (2000). *Introduction to total quality management, quality management for production, processing, and services* (3rd ed.). Englewood, Cliffs, NJ: Prentice-Hall.

Goldratt, E. M. (1977). *Critical chain.* Great Barrington, MA: North River Press.

Gupta, R. (1996, July). Everything in the garden's lovely, *Economist, 340*(7976), 56.

Harry, M., & Schroeder, R. (2000). *Six Sigma: The breakthrough management strategy revolutionizing the world's top corporations.* New York, NY: Doubleday Currency.

Heard (Ed.) (1987). Short cycle manufacturing: The route to JIT. *Target, 2*(3), 22–24.

Henderson, K. M., & Evans, J. R. (2000). Successful implementation of Six Sigma: Benchmarking General Electric Company. *Benchmarking: An International Journal, 7*(4), 260–281.

Jabnoun, N. (2002). Control process for total quality management and quality assurance. *Work The study, 51*(4), 182–190.

James, R. E. (2003). *Total quality: Management, organization and strategy.* Thomson/South-Western, 2005—Business & Economics.

Juran, J. M. (1989). *Juran on leadership for quality.* New York: The Free Press.

Kerzner, H. (1995). *Project management: A systems approach to planning, scheduling, and controlling.* Van Nostrand Reinhold.

Kubiak, T. M., & Benbow, D. W. (2010). *The certified Six Sigma Black Belt Handbook* (2nd ed.). Dorling Kindersley (India) Pvt. Ltd.

Lee, S. S., Dugger, J. C., & Chen, J. C. (2000). Kaizen: An essential tools for inclusion in industrial technology curricula. *Journal of Industrial Technology, 16*(1), 1–7.

Michels, T. B. (2007). *Application of Shingo's single minute exchange of dies (SMED) methodology to reduce punch press changeover times at Krueger International* (Thesis), The University of Wisconsin-Stout.

Muir, A. (2006). *Lean Six Sigma way.* New York: McGraw Hill.

Muralidharan, K. (2008a). A quality philosophy called Six Sigma. *University News, 46*(26), 6–8.

Muralidharan, K. (2008b). Six Sigma: A success story of quality management. *University Today, XXVIII*(22), New Delhi.

Muralidharan, K. (2010). Data mining: A subject to explore for management. *Quality Council Forum of India, 45,* 1–2.

Muralidharan, K. (2015). *Six Sigma for organizational excellence: A statistical approach.* India: Springer Nature.

Muralidharan, K., & Syamsundar, A. (2012). *Statistical methods for quality, reliability and maintainability*, PHI Learning in PVT, India.

Nakajima, S. (1988). *Introduction to TPM: Total productive maintenance.* Cambridge, MA: Productivity Press.

Oakland, J. S. (1997). Interdependence and Cooperation: The essentials of total quality management. *Total Quality Management, 8*(2 & 3), 31–35.

Pande, P. S., Newuman, R. P., & Cavanagh, R. R. (2003). *The Six Sigma way.* New Delhi: Tata McGraw Hill.

Parasuraman, A., Zeithml, V. A., & Berry, L. L. (1985). A conceptual model for service quality and its implication for future research. *Journal of Marketing, 49*, 41–50.

Prabhuswamy, M., Nagesh, P., & Ravikumar, K (2013, February). Statistical analysis and reliability estimation of total productive maintenance. *IUP Journal of Operations Management, XII*(1), 7–20.

Rampersad, H. K., & El-Homsi, A. (2008). *TPS-lean Six Sigma.* New Delhi: SARA Books, Private Limited.

Roebuck, K. (2011). Business process modeling: High-impact emerging technology—What You need to know: Definitions, adoptions, impact, benefits, maturity, vendors (p. 32). Tebbs.

Rubrich, I., & Watson, M. (2004). Implementing world-class manufacturing, Fort Wayne, IN: WCM Associates.

Seiling, J. G. (1999). Reaping the rewards and rewarding work. *The Journal of Quality and Participation, 22*(2), 16–20.

Shingo, S. (1985). A revolution in manufacturing: The SMED system. Cambridge, MA: Productivity Press.

Stamaitis, D. H. (1996). *TQM in healthcare.* Burr Ridge, IL: Irwin.

Steyn, G. M. (1999). Out of the crisis: Transforming schools through TQM. *African Journal of Education, 19*(4), 457–461.

Wienclaw, R (2008). *Operations & Business Process Management*, Research Starters B2B models; 4/1/2018, p1, Essay.

Zeitz, G., Johannesson, R., & Richie, J. E., Jr. (1997). An employee survey measuring total quality management practices and culture. *Group and Organization Management, 22*(4), 414–444.

http://seir.sei.cmu.edu/seir

http://www.sei.cmu.edu/cmmi

# Chapter 5
# Lean, Green, and Clean Quality Assessment Models

*Modern organizations can't succeed unless the people they employ agree to contribute to their mission and survival.*
—Denise M. Rousseau (2004)

## 5.1 Introduction

All business projects must be certified for quality and environmental standards by some agencies. Generally, the lean projects are evaluated through Lean Six Sigma organizations or equivalent certifying bodies. There are many agencies offer services for quality certifications and compliance adherence. However, most of the companies get certified through international agencies like international standards of the organization (ISO), American Society for Quality (ASQ), quality councils (QC), or bureaus of standards of respective nations. These organizations' objectives are to cover the promotion of standards in the world to facilitate the international exchange of goods and services and to develop cooperation in the sphere of intellectual, scientific, technological, and economic activity, including the transfer of technology to developing countries. It also aims at harmonization of standards at the international level to minimize trade and technical barriers. That is to eliminate country-to-country differences, reduce terminology confusion, increase quality awareness, etc. Business organizations practicing the creative and innovative way to win both the confidence and trust of their customers are required to adhere to the quality guidelines for consistent services and sustainable products. Therefore, we believe that the LGC concept is the best option for those organizations working for the more significant interest of their process, product, people, and community.

In this chapter, we describe various assessment models to study the performance standards of projects and processes. The rules, regulations, and compliance issues impacting environmental concerns and sustainable problems are elaborated. They range from operational (principles and tools to provide consistent and sustainable products) to tactical (principles to link current state of affairs) to strategic (policies to coordinate business vision with objectives and values) way of approaching the concerns to promote the necessity of innovation and productivity in managing the world free of wastes and problems. The personal aspects of business are integrated into all the three components. The working philosophy of all standards described below is the famous *Plan-Do-Check* or *Study-Act* way (See also Fig. 3.2).

- *Plan*: Establish objectives and processes required; gather data on critical problems; identify and target root causes of problems; devise possible solutions; and plan a Greening and Turban (2000) test to implement the highest potential solution.
- *Do*: Implement the processes by identifying the resources required, and assign responsibilities to those members of the organization who are responsible for the improvements and performances. Also, prepare an environmental management system (EMS) manual for implementation and control.
- *Check or study*: Measure and monitor the processes and report results. If problems arise, look into the barriers that are obstructing your improvement efforts.
- *Act*: Take action to improve the performance of EMS based on results.

## 5.2  ISO 9000 Series

ISO is the International Organization for Standardization based at Geneva, Switzerland, and was formed in 1947. The objectives of ISO cover the promotion of standards in the world to facilitate the international exchange of goods and services and to develop cooperation in the sphere of intellectual, scientific, technological, and economic activity, including the transfer of technology to developing countries. ISO 9000:2000 consists of four core areas:

1. ISO 9000:2000-Quality Management Systems: fundamentals and vocabulary
2. ISO 9001:2000-Quality Management Systems—requirements (required for certification)
3. ISO 9004–2000-Quality Management Systems—guidelines for performance improvement
4. ISO 19011-2002-Guidelines on quality and environmental management auditing.

Apart from these, there are other international standards like ISO 10012–2001 (quality assurance requirements for measuring equipment and ISO 10015–1999 (quality management guidelines for training). The technical reports include

- ISO 10006-1997-Guidelines to Quality in Project Management
- ISO 10007-1995-Guidelines for Configuration Management
- ISO 10013-1995-Guidelines for Developing Quality Management System Documentation
- ISO 10014-1998-Guidelines for Managing the Economics of Quality
- ISO/TR 10017-1999-Guidance on Statistical Techniques for ISO 9001.

The only standard in this family that may be used for registration, or certification, of a quality management system is ISO 9001. ISO 9000-2000 lists the following steps that should be used to develop a quality management system (QMS):

1. Determine the needs and expectations of customers and other interested parties.
2. Establish the quality policy and quality objectives of the organization.
3. Determine the processes and responsibilities necessary to attain the objectives.
4. Determine and provide the resources necessary to attain the quality objectives.
5. Establish methods to measure the effectiveness and efficiency of each process.
6. Apply these measures to determine the effectiveness and efficiency of each process.
7. Determine the means to prevent nonconformities and eliminate their causes.
8. Establish and apply a process for continual improvement.

The detailed requirements of the standard are stated in various clauses like management responsibility, resource management, product realization, and measurement, analysis, and improvement. Among all the clauses, the management responsibility stated under clause 5 is the most important. It is the responsibility of management to ensure that its quality policy is understood, implemented, and maintained. This clause also requires the appointment of a management representative charged with the operation and maintenance of the quality system. This person must have different access to top management. The person also monitors the formulation and dissemination of a quality policy by senior management with outreach to all levels of the organization, the identification of authority for all work affecting quality, and management's review of the system.

The ISO 14001 standard is part of the ISO 14000 series, which has been developed to provide organizations with a structure for handling environmental impact. ISO 14000:2004 sets out the criteria for an environmental management system (EMS) and can be certified to. It does not state requirements for the environmental recital but maps out a framework that a company or organization can follow to set up an effective environmental management system. It can be used by any organization irrespective of its activity or sector. Using ISO 14000:2004, one can assure company management and employees as well as external stakeholders that environmental influence is being measured and improved. The guidelines for measuring, quantifying, and reducing greenhouse gas emissions are based on ISO 14064 series.

The impact of ISO 14001 includes (Curkovic & Sroufe, 2011)

- more proactive environmental management
- higher levels of communication
- higher levels of waste reduction and cost efficiency
- better return of investment
- higher levels of customer relationship management
- fewer issues with employee health, and
- reduced number of safety incidents.

ISO 26000:2010 guides how businesses and organizations can operate in a socially responsible way. This means acting ethically and transparently that contributes to the health and welfare of society. ISO 26000 provides guidance rather than requirements, so it cannot be certified to unlike the other standards mentioned above. Instead, it helps clarify what social responsibility is, helps businesses and organizations translate principles into practical actions, and shares best practices relating to social responsibility globally (Hahn, 2012). It is aimed at all types of organizations, regardless of their activity, size, or location (ISO 2013). The standard is based on seven fundamental principles: accountability, transparency, ethical behavior, respect for stakeholder interests, respect for the rule of law, respect for international norms and practice, and respect for human rights (ISO 2010) (www.iso.org).

As mentioned above, ISO management standards follow the PDCA cycle that most businesses and health–safety–climate–environment (HSCE) managers will already be familiar with. The other famous series in this is the OHSAS 18001 meant for Operational Health and Safety Assessment Systems (discussed in detail in Sect. 5.6 below). The latest ISO series is the ISO 45001:2018, the first internationally agreed standard for occupational health and safety management. Since most businesses worldwide are small- and medium-sized enterprises (SMEs), their contribution in terms of production, services, and processing is very significant. SME's key role is to promote entrepreneurial focus and innovation, thereby ensuring competitiveness. Therefore, their contribution to the world economy is substantial. There are many layers of categories of SMEs to simplify the type of business they are into. One such relevant group aligned with SMEs is the inclusion of micro, small, and medium enterprises (MSMEs). This is due to continuous research efforts in recent years into the nature and functioning of enterprises.

SMEs account for 90% of businesses worldwide and 50–60% in terms of employment (Sannajust, 2014). SMEs are significant contributors to Poland's social and economic development since Poland has SMEs employing over 6 million people and generating 50% of its GDP. In India, the growth of MSME has consistently registered a higher growth rate compared to the other industrial sectors for the last few years. The industry contributes 8% to India's GDP while accounting for 50% of the total manufacturing output and 45% of India's exports (Nayak et al., 2014). The Micro, Small, and Medium Enterprises Development Act were framed by India's government to address policy issues that affect SMEs' financial performance and thereby enhance their competitiveness. As per the new MSME Act 2006, the number of Indian SMEs ranges from 7.8 million to 13 million. The total share in the GDP of SMEs in India is more than 80%, and more than 90% of all the enterprises are SMEs. The notable characteristics of SMEs are low start-up costs, portability, leadership, management structure, planning, systems and procedures, market and customer focus, operational improvement, innovation, networking, revenue and profitability, ownerships and taxes, and locations. ISO 45001 has been developed to promote all types of business activities that impact the overall growth of a nation, though it is also applicable to larger and more complex businesses.

In India, the MSME contributes to lots of employment generation through start-up companies and contributes to the growth of the economy. MSMEs are under the surveillance of the industrial and socio-economic policies of the government. Some of the objectives of MSME are to provide competitive, cost-effective, and fast services to the industries; to promote e-governance for empowering citizens; promoting the inclusive and sustainable growth of the electronics, IT and IT-enabled service industries; enhancing India's role in Internet Governance, adopting a multi-pronged approach that includes the development of human resources; promoting R&D and innovation; enhancing efficiency through digital services; and ensuring secure cyberspace (https://msme.gov.in/about-us/mission-objectives, last accessed on 19.02.2019). The quality management system adopted by MSME also follows the adherence of ISO standards and guidelines.

Since ISO 9000 series is a family of standards, it is subject to periodic review and revisions. Hence, developing a quality system needs input from various quality improvement programs that are tried and tested. Some of these are Kaizen, Kanban, Poka-Yoke, 5S, TPM, JIT, and so on, discussed in detail in the previous chapter. Most of these concepts are not an integral part of the ISO 9000 series, and none of these terms, ideas, and techniques represent quality system assessment or registration. The strengths of ISO 9001-2000 lie in the structure that sets forth a uniform, consistent set of requirements that can be applied universally, within limitations of interpretations and individual implementation. It provides a basis for designing, implementing, evaluating, specifying, and certifying a quality management system (Wadsworth et al., 2004). The ten steps to ISO 9000 registration are as follows:

1.   Set the registration objective.
2.   Select the appropriate standard.
3.   Develop and implement the quality system.
4.   Select a third-party registrar and apply.
5.   Perform self-analysis audit.
6.   Submit quality manual for approval.
7.   Pre-assessment by registrar
8.   Take corrective actions.
9.   The final assessment by a registrar
10.  Registration!

Bureau of Indian Standards (BIS), the national standard body of India, is responsible for the harmonious development of the activities of standardization and quality certification of goods and products. This benefits the national economy by providing safe, reliable, and quality goods, minimizing health hazards to consumers, protecting the environment, promoting exports and import substitutes, controlling over the proliferation of varieties, etc. The standards and certification scheme of BIS, apart from benefitting the consumers and industry, also support various public policies, especially in areas of product safety, consumer protection, food safety, environment protection, building and construction, etc. (https://consumeraffairs.nic.in/organisat ion-and-units/division/bureau-indianstandards: last accessed on 21.06.2020).

## 5.3  Environmental Management System

The environmental management system (EMS) was first developed to respond to new environmental regulations being imposed on companies. Environmental management has short-term and long-term consequences, affecting the current performance and long-term sustainability of businesses. Carbon management is a relatively new part of this process, gaining a lot of significance in the light of the climate change threat (Walley & Whiltehead, 1994). It is essential to create one comprehensive environmental strategy and understand the potential trade-offs between its constituent parts. The EMS is a structured framework for managing an organization's significant impact on the environment. These impacts can include waste, emissions, energy use, transport and consumption of materials, and increasingly, climate change factors (Yuri et al., 2019).

In response to climate change, the global community adopted two chief mechanisms, mitigation and adaptation (See TEDDY, 2010, page 398). *Mitigation* refers to the policies and measures designed to reduce greenhouse gas (GHG) emissions. At the same time, *adaptation* aims to minimize the adverse impacts of a wide range of system-specific actions. According to IPCC (2007), mitigation refers to the technological change and substitution that reduce resource inputs and emissions per unit of output, and adaptation refers to the adjustment in natural or human systems to a new or changing environment. While mitigation strategies are essential, adaptation strategies are indispensable because even a drastic and immediate cut in global GHG emissions would not entirely prevent the adverse impacts of climate change.

For companies interested in certification of their environmental credentials, there exists a series of international standards like the ISO 14000–14001 series. As mentioned above, these are a set of voluntary rules and guidelines for companies aiming to minimize their environmental impact. This series was published in the year 1996 and is the only standard in the ISO 14000 series for which certification by an external authority is available. It concerns the specification of requirements for a company's environmental management system.

ISO 14001 encourages a company to improve its environmental performance continually. Apart from the obvious—the reduction in actual and possible negative environmental impacts—this is achieved in three ways (Gasti, 2009):

- *Expansion*: Business areas increasingly get covered by the implemented EMS.
- *Enrichment*: Activities, products, processes, emissions, resources, etc., increasingly get managed by the implemented EMS.
- *Upgrading*: The structural and organizational framework of the EMS, as well as an accumulation of knowledge in dealing with business-environmental issues, is improved.

Improving quality is the core of all standards. The customer's perception about quality is that the product or service should have a proper design with a better look, feel, and function apart from its purpose of existence and longevity. According to Deming's quality philosophy, companies that sought to improve their quality have often found they had to change their entire quality culture and approach to problem-solving. An EMS supplies the framework to do so by creating a quality structure (Darnall et al., 2000) to

- adopt a written environmental policy
- identify the environmental aspects and impacts of operations
- set priorities, goals, and targets for continuous improvement in environmental performance
- assign clear responsibilities for implementation, training, monitoring, and corrective actions; and
- evaluate and refine application over time to achieve continuous improvement in the implementation of environmental goals and targets and in the EMS itself.

## 5.4  Eco-Management and Audit Scheme

The Eco-Management and Audit Scheme (EMAS) is a voluntary European-wide standard introduced by the European Union and applied to all European countries. It was formally introduced into the UK in April 1995. According to the Institute of Environmental Management Assessment (IEMA, 2011, last accessed online: iema.net/iema on 14.02.2020), the aim of EMAS is "to recognize and reward those organizations that go beyond minimum legal compliance and continuously improve their environmental performance." Participating organizations must regularly produce a public environmental statement, checked by an independent ecological verifier that reports on their environmental performance.

Among the ISO standards, ISO 50001 supports organizations in all sectors to use energy resourcefully, through the development of an energy management system (EnMS) and can be certified. An EnMS is a set of consistent or interrelating elements to establish an energy policy and energy objectives, and processes and procedures to achieve those objectives. ISO 50001 is also based on the PDCA cycle that makes it easy to assimilate the EnMS into other management systems. The EnMS requirements are divided under the headings: energy policy, planning, implementation and operation, checking, and management appraisal. The issues under each title are the same as for ISO 14001 with the exclusion that the energy policy is a statement of the overall purposes and direction related to the energy performance and that the system is based on the energy aspects of environmental ones.

Among ISO 14000 series, the series ISO 14040–14049 details the life-cycle assessment (LCA), which includes the guidelines for preproduction planning and environment goals setting. The ISO 19011 specifies one audit protocol for both ISO 14000 and 9000 series standards together. This replaces ISO 14011—how to tell if your intended regulatory tools worked. Using ISO 19011 is now the only recommended way to determine this. Thus, ISO 14001:2004 specifies the requirements for every aspect of EMS. Fulfilling these requirements demands objective evidence that can be audited to demonstrate that the EMS is operating effectively in conforming to the standard. In the future, customers will require suppliers to show that they are actively concerned about the environment and that the products or services are produced free from environmental hazards. This will place the importance of these standards on a par with quality standards. As more customers demand confidence in an organization's ability to prove that the organization is worthy, the pressure will be placed on the organizations to be certified in ISO 14000 standards (Juran & De Feo, 2010).

## 5.5    Corporate Sustainable Business Practices

Corporate sustainability is an approach that creates long-term stakeholder value by implementing a business strategy that considers every dimension of how a business operates in the ethical, social, environmental, cultural, and economic spheres. It also formulates strategies to build a company that fosters longevity through transparency and proper employee development (https://en.wikipedia.org/wiki/Corpor ate_sustainability: last accessed on 14.04.20). Corporate sustainability mainly refers to the role that companies can play in meeting the agenda of sustainable development and entails a balanced approach to economic progress, social progress, and environmental stewardship (Atkinson, 2000; Marrewijk, 2003). Sustainable business and corporate social responsibility (CSR) are some of the other terms inadvertently used for describing this concept. According to CSR's various perceptions, the CSR approach is holistic and integrated with the core business strategy for addressing the social and environmental impacts of businesses. Also, CSR needs to address all stakeholders' well-being and not just the company's shareholders (Shukla et al., 2013).

CSR in India has traditionally been seen as a philanthropic activity. And in keeping with the Indian tradition, it was an activity that was performed but not deliberated. As a result, there is limited documentation of specific activities related to this concept. However, what was evident that much of this had a national character encapsulated within it, whether it was endowing institutions to actively participating in India's freedom movement, and embedded in the idea of trusteeship (Shukla et al., 2013). As some observers have pointed out, the practice of CSR in India remains within the generous space but has moved from an institutional building (educational, research, and cultural) to community development through various projects. With global influences and with communities becoming more active and demanding, there appears to be a discernible trend that while CSR remains mostly restricted to community development, it is getting more strategic (that is, getting linked with business) than philanthropic. A large number of companies report the activities they are undertaking in this space in their official Web sites, annual reports, sustainability reports, and even publishing CSR reports.

According to the Brundtland Commission Report (1987), corporate sustainability is defined as *"development that meets the needs of the present without compromising the ability of future generations to meet their own needs."* Globally, the notion of CSR and sustainability seem to be converging, as is evident from the various definitions of CSR put forth by global organizations. CSR, according to Greening and Turban (2000), is used as a framework for measuring an organization's performance against economic, social, and environmental parameters. The standards most relevant for CSR are ISO 14001, ISO 50001, OHSAS 18001, ISO 9001, SA 8000, and ISO 26000.

ISO 14001 is the current environmental management system standard, providing a model for CSR to follow when creating and applying your environmental policy. Focusing on the issues that matter, ISO 14001 is designed to help you achieve consistent regulatory compliance while embedding the concept of continuous improvement at the heart of everything. Essential requirements of the standard include

- Creation of an environmental policy
- Appreciation of your environmental issues and their potential impact
- Formalization of control systems and monitoring processes
- Communication of your policy and approach to staff and stakeholders
- The setting of targets and objectives
- Compliance with relevant legislation.

Since CSR's impact is enormous, an increasing number of companies are seeking to back up their processes with formal ISO certification because of the following potential benefits:

- Operational efficiencies and cost savings are high.
- Supports legislative compliance, significantly reducing the potential for failure
- Opens the door to new business, as certification is often a condition of the supply
- Sends a clear message about your commitment to the sustainability agenda
- It provides a framework for continuous improvement, increasing staff morale, etc.

Thus, CSR works more proactively to environmental issues, climatic conditions, the people's well-being, and the elevation of socially and economically deprived sections of people. According to many authors like Marrewijk (2003), Waddock and Graves (1997), Andersen and Skjoett-Larsen (2009), and Peto (2012), the standard features of lean and CSR can be described as follows:

- Lean creates value for people, community, environment, and economy, whereas CSR invests in long-term prosperity for people, society, and environment.
- Lean creates the right process to produce the correct outputs, whereas CSR ensures that the ecosystem is in balance, if necessary, to intervene in the system.
- Lean adds value by developing people and partners, whereas CSR invests in people and processes.
- Lean continuously improves the processes, whereas CSR encourages the detection of problems in the products and services.
- Lean eliminates waste and variations in the process, whereas CSR educates the people for the adverse effects of waste and variations.

Thus, lean is relevant locally, and CSR is applicable globally, but sustainability is common to both. If lean focuses on professional development, CSR focuses on social development and professional prosperity (Agrogiannis & Agrogiannis, 2015).

Therefore, integrating lean concepts into CSR can boost management activities to a higher level of sustainability, which can prove exciting and rewarding.

## 5.6  Occupational Health and Safety Assessment Systems

OHSAS 18001 (officially "BS OHSAS 18001:2007") is an internationally applied British Standard for occupational health and safety management systems. An OHSAS endorses a safe and healthy working environment by providing a framework that allows an organization to reliably identify and control its health and safety risks and improve overall show (BSI, 2013). BS OHSAS 18001:2007 is the most common standard for OHSAS in the world, and it has a structure that corresponds well with other ISO standards.

The requirements for certifications are divided under the headings:

- Policy and planning
- Hazard identification, risk valuation, and risk control
- Legal and other requirements
- Objectives, targets, and management programs
- Implementation and operation
- Checking and corrective action; and
- Management review.

Looking at the growth and demand of MSEs and MSMEs, the existing standard was revised, keeping the fundamentals intact. And the new standard to join in this series is the ISO 45001, which takes care of health and safety issues, as it is the first internationally agreed standard for occupational health and safety management. The new rule also sparks conversations that focus on business impact, business risk, and conducting business morally and ethically. ISO 45001 will also ensure that companies recognize and embrace the importance of worker involvement and worker consultation in improving working conditions through better engagement of the workforce (Quest Magazine, 2018). An ISO 45001-certified system will bring global recognition to its risk management approach, which attests to the business's moral and ethical credentials as a whole and will give them an edge when competing for international contracts. Among many changes affected, the main difference is that ISO 45001 concentrates on the interaction between an organization and its business environment. At the same time, OHSAS 18001 was focused on managing OH&S hazards and other internal issues.

The benefits of the increasing popularity of OHSAS will be

- Improve business performance by reducing absenteeism due to workplace illness and injury.
- Facilitate continual improvement in employee morale while decreasing insurance costs and liability.
- Transform work operation from detection mode to prevention mode.

## 5.7 Emission Auditing

Another relatively simple way of addressing the green concern and assessing environmental pollution is through the use of emission auditing. Emission auditing calculates the amount of greenhouse gas (GHG) or other pollutants released into the atmosphere by a given activity. This can be due to vehicles, machines, production, maintenance, etc. When estimating vehicle emissions, a variety of factors can be taken into account, including load weight and distribution, vehicle age, vehicle size, vehicle design, driving style, road gradient, and speed. Fleet congestion, road congestion also should be taken into consideration while estimating the emissions. It is also worth mentioning the Leadership in Energy and Environmental Design (LEED) certification process developed by the US Green Building Council to propel green building design in the United States (Boeing, 2014). LEED certification is a very prestigious title and can be attained through compliance with all environmental laws and regulations, occupancy scenarios, building permanence, pre-rating completion, site boundaries, area-to-site ratios, and obligatory five-year sharing of whole-building energy and water use data from the start of occupancy (for new construction) or date of certification (for existing buildings).

## 5.8 Carbon Management

Carbon profile or footprint refers to "*total greenhouse gas emissions caused directly and indirectly by a person, organization, event or product, gases that include non-carbon compounds.*" The greenhouse gases (GHGs) are carbon dioxide ($CO_2$), methane ($CH_4$), nitrous oxide ($N_2O$), sulfur hexafluoride ($SF_6$), and a class of compounds known as fluorocarbons. To calculate a carbon footprint, the amount of GHG emissions that a specific process causes is converted to a carbon dioxide equivalent. The class of greenhouse gases known as fluorocarbons contains over a hundred compounds, some of which are included in specific carbon footprinting databases and not in others. Compared with fossil fuels like natural gas and petroleum, coal emits the highest amount of GHGs per calorie of energy produced. Poor carbon management can contribute to risks like business risks, financial risks, and regulatory risks. Reducing carbon footprint could reduce costs and manage future carbon quotas (Juran's EcoQuality, www.juran.com).

A carbon management plan is a documented strategy and a set of actions to help an organization meet its carbon reduction objectives. It will typically include a summary of previous carbon footprint assessments, carefully identified carbon reduction targets, and defined actions to achieve the organization's goals. According to an estimate by Shah and Venkatramanan (2019), the GHG emission from cultivation in India has increased by 161% from 14.81 TgCE/year in 1960 to 38.71 TgCE/year by 2010 over the past 50 years. This is due to an increase in the use of inputs primarily chemical nitrogen fertilizer, shifting from conventional animal and human energy sources to carbon-intensive diesel and electricity-dependent machinery. To make matters worse, there are many other sources and estimates, which are not accounted for GHG emissions in India. One of the severe concerns is the increasing usage of the Internet and its growing demand for information and communication technology (ICT) and other enabled services.

Much literature supports the findings that the Internet consumes a considerable amount of GHG-emitting energy, which can be divided into two parts. Technology companies must manufacture (by extracting raw materials and assembling in factories) and ship (using airplanes, trucks, etc.) the internet hardware: servers, personal computers, iPhones, etc. Those devices must then be powered and cooled, drawing electricity from their local grids and servers to enable the energy generation in its own ways with varying GHG emissions (Gombiner, 2011). According to Harvard physicist Alex Wissner-Gross, a Google search accounts for seven grams of carbon dioxide emissions. Seven grams of $CO_2$ emissions are equivalent to boiling a pot of tea or driving a car 52 feet. With one billion Google searches occurring every day, there are one billion grams of $CO_2$ being emitted into the atmosphere due to Google searches alone. This is the same as driving a car 2,375,000 miles (or 80,000 people commuting to work 15 miles each way). Wissner-Gross estimates that looking at a webpage with pictures or videos emits 0.2 g of $CO_2$ per second. Secondly, YouTube users watch two billion videos every day and upload hundreds of thousands more (YouTube factsheet). If the average time a user views a video is ten seconds (a random guess), watching videos on YouTube accounts for another four billion grams of $CO_2$ emissions or an additional 320,000 car commuters (Hölzle, Urs, 2009).

According to Andrew (2018), India's $CO_2$ emissions grew by an estimated 4.6% in 2017, despite a turbulent year. Measured per person, India's emissions are still very low—at only 1.8 tones of $CO_2$ per capita—which is much lower than the world average of 4.2 tones. But those emissions have been growing steadily, with an average growth rate over the past decade of 6%. Many studies attribute electricity, car, and air travel as the primary contributor to $CO_2$ emissions (Baliga et al., 2009). Air travel is believed to be a significant source of carbon emissions (Penner et al., 1999). A reduction in business travel would reduce greenhouse gas emissions (Nairn, 2007). However, the Internet's capacity would need to be significantly increased to support good quality video conferencing. If Internet capacity is increased, energy consumption and consequently the greenhouse footprint of the Internet will also increase (Baliga et al., 2009).

## 5.9  Environmental Stress Screening

*Environmental stress screening* (ESS) is a reliability technique with universal application to improve product reliability. ESS is subjecting the new product to a period of use and external stress that precipitates instantaneous or early time failures (ETF) within the product before shipment to the customer. The two main reasons for performing

ESS are that (i) product development engineers will quickly identify reliability problems and initiate corrective actions by obtaining data about ETFs. Secondly, the product shipped to the customer after ESS has improved reliability because of ETFs' elimination. According to Harrington and Anderson (1999), ESS is a technique that has provided a significant improvement in product reliability, allowing it to meet required, specified reliability goals. The four predominant reasons for employing ESS are

1. ESS is a tool to improve product reliability. It is the best way of approaching the best MTBF based on the flat or exponential modeled part of the failure rate curve.
2. ESS detects and eliminates lousy product lots from being shipped to the customer. The period to control deviation during sampling is also shortened, and hence, it helps determine the root cause of the product reliability and functional problems.
3. ESS shortens the development cycle in two ways. First, it provides a quick feedback loop for identifying and corrective action required for reducing and eliminating product problems. Secondly, ESS provides an accurate measure of both the functional and reliable performances needed to specify the next product.
4. ESS is a monitor of product performance. The manufactured product is evaluated continuously, providing a measure of effect for engineering changes during manufacturing, the variation in parts supplied to construction, and the stability of the manufacturing process.

Thus, ESS will identify most of these problems and provide an evaluation of change interactions. The word *screening* in ESS is appropriate in that it indicates that potential defective products will be screened from the shipped product. Some other types of screens are vibration, mechanical shock, acceleration, humidity, chemicals, altitude, radiation, etc. A concept very close to ESS is the *accelerated life test* (ALT) plan to obtain new information about the failure of items or components, as estimating the failure time distribution or long-term performance of parts of high-reliability products is particularly tricky. Most new products are designed to operate without failure for years, decades, or longer. Thus, few units will fail or degrade appreciably in a test of reasonable length at normal use conditions. ALT is used widely in manufacturing industries, mainly to obtain timely information on the reliability of pure components and materials.

According to Meeker and Escobar (1998), there are three methods of accelerating a reliability test:

- Increase the use-rate of the product, by increasing the frequency of use, items life decreases and eventually fails.
- Increase the aging rate of the product. For example, increasing experimental variables like temperature or humidity can accelerate the chemical processes of specific failure mechanisms, such as chemical degradation of an adhesive, mechanical bond, or the growth of conducting filament across an insulator (eventually causing a short circuit).

- Increase the level of stress under which test units operate. Units fail when their strength below applied stress. The stress level can be increased using temperature cycling, voltage, or pressure.

Combinations of these methods also work in many situations. Today's manufacturers face intense pressure to develop new, higher technology products in record time while improving productivity, product field reliability, and overall quality. This has motivated the development of methods like concurrent engineering and encouraged extensive use of designed experiments for product and process improvement. This is in line with the modern quality philosophy for producing high-reliability products. Therefore, methods like ESS and ALT are proving to be methods to assess or demonstrate component and sub-system reliability, to certify components, to detect failure modes so that they can be corrected, to compare different manufactures, and so forth (Meeker & Escobar, 1998; Lawless, 1982).

## 5.10  Climate Models

Climate models are mathematical representations of the interactions between the atmosphere, oceans, land surface, ice, and the sun. Climate models are used to assess the $CO_2$-global warming hypothesis and to quantify the human-caused $CO_2$ "fingerprint."

A new study published in a peer-reviewed journal by Nic Lewis and Judith Curry (2018) finds that climate models exaggerate global warming from $CO_2$ emissions by as much as 45%. In the study, the authors looked at actual temperature records and compared them with climate change computer models. They found that the planet has shown itself to be far less sensitive to increases in $CO_2$ than the climate models. Randal et al. (2003) provide various climate models based on measurement-based calculations. Authors argue that climate models are being subjected to more comprehensive tests, including, for example, evaluations of forecasts on time scales from days to a year for projections and predictions. Among various models proposed, the Atmosphere–Ocean General Circulation Models (Randall et al., 2007) can simulate extreme warm temperatures, cold air outbreaks, and frost days reasonably well.

According to the Times of India daily newspaper report dated 14.02.2019, the climate change issues are so severe that it could wipe out Bengal tigers in 50 years. Sundarbans, the iconic Bengal tiger's last coastal stronghold and the world's biggest mangrove forest, could be destroyed by climate change and rising sea levels in the next 50 years. According to scientists from James Cook University in Australia, the Sundarbans are under growing pressure from industrial developments, new roads, and higher poaching beyond climate change. Many endangered species of animals and plants face similar extinction due to the adverse effects of climate change. This is costing the natural habitat to a great extent.

The United Nations Framework Convention on Climate Change (UNFCCC) sets an overall framework for international efforts to address climate change issues.

Concerning emissions, the parties to the convention are required to furnish information on GHG emissions, describing the steps taken and strategies planned to implement the objectives. India submitted its initial National Commission (NATCOM) plan to the UNFCCC in June 2004 (MoEF, 2004). In 2008, the Government of India announced the National Action Plan on Climate Change (NAPCC) to adapt to and mitigate climate change impacts in eight key sectors. The NAPCC has eight national missions, which deal with topics like solar energy, the Himalayan ecosystem, water resources, sustainable agriculture, forests, energy efficiency, sustainable habitat, and strategic knowledge. In August 2009, the state governments were urged to make state-level action plans consistent with the objectives of the NAPCC (Teddy, 2010).

## 5.11  Life-Cycle Assessment

*Life-cycle assessment* (LCA), also known as life-cycle analysis, eco-balance, and cradle-to-grave analysis, is a technique to assess environmental impacts associated with all the stages of a product's life from raw material extraction through materials processing, manufacture, distribution, use, repair and maintenance, and disposal or recycling (US Environmental Protection Agency, last accessed on 11.01.2020). Cradle-to-grave is the full LCA from resource extraction ("cradle") to the use phase and disposal phase ("grave"). The most critical applications of LCA are analysis of the contribution of the life-cycle stages to the overall environmental load and to prioritize improvements on products or processes and comparison between products for internal use (Muralikrishna & Manickam, 2017; Widheden & Ringström, 2007).

There are four linked components of LCA:

- *Goal definition and scoping*: Identify the LCA's purpose and the expected products of the study, and determine the boundaries and assumptions based upon the goal definition.
- *Life-cycle inventory*: quantifying the energy and raw material inputs and environmental releases associated with each stage of production
- *Impact analysis*: assessing the impacts on human health and the environment associated with energy and raw material inputs and environmental releases quantified by the inventory
- *Improvement analysis*: evaluating opportunities to reduce energy, material inputs, or environmental impacts at each stage of the product life cycle.

The technological advancements in manufacturing processes and the steadily increasing variety of materials and products have led to a continuous rise in the amounts of waste generated. The cradle-to-grave flow of articles has proven to be just enough to protect the environment but inefficient due to the depletion of natural resources. The ISO standards developed for assessing LCA are ISO14040 (LCA—Principles and Guidelines), ISO14041 (LCA—Life Inventory Analysis), ISO14042 (LCA—Impact Assessment), and ISO14043 (LCA—Interpretation).

According to El-Haggar (2007), LCAs can be used in the following ways:

1. Consumers wishing to make the best choice for the environment when purchasing a product can use the analyses from an LCA to compare the environmental "preferability" of different products.
2. Manufacturers wishing to minimize their products' environmental impacts can use LCAs to determine where to make changes in their products' life cycles to reduce significant environmental impacts.
3. Governments can use LCAs to pinpoint particular steps in a product's life cycle, where regulations will be the most effective at reducing environmental degradation and conserving natural resources.

A closed variant of LCA is the economic input–output LCA (EIOLCA), which involves the use of aggregate sector-level data on how much environmental impact can be attributed to each sector of the economy and how much each sector purchases from other industries (Hendrickson et al., 2005). Such analysis can account for long chains (e.g., building an automobile requires energy, but producing energy requires vehicles, and making those vehicles requires energy, etc.), which somewhat alleviates the scoping problem of process LCA.

## 5.12  Cleaner Production Assessment

Cleaner production (CP) means the continuous application of an integrated, preventive environmental strategy to processes and products to reduce risks to humans and the environment. Cleaner production is a creative way of thinking about products and processes that makes them. It is realized through continuous applications of business and production strategies to minimize the generation of wastes and emissions. For operations, CP involves conserving raw materials and energy, eliminating toxic substances, reducing quantity and toxicity of emissions, and debris before they leave the process. Whereas, for products, CP is nothing but reducing the environmental impact during the entire life cycle from raw material extraction until its disposal.

For CP to be active and sustaining, a strong organizational commitment must be formulated and adopted. The promise should include taking responsibility, fixing targets, reviewing progress, and timely implementation of production programs and plans. The technique includes

- *Source reduction.* This is managed through new approaches to the production process, where corrective and preventive actions are carried out to minimize resources and technology use. The strict adherence of SOPs is encouraged to gain the material and equipment usages.
- *On-site recycling.* It is the process of recovering useful materials from wastes and recycled materials.
- *Product modification.* This is the process of minimizing the environmental impacts of production or products through appropriate product reformulation or change in product composition.

The theory behind a cleaner production assessment (CPA) aims to identify, evaluate, and implement CP opportunities in a company. Therefore, CPA helps track the flow of materials through the industry to reduce or eliminate waste. For this, the involvement of the production- and non-production-related activities in the company is ensured to focus on targeted goals and aims. The working principle of CP consists of the following steps:

- Preparation stage: It involves planning and organization of the CP assessment, including the establishment of a project team, performing a preliminary survey, and the selection of audit items.
- Analyzing process steps: It involves evaluating the unit operations relevant to the selected audit item and preparing materials to quantify waste generation, its costs, and its causes. Descriptive and inferential statistics are predominantly used for analyzing the process steps.
- Generating cleaner production opportunities: Here, the team members get into action mode to identify the methods for identifying and eliminating wastes. Inputs from management and expertise are critically evaluated for each opportunity.
- Selecting cleaner production solutions: A sustainable and environmentally friendly solution will be suggested, recommended, and implemented.
- Implementing cleaner production solutions: It involves the preparation of checklists of tasks completed, agencies and departments consulted, and help for other production areas.
- Sustaining cleaner production: It involves integrating all stakeholder views for good housekeeping, process optimization, and best production practices for maintaining the company's objectives.

In their review, Bonilla et al. (2010) point out that the adoption of environmentally sound technological solutions based upon scientific research and societal testing is necessary for understanding the role of cleaner production in sustainable development. Fortunately, some companies are increasingly done via the implementation of environmentally oriented, socially responsible economically sound, management programs; such approaches provide the benefit of helping them make environmental, social, and economic progress simultaneously. This win-win approach is sometimes called the *triple bottom line* (TBL) approach. It has become clear that new legislation, which requires that companies engage in waste minimization and extended producer responsibility from the product design phase to the production phase, to the consumer phase, to the end-of-life management of the 'dead' products phase, can be an increasingly important catalyst for changes in corporate and societal attitudes, values, paradigms, and practices.

## 5.13 Other Assessment Models

The green methods of evaluation are mainly done for air and climate change-related environmental problems. There are many other green methods of assessments as well to support the business and trade. Two such ways, where the involvement of data-based management techniques is generally used, are as follows:

- Willingness to pay (WTP) assessment is done to make customers aware of the pricing and product development details about environmental impacts and influences. WTP is also the maximum amount that the homeowner would pay for the right to clean air.
- Willingness to accept (WTA) assessments are the minimum amount a customer pays in acceptance of good or clean air. Both WTP and WTA are essential for public policy (Horowitz and Mcconnee, 2003; Ju et al., 2007).

The climate change issues are generally propagated through the use of climate models to predict temperature rise, assess future damage to human activity/ecosystems, the cost of the damage, calculation of marginal abetment costs, etc. There are many other programs also practiced by governments and NGOs globally. Some of them are as follows: Design for Environment Program (www.epa.gov/dfe); Green Engineering Program (www.epa.gov/oppt/greenengineering); Green Suppliers Network Program (www.greensuppliers.gov); Lean and Environment Initiative (www.epa.gov/lean); National Partnership for Environmental Priorities (www.epa.gov/npep); Sector Strategies Program (www.epa.gov/sectors); and Sustainable Futures Program (www.epa.gov/oppt/sf), etc.

## 5.14  Exercises

5.1.  What is the significance of ISO 9000 certifications?

5.2.  Distinguish between SME and MSME.

5.3.  Discuss how EMS helps improve climate change issues. How does it differ from EMAS?

5.4.  Discuss the importance of CSR activities. How is it different from sustainable practices?

5.5.  Discuss various features of CSR and lean activities. State how they can be compared to multiple standards of ISO.

5.6.  What is the importance of OHSAS in safety management systems? What are the particular standards of the system?

5.7.  Write short notes on

   (i)    Emission auditing
   (ii)   Carbon management
   (iii)  Climate models

5.8.  Distinguish between environmental stress screening life-cycle assessments.

5.9.  State all features and importance of cleaner production.

# References

Agrogiannis, S., & Agrogiannis, C. (2015). *A critical review and evaluation of the lean concept and corporate social responsibility/sustainability: Investigating their interrelation and contribution in terms of business competitive positioning.* KTH Royal Institute of Technology, Industrial Engineering, and Management.

Andersen, M., & Skjoett-Larsen, T. (2009). Corporate social responsibility in global supply chains. *Supply Chain Management: An International Journal, 14*(2), 75–86.

Andrew, R. (2018). Why India's $CO_2$ emissions grew strongly in 2017? *Guest Post.* CICERO Center for International Climate Research.

Atkinson, G. (2000). Measuring corporate sustainability. *Journal of Environmental Planning and Management*, 235–252. https://doi.org/10.1080/09640560010694

Baliga, J., Hinton, K., Ayre, R., & Tuckder, R. S. (2009). The carbon footprint of the Internet. *Telecommunications Journal of Australia,* Monash University e-press.

Boeing, J. S., Barizão, E. O., Costa e Silva, B., Montanher, P. F., Almeida, V. C., & Visentainer, J. V. (2014). Evaluation of the solvent effect on the extraction of phenolic compounds and antioxidant capacities from the berries: Application of principal component analysis, *Chemistry Central Journal, 8*, 48. https://doi.org/10.1186/s13065-014-0048-1.

Bonilla, S. H., Almeida, C. M. V. B., Giannetti, B. F., & Huisingh, D. (2010). The roles of cleaner production in the sustainable development of modern societies: An introduction to this special issue. *Journal of Cleaner Production, 18*(1), 5.

Brundtland Commission's Report. (1987). *Report of the World Commission on environment and development, General Assembly resolution 42/187*, 11 Dec 1987.

BSI (2013). *ISO standards, certification news.*

Curkovic, S., & Sroufe, R. (2011). Using ISO 14001 to promote a sustainable supply chain strategy. *Business Strategy and the Environment, 20*(2), 71–93. https://doi.org/10.1002/bse.67

Darnall, N., Gallagher, D. R., Andrews, R. N. L., & Amaral, D. (2000). Environmental management system. *Environmental quality management*, 1–9.

El-Haggar, S. M. (2007). Current practice and future sustainability. *Sustainable Industrial Design and Waste Management,* 1–19.

Gastl, R. (2009). *Kontinuierliche Verbesserung im Umweltmanagement: Die KVP-Forderung der ISO 14001 in Theorie und Unternehmenspraxis.* vdf Hochschulverlag AG. p. 336. https://doi.org/10.3218/3231-4.

Gombiner, J. (2011). Consilience. *The Journal of Sustainable Development, 5*(1), 119–124.

Greening, D. W., & Turban, D. B. (2000). Corporate social performance as a competitive advantage in attracting a quality workforce. *Business & Society, 39*(3), 254–280.

Harrington, H. J., & Anderson, L. C. (1999). *Reliability simplified.* McGraw-Hill.

Hahn, R. (2012). Transnational governance, deliberative democracy, and the legitimacy of ISO 26000. *International Journal of Business and Society.* https://doi.org/10.1177/0007650312462666

Hendrickson, C. T., Lave, L. B., & Matthews, H. S. (2005). *Environmental life cycle assessment of goods and services: An input-output approach.* Resources for the Future Press ISBN 1-933115-24-6.

Hölzle, Urs. (January 11, 2009). Powering a Google search. *The Official Google Blog.* http://www.youtube.com/t/fact_sheet.

Horowitz, J. K., & Mcconnell, K. (2003). Willingness to accept, willingness to pay and the income effect. *Journal of Economic Behavior & Organization, 51*(4), 537–545.

https://www.carbonfootprint.com/carbonmanagement.html.

IPCC (2007). *Climate change 2007: Synthesis report.* http://www.ipcc.ch/pdf/assessment-report/ar4/syr/ar4_syr.pdf.

ISO (2010). *Guidance on social responsibility.* ISO/TMBG Technical Management Board.

ISO (2013). *Information technology-security techniques: Information security management systems requirements*, ISO/IEC JTC 1/SC 27, Information security, cybersecurity and privacy protection.

Ju, Y.-J., Zhang, Q., & Hu, C.-M. (2007). Assessment of the disparity between Willingness to Accept (WTA) and Willingness to Pay (WTP) by Value. *International Conference on Management Science and Engineering, 20–22*, 1184–1189.

Juran, J. M., & De Feo, J. A. (2010). *Juran's quality handbook* (6th ed.). Tata McGRAW-HILL.

Lawless, J. F. (1982). *Statistical models and methods for lifetime data.* John Wiley & Sons.

Lewis, N., & Curry, J. (2018). The impact of recent forcing and ocean heat uptake data on estimates of climate sensitivity. *Journal of Climate.*

Marrewijk, M. V. (2003). Concepts and definitions of CSR and corporate sustainability. *Journal of Business Ethics, 44*, 95–105.

Meeker, Q. W., & Esobar, L. A. (1998). *Statistical methods for reliability data.* Wiley.

MoEF(2004). *India's initial national communication to the United Nations framework convention on climate change.* Ministry of Environment and Forests, Government of India.

Muralikrishna, I. V., & Manickam, V. (2017). *Environmental management: Science and engineering for industry.* Elsevier.

Nairn, R J. 2007. Broadband telecommunications and urban travel. *Telecommunications Journal of Australia, 57*(2–3): 26.1–26.9. https://doi.org/10.2104/tja07026.

Nayak, R., Kumar, A., & Sengupta, R. (2014). Barriers affecting the implementation of Technology Transfer (TT) in apparel manufacturing Indian SMEs. *International Journal of Applied Sciences and Engineering Research, 4*(4), 417–426.

Penner, J. E., Lister, D. H., Griggs, D. J., Dokken, M., & Mcfarland (eds.). (1999). Aviation and the global atmosphere.—A special report of IPCC working groups I and III. Intergovernmental Panel on Climate Change (p. 365). Cambridge University Press.

Peto, J. (2012). That the effects of smoking should be measured in pack-years: Misconceptions. *British Journal of Cancer, 107*(3), 406–407. https://doi.org/10.1038/bjc.2012.97.

Quest Magazine (2018). *ISO 45001:2018 occupational health & safety management.* QuEST Management Solutions.

Randall, D. A., Wood, R. A., Colman, R., & Fichefet, T., et al. (2007). Climate models and their evaluation. In Solomon (Ed.), *Climate change 2007: The physical basis* (pp. 589–662). Cambridge University Press.

Randall, D. A., et al. (2003). Confronting models with data: The GEWEX cloud systems study. *Bulletin of the American Meteorological Society, 84*, 455–469.

Sannajust, A. (2014). Impact of the world financial crisis on SMEs: The determinants of bank loan rejection in Europe and the USA. Working paper, 327.

Shah, S., & Venkatramanan, V. (2019). Advances in microbial technology for upscaling sustainable biofuel production. In V. K. Gupta, & A. Pandey (Eds.), *New and future developments in microbial biotechnology and bioengineering* (pp. 69–76). Netherlands, Europe: Elsevier. https://doi.org/10.1016/b978-0-444-63504-4.00005-0.

Shukla, S., Jawakar, K., & Chakravarthy, S. (2013). *Handbook on Corporate Social Responsibility in India.* CII and PwC India.

TEDDY. (2010). *TERI energy data directory and yearbook.* New Delhi.

Waddock, S. A., & Graves, S. B. (1997). The corporate social performance-financial performance link. *Strategic Management Journal, 18*(4), 303–319.

Wadsworth, M. E., Raviv, T., Compas, B. E., Jennifer, K., & Connor-Smith. (2004). Parent and adolescent responses to poverty related stress: Tests of mediated and moderated coping models. *Journal of Child and Family Studies, 14*, 283–298.

Walley, N., & Whitehead, B. (1994). It is not easy being green. *Harvard Business Review, 72*(3), 54–64.

Widheden, J., & Ringström, E. (2007). *Handbook for cleaning/decontamination of surfaces.*

Yuri, K., Albina, G., & Muralidharan, K. (2019). Lean six sigma for sustainable business practices: A case study and standardization. *International Journal for Quality Research, 13*(1), 47–74. ISSN 1800-7473.

# Chapter 6
# Green Statistics: Essence of Lean, Green, and Clean Sciences

*An organization must constantly measure the effectiveness of its processes and strive to meet more difficult objectives to satisfy customers.*
—Taiichi Ohno: Toyota Production System

## 6.1 Introduction

Quality management is increasingly becoming a process/product improvement tool for many corporate institutions and organizations. A quality concept creates constancy of purpose toward improvement of product and service to become competitive, to stay in business, and to provide jobs. This increases a common interest in achieving better goals in product and service quality along with a long-term relationship of loyalty and trust. A quality concept also institutes leadership and practice. Such a business environment eliminates management by objective; instead, learn the capabilities of the system and make way for improvement (Muralidharan, 2018a, b).

As we know, there are many activities in a company, irrespective of whether they are into manufacturing, production, information technology, research, and development, or automobile or service industries. Uncertainties and probabilities are pervasive in all these institutions. They are due to man, machine, management, materials, methods, or any other valid causes and problems. Therefore, cooperation within and between organizations is vital to minimize the logistical impact of workforce utilization, product movement, material flows, sales and marketing of products, etc. Ideally, the entire supply chain needs to be statistically estimated and validated for suitable conclusions and interpretations regularly for the business's long-term sustainability. It borders the firm's environmental impact and corporate boundaries within which the organization functions. Once the inventories are known, it becomes easy to propose necessary amendments and the number of greening activities required in the organization.

A variety of business models exists for evaluating the performance of the organization. Specifically, for assessing an organization's environmental performance,

K. Muralidharan, *Sustainable Development and Quality of Life*,
https://doi.org/10.1007/978-981-16-1835-2_6

we have models like EMS, EMAS, corporate sustainable business practices, climate models, and others (already discussed in the previous chapter). To assess and evaluate the performance of the organizational supply chain, we have a new concept called the *"Green Statistics"* introduced by Muralidharan and Ramanathan (2013). The Green Statistics concept provides measurement-based performance checks along with quality guidelines for every activity in the organization. They are used to assess business activities' economic and financial performance and improve the environmental concerns which impact climate change. This concept connects the organization's supply chain dynamics, as it involves material supplies, production facilities, distribution services, and customer-linked activities. One should understand that uncertainty is a common feature in all technical processes, and data plays a significant role in managing that information. Therefore, Green Statistics encourages statistical analysis for performance evaluation and strategic decision making efficiently. In different sections below, we present the necessity of Green Statistics, its relevance in managerial decision making, and various methods of green evaluation without inviting much of technical and computational aspects.

## 6.2  Green Statistics

With the advent of quality improvement techniques like TQM, Kaizen, Taguchi methods, Six Sigma, CMMI, etc., management accountability and customer expectations have also increased manifold. This has put a considerable amount of pressure on management to deliver a superior quality product to the stakeholders. Besides, the product has to be environmentally friendly and produced in a safe environment with fewer waste and carbon emissions. Therefore, a data-based management decision is necessary to bring transparency in the system for the business's long-term sustainability. The Green Statistics concept facilitates this necessity.

Green Statistics (GS) could be defined as *"the systematic collection and analysis of data that could lead to assessing environmental impacts of the operations of a firm. These statistics can be qualitative and quantitative"* (Muralidharan, 2015; Muralidharan & Ramanathan, 2013). From the definition of GS, it is apparent that the stakeholders have lots of responsibility and accountability toward the quality of their products and services. Hence, companies should make available the proper use of statistical (data or measurement-based) information and its impacts on the spot assessment about their functional activities.

The component *qualitative statistics* of GS can be anything like:

- The source of information regarding the raw materials and machinery
- Strength of the company in terms of its size, human resources, machinery, and methods
- Display of breakthrough achievements in the past and present
- The complete information on innovations carried out by the companies
- The green initiatives carried out by the company

- Waste disposal schemes employed
- Sources of energy conservation and fuel consumption
- Methods to minimize noise pollution in the organization
- Employee retention and hiring
- Performance measurement
- Statistical process/quality control status
- Tax paid and company's liabilities
- State of reactive and proactive maintenance activities
- Status of energy recovery processes
- Status of renewable and recycling processes.

and the component *quantitative statistics* of GS can be anything like:

- Number of significant accidents/incidents, catastrophes happened in the organization
- Number of bugs reported in computer programs and software
- Number of training programs held
- Number of projects on quality improvements
- Number of collaborations between companies and governments
- Number of corporate social responsibility programs carried out by the company
- Number of significant conference/seminars/workshops held
- Amount of carbon released by the company
- Amount of level of greenhouse gas emissions
- Amount of hazardous wastes collected and disposed of.

The above statistical information leads to a robust database, which can be further analyzed and used for appropriate decision making. If necessary, the data collection plan may be modified and prepared according to the customer specifications and requirements. The flexibility for modifications and amendments should be suitably allowed as per the objectives of the institution. It is considered a good practice to introduce and explain the project's goals to the stakeholders and, hopefully, get their buy-in and active support. When data is outside the organization, it is useful to have designated people in all companies involved to internally coordinate and manage the data collection process. At this stage, an expert's (maybe a statistician, lean consultant, Six Sigma professional, quality engineer, or an operations manager) involvement may be made available for initial assessment analysis and action plan thereof.

There are many driving forces behind the information-oriented concepts like GS, as discussed above. They can be operational, strategic, tactical, or even personal. Some of them are:

- Improve public relations.
- Improve customer relations.
- Improve investor relations.
- Improve corporate relations.
- Understand competitors' potential.
- Understand product variants.

- Understand market variants.
- Reduce logistic costs.
- Satisfy customer requirements.
- Bound to use government compliance for all sustainable activities.

As per the theoretical constructs, the GS necessitates the use of lean and green principles to identify and quantify the wastes (variations, errors, redundancies, randomness, clutters, etc.) generated from all the process activities. Since the environmental concern is at its core, GS also facilitates awakening in the collective responsibility of people associated with each process. Therefore, the benefits of GS are tremendous and spontaneous. Some of these are:

- Reduce overall business costs and thereby improve profits.
- Improve visibility of green drivers.
- Improve brand image.
- Reduce greenhouse gas (GHG) emissions.
- Reduce wastage.
- Improve process and product quality.
- Improve customer satisfaction.
- Improve employee confidence.

To coordinate and integrate the flows of information both within and among companies, one should understand GS's importance in the supply chain management system. The ultimate goal of any effective supply chain management system is to reduce inventory. In the next section, we discuss the importance of strategic decisions necessary to contain the uncertainty in the green movement of stock and dynamics.

## 6.3   Green Statistics Supply Chain Dynamics

Traditionally, a supply chain has been defined as *a system whose constituent parts include material suppliers, production facilities, distribution services, and customers linked together by the feed-forward flow of materials and the feedback flow of information* (Stevens, 1989). When designing a supply chain network, different levels of decisions need to be considered, from strategic to operational. Strategic decisions typically have a planning period of many years and long-lasting effects. Identifying the number of products, locations, and capacities of serving facilities, such as distribution centers and warehouses, would generally be a part of strategic planning. Tactical activities include selecting suppliers, assigning products to distribution centers, determining the distribution channel, and the type of transportation mode. Finally, operational decisions, such as scheduling and routing activities, consider the day-to-day flow of products through the network, the amount of the inventory to be held by the facilities, and so on. All this can be achieved through some quantitative analysis and operations research techniques. The flow of information established through this analysis will get you to identify the variations, wastes, and their

**Fig. 6.1** Green Statistics
Supply Chain Management
system

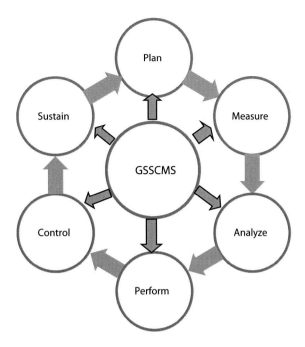

management. Eventually, through appropriate techniques applied to resources like man, machine, materials, supervision, methods, etc., one can plan and utilize the resources effectively, to reduce the wastes.

Integrating GS into SCM leads to a productive working model for any organization committed to quality and best business practices. We call this model as Green Statistics Supply Chain Management (GSSCM) system model, as shown in Fig. 6.1. The component of each is explained as follows:

- *Plan*—Plan all the aspects of organizational transactions, including sales and operations, including supplies, procurement, etc.
- *Measure*—Measure the performance parameters related to all transactional processes, as mentioned in the previous stage. Identify and eliminate all the non-value-added activities in the supply chain process to remove all types of variations and wastes.
- *Analyze*—Analyze the pattern of variations and its root causes and test the supply chain metrics for concurrent realization and delivery. Suggest the optimum level of inventories to avoid excess wastage.
- *Perform*—Perform capability analysis, prepare SOPs for implementing the solutions, and develop an efficient quality improvement program for the ongoing supply chain.
- *Control*—Control and monitor the improvement of the ongoing supply chain. Ensure that the improvement meets customer specifications and satisfaction.

- *Sustain*—Sustain the new supply chain's development and be aware of the need for further chain.

According to Towill (1991), individual inventory control of any supply chain has the following characteristics:

- Perceived demand for products, which may be firm orders or only forecasts
- A production or "value-added" process
- Information on current performance
- Information on "disturbances," for example, due to breakdown, delays, absenteeism
- Decision points where information is brought together
- Transmission lags for both value-added and other activities
- Decision rules based on company procedures, for example, changing stock levels, placing new orders, and production requirements.

The above characteristics are bound to have uncertainties at any point in time, requiring special attention and consideration. One has to think stochastic or dynamic models instead of deterministic models to encounter those variations. Once again, there is a call for robust statistical methods and technical analysis to counter the necessity. From the above discussions, it is evident that uncertainty (a measure of probability) can influence any managerial decision. These problems can only be treated with proper quantification of measurements and expertise modeling of data (Muralidharan, 2018b). Thus, statistical information helps business people to plan various courses of action according to the tastes and preferences of the costumers, societal demands, and the environmental issues associated with a product and services. Besides, the quality of the products and services can also be checked more efficiently by using statistical methods.

Davis (1993) was the first author to explicitly consider uncertainty as a strategic issue for supply chain performance when he stated, "there are three distinct sources of uncertainty that plague supply chains: suppliers, manufacturers, and customers." To understand the impact on customer service fully and to be able to improve performance, each characteristic must be measured and analyzed. Mason-Jones and Towill (1999) have further modified the above model incorporating one more source of uncertainty called the control systems. Thus, a lean-enabled supply chain process uncertainty model can be described as in Fig. 6.2.

According to Geary and Pand (2002), the uncertainty from the supply side can be due to short-notice amendments to suppliers, excessive supplier delivery lead time, adversarial supplier relationships, no vender measure of performance, etc, whereas from the demand side, it can be due to no customer stock visibility, no synchronization between processes, adversarial customer relationships, large, infrequent deliveries to the customer, etc. The most important sources of uncertainty from processes are due to poor value stream mapping of the processes, different measures of performance measures, reactive rather than proactive maintenance, random shop floor layout, etc. All these can contribute to the uncertainty in the control systems. Specifically, poor stock auditing, incorrect supplier lead time and continuous product modifications,

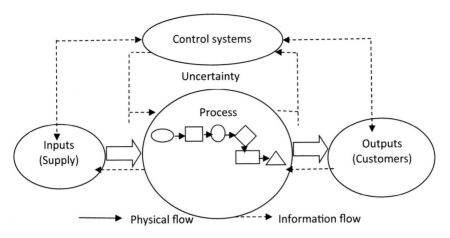

**Fig. 6.2**  Lean-enabled process uncertainty model

high obsolescence, government regulations, etc., can contribute a high amount of uncertainty in the model. Uncertainty also leads to variations, variations lead to redundancies, redundancies lead to wastes, and this chain continues. Therefore, the best part of the GSSCM model is that it addresses the uncertainty part of the process in the model and helps monitor whether the process is statistically in control.

As we know, uncertainty is bound to happen in any organizational process and decision making. Beyond the factors selected for consideration, many uncontrollable variables may also affect the outcome of the experiments and supply chains. These variables are generally called "noise," "experimental error," "nuisance variables," etc. Prior knowledge of these variables facilitates the identification of the significance of the main effects and interaction effect. A planned experiment subject to statistical investigation promotes the identification of homogeneous blocks and sections, which can minimize the effect of unwanted variables and accentuate the effect of the variables of interest. One of the best and statistically proved improvement tools for identifying variables of interest is the design of experiments (DOE). The DOE is the process of planning experiments so that appropriate data will be collected. The minimum number of experiments will be performed to acquire the necessary technical information, and suitable statistical methods will be used to analyze the collected data. A DOE is carried out by researchers or engineers in all fields of study to compare the effects of several conditions or to discover something new. If an experiment is to be performed most efficiently, then a scientific approach to planning must be considered. See Sect. 8.3.7 for more discussion on this concept and other tools for control and monitoring the process variation.

## 6.4  Key Performance Measures Associated with Green Statistics

Hervani et al. (2005) give a review of performance measurement systems and metrics under development for green supply chains. The selected list of parameters ranges from atmospheric emissions to energy recovery. They include measures for on-site and off-site energy recovery, recycling and treatment, spill and leak prevention, and pollution prevention. Additional energy measures include total energy use, aggregate electricity use, total fuel use, other energy use, total water use, habitat improvements, and damages due to enterprise operations, the cost associated with environmental compliance, and others. It is also necessary that the organizations may choose their environmental performance measurements specifically to meet new government regulations on emissions, energy consumption, or the disposal of hazardous waste. According to the authors, there is a need for industry-specific research to address which performance measurement systems work best; there needs to be interorganizational agreement on performance management and measurement. There is a need for tools to promote the development and improvement of green performance measures and supply chain management and data and information use relating to green supply chain management (GSCM). Singh et al. (2009) point out that even though there is an international effort to measure sustainability, relatively few approaches consider environmental, economic, and social aspects in an integrated way.

Apart from the above, some of the key performance indicators (KPIs) associated with the greening of business development and management specific to business areas are given below. These indicators provide leading information on future performance, and it helps to establish the organization's benchmarks. Benchmarks also help in checking what other successful organizations see as crucial in building and maintaining competitive advantage. They are central to any type of competitive analysis (http://www.supplychainmetric.com/: last accessed on 21.12.2019).

- *Administration*: For any administrative setup, strong leadership is required to plan and monitor business activities. The correct identification of standard operating procedures (SOPs) and responsible personals are prioritized for better corporate governance. This will help management use better capacity utilization by suitably adjusting the time and money involved with the premises' activities. Some of the performance indicators measured for administrative efficiency are budget allocation, strategic risk assessment, portfolio risk levels, insurance costs/sales, productive hours percentage, utility cost, percentage outsourcing, complaint resolution speed, complaint resolution cost, average meetings/month, utility cost/market cost ratio, premises cost/market cost ratio, space utilization, health and safety breaches, security breaches, certification, litigation, internal service satisfaction levels, percentage mentoring, etc. Although the indicators mentioned above are exclusively used for performance evaluation, they can also be used as green statistical measures, once they are quantified. This can bring economic sustainability freedom in the organization.

- *Finance*: Cash management, currency management, tax/depreciation, funding options, and other creative payment activities need better decision making in the command. To bring synergy between the auditing and reporting of financial activities in the organization, it is essential to have a healthy cash management expert for any management. Some other performance indicators need special attention for organizational coordination is the cost of finance, budget ratio, capital allocation ratio, gross yield, overdue accounts, productive hours, percentage of market dynamics capital allocation, sales tax rate percentage, cash interest rate percentage, depreciation percentage, internal service satisfaction levels, active headcount, percentage mentoring, risk profile, average project return, etc. The control over these measures brings plenty of transparency in the financial process of the organization.
- *Sales and marketing*: Statistical analysis like trend calculation, forecasting, estimation, and process control techniques are the critical tools of performance evaluation generally used by this department. The other indicators may be market share by segment, budget allocations, competitive score, sales by channel, percentage repeat purchase, average sales value, sales productivity, market share, advertising productivity by channel, cost per lead, cost per converted lead, bid success rates, average discount, service call-out times, productive hours percentage, inquiry response time, seasonality ratio, price index, customer satisfaction, advertising awareness, branding percentage, customer investment review, customer transition rate, customer investment return, customer complaints, warranty claims, project success, pricing, price elasticity, product spread, product age spread ratios, and internal service satisfaction levels. These indicators help the management to assess the customer behavior and the market potential and establish a trade-off between the customer–producer relationships. They also help to identify the market drivers, competitive drivers, the brand image, and the customer spread of a product and service.
- *Production and logistics*: To understand various factors that influence the organizational logistics and to identify the weaknesses and areas of improvement in the current production processes, the necessary logistical measures need to be statistically or numerically validated from time to time. Some of these performance measures are production cycle times, order processing cycle, downtime, percentage outsourcing, capacity utilization, logistics cost, load utilization, failure rates, space utilization, setup time, waste rates, pollution levels, out of stock percentage, obsolescent stock percentage, recycling percentage, back order percentage, just-in-time energy efficiency ratio, peak capacity percentage, supplier ratio, partnering, economic order quantity, number of suppliers, number of components, delivery failures, productive hours percentage, e-enablement, vendor rating, internal service satisfaction levels, percentage mentoring, etc. As mentioned above, the uncertainties associated with these areas need to be addressed efficiently using measurement-based methods so that recurrence will be minimum.
- *Personal*: For any improvement, we need to set goals not only in technical activities but also for the personals associated with every operation. These goals should be aggressive, but yet obtainable. Some of the green areas include productivity, absenteeism, staff alignment for each process, labor cost percentage, wages ratio,

employee satisfaction levels, overtime percentage, skills, training, discipline, timekeeping ratio, apprenticeship, recruitment costs, training days, productive hours percentage, appraisals, project success, internal service satisfaction levels, percentage mentoring, risk profile, etc.

- *Information Technology*: It is not only management, materials, and department matter; computing technology also matters for the performance of a firm. Quite often, the measurements make the areas more effective in communication and setting the goals. The measures that need some attention in IT department are management information system functionality, Web hits, access speed, site downtime, productive hours percentage, Intranet, Extranet, percentage outsourcing, security breaches, data storage, quality of data, information overload, project success, bugs reported, internal service satisfaction levels, software alignment ratio, e-enablement ratio, etc.

- *Service development and product delivery*: This is one of the most critical areas of the organization, where there is a lot of scope for greening the business. Customer satisfaction, brand image, and resource mobilization entirely depend on this area. Some specific indicators are R&D allocation, product age spread, ideas generated, strategic fit, total cycle time, project review, team creation, testing, percentage outsourcing, license fees, advertisement costs, patents, intellectual property right (IPR) infringements, IPR maintenance costs, royalty rate percentage, productive hour's percentage, specification limits, internal service satisfaction levels, reverse engineering, reverse logistics, etc.

Note that reverse engineering and reverse logistics (see Sect. 6.7 of this chapter for details) are the two emerging concepts in GSSCM logistics that can facilitate the coordination of all the above departments. The indicators discussed above are all critical for better planning and resource utilization in the organization. They help organizations speed up the processes, retain the green characteristics in the structure, and maintain a long-term impact on performance evaluation. However, it is essential to evaluate all the metrics from a statistical analysis point of view. This may include descriptive analysis, probability-based assessments, product and process control, multivariate techniques, forecasting and confirmatory analysis using hypothesis testing, etc. Thus, the combination of GSSCM and GSCM can play a vital role in developing viable business models for future applications and use.

In the next section, we discuss some important methods of green evaluations. They include all aspects of modern-day business activities like e-commerce, e-governance, and Industry 4.0. The growing importance of e-governance in information and communication technology (ICT) and big data analytics are also given space for consumer's concern. Each of the methods is critically examined, and their benefits are clearly stated.

## 6.5  Green Methods of Evaluation

The green methods of evaluation are generally done for air and climate change-related environmental problems. There are many green methods of evaluations and assessments practiced by institutions and organizations. In Chap. 5, we have already discussed this to some extent. Most of those methods involve statistical and numerical evaluations of information. For instance, *recursive partitioning* (RP) analysis is an analytical method for multivariable analysis, where it creates a decision tree that strives to correctly classify members of the population based on several dichotomous-dependent variables. Similarly, Breiman's (1984) estimate of soundproofing, etc., is all done for air and noise pollution through statistical analysis. For climate change issues, it is achieved through the use of climate models to predict temperature rise, assess future damage to human activity/ecosystems, the cost of the damage, calculation of marginal abetment costs, etc.

Even the *contingent valuation method* (CVM) is used to assess the preferences for environmental quality these days (Aldred, 1994; Hanemann et al., 1991). The CVM is an interview technique used to estimate the value people attach to certain environmental goods. In principle, there are two main fields of application for the CVM. One is the economic valuation of ecological projects, i.e., of projects meant to improve environmental quality. The second relevant field of application for the CVM is damage assessment after environmental accidents, i.e., after incidents that deteriorate environmental quality. Comparatively, one of the best methods of pollution accounting can be achieved through emission auditing. Emission measurement and auditing (a part of emission trading) are becoming increasingly critical to successful business operations. Companies are required to prove the integrity of their measurement procedures through independent audits (Montgomery, 1972; Rosen et al., 2008).

We now discuss some essential aspects of GS in business: e-business, big data analytics, reverse logistics, and so on. All these methods address the relevance of cost-effective ways of running a business. We connect the objective of lean, green, and clean principle to the notions of these new ideas for sustaining quality life for the future generation.

## 6.6  Green Statistics for Emerging Aspects of Business

### 6.6.1  E-Commerce and E-Governance

In the current dynamic, challenging, and data and information explosion environment, e-governance and adoption of statistical methods and LSS implementation have become vital for higher productivity, improved quality, reduced cost, and waste control in manufacturing, services sectors, and administrative systems. Skills in statistical and mathematical methods have to be developed at all levels to achieve efficiency and effectiveness. These analytical tools and techniques are widely adopted in the West, but unfortunately, it is not geared up in India. Knowledge of statistical methods and Six Sigma have still largely remained at academic levels and not yet percolated to the application level. Executives and professionals acquire certificates through short-term crash courses but do not translate their knowledge into operational areas for critical improvements (Gupta, 2008; Muralidharan & Gupta, 2016).

The need and urgency of data-based e-governance for efficient and effective administration are increasingly felt. Large-scale delay in delivery of products and services has been hurting the common man in several ways. The backlog/pending court cases for decades is a classic case of delay. A large amount of paperwork and their abnormally slow processing/movement and physical filing system has been a matter of concern in government offices. The information and computing technology have not been fully adopted through e-governance. The paperwork and the current method have been choking the systems. However, the impact and efficiency of e-government have been partially demonstrated in railways, banking, and insurance systems. But to a large extent, the traditional method of decision making and the office-keeping processes continue in these segments as a means for environmental accounting and auditing purposes. While e-commerce, e-marketing, e-retailing, e-teaching, e-banking-post, etc., are on the way, these initiatives have touched only the surface. The word "electronic" in the term e-commerce or e-governance implies technology-driven activity. There is vast scope in other areas of public concerns as well. The advent of big data and data analytics concepts has further added the necessity of fast processing and analysis of information with the help of information technology.

E-commerce or e-business refers to the transformation of key business processes through the use of Internet technologies. It relates to the use of information and communication technologies (ICT) to buy and sell goods and services, collaborate with business partners, and conduct electronic transactions within an organization (Turban et al., 2007). The impact of e-business covers both business-to-business (B2B) and business-to-customer (B2C) transactions (Edwards et al., 2010). E-commerce became popular during the 1990s with the advent of information technology and the widespread use of the Internet.

Daniel et al. (2004) describe the B2B concept close to the electronic marketplace. Business systems are attributed to Web-based systems that enable the automated transaction, trading, or collaboration between business partners. Using the Web, companies can reduce the complexity and cost of implementation and ease the integration with other systems. Electronic management systems and cloud systems are emerging quickly as a viable alternative to large-scale client–server solutions (Horvath, 2001).

B2C e-commerce (Nielsen, 2008) is simply electronic retailing using the Internet as a medium to place orders for goods, with the consumer interacting directly with the supplier's system. With the advent of Internet shopping, the "store-reach" boundaries of the traditional retailer have been extended to a global scale. Retailers can offer goods via the Internet to the final consumer anywhere in the world, provided each has access to a computer. Numerous factors (economic, infrastructural, cultural, and political) determine both businesses' and consumers' engagement with e-tailing.

E-governance is the application of ICT for delivering government services, exchange of information, communication transactions, integration of various stand-alone systems and services between government-to-citizen (G2C), government-to-business (G2B), government-to-government (G2G), government-to-employees (G2E) as well as back-office processes and interactions within the entire government framework (Marche & McNiven, 2009; Saugata & Masud, 2007).

The goal of government-to-citizen (G2C) e-governance is to offer a variety of ICT services to citizens efficiently and economically and to strengthen the relationship between government and citizens using technology. E-government-to-employee (G2E) partnership is one of four main primary interactions in the delivery model of e-governance. It is the relationship between online tools, sources, and articles that help employees to maintain communication with the government and their own companies. E-governance relationship with employees allows new learning technology in one single place as the computer. Documents can now be stored and shared with other colleagues online (Zhiyuan, 2002).

Government-to-government (G2G) is the online non-commercial interaction between government organizations, departments, and authorities, and other government organizations, departments, and bodies. Its use is every day in many countries, along with G2C, the online non-commercial interaction of local and central government and private individuals, and G2B, the online non-commercial cooperation of local and central government and the commercial business sector. G2B refers to the conduction through the Internet between government agencies and trading companies. B2G is the professional transaction between the company and the district, city, or federal regulatory agencies. B2G usually includes recommendations to complete the measurement and evaluation of books and contracts.

E-governance can only be possible if the government is ready for transacting business through the Internet. It is not a one-day task, and so the government has to make plans and implement them before switching to it. Some of the measures include investment in telecommunication infrastructure, budget resources, ensure security, monitor assessment, Internet connection speed, promote awareness among the public regarding the importance, support from all government departments, and so forth (https://businessjargons.com/e-governance.html, last accessed on 11.01.2019). Benefits of e-governance include:

- High transparency in governance
- Increased convenience in transacting business
- Growth in gross domestic product (GDP)
- Growth in gross national product (GNP)
- Direct participation of constituents and stakeholders
- Reduction in the overall cost of business
- Expanded reach of government and people
- Reduced corruption.

Scholl et al. (2009) study the importance of e-commerce information systems and e-governance information systems, as they can facilitate an increased opportunity for and engagement in institutional collaboration. In G2G, mainly, collaboration was found to be motivated through citizen and business demand as well as through expected economies of scale and scope. Creating a supportive culture for collaboration strongly depends on personal relationships between the decision-makers of the collaborating entities. According to the authors, the associations and correlations with e-commerce and e-government can be found regarding (i) process improvements, (ii) back-end (process) integration, (iii) cost savings, (iv) information sharing, (v) vertical and horizontal system integration, (vi) increased responsiveness and service quality, (vii) standardization efforts, (viii) the criticality of senior leadership support, and so on.

As expected, the current drive for e-governance will boost the importance of data management for obtaining the right data for the right decisions besides planning, implementation, monitoring, control, and evaluations of ongoing short-term and long-term operations and projects. This will further improve the quality of the information passed onto the stakeholders, leading to stakeholder confidence and morale. Hans et al. (2009) also report several similarities and differences between e-commerce and e-government from the stakeholder's perspective. According to them, the sheer volume of information, old and new, as well as accuracy and timeliness of the information it was said required new presentation formats, exchange strategies, and operations in public and private sectors. In process management, although the transaction volumes in e-commerce were found far higher than in e-government, in e-government, online transactions play an increasingly important role. Differences were found in the extent and sophistication of process redesign between the sectors. E-commerce redesigns were seen as far more advanced than those in e-government. Further, while the speed of transaction was found an essential element of e-commerce, less emphasis was put on this aspect in e-government (Ngai et al. 2002; Wang, et al., 2007; Muralidharan & Gupta, 2016).

Interestingly, leadership in government appeared to be more supportive of (in particular, collaborative) e-projects than their commercial counterparts. It also appeared that collaborative structures in the public sector were markedly more robust than those in the private sector. Hence, a cooperative arrangement between two or more public and private sectors is necessary for private entity financing, constructing, or managing a project in return for a promised stream of payments directly from the government or indirectly from users over the projected life of the project or some other specified period (Hodge & Greve, 2016). It is worth noting that the public–private partnership (PPP)-based e-governance projects are hugely successful in India. Many countries implement e-government policies in an attempt to build a corruption-free government.

## 6.6.2   E-Governance in Information and Communication Technology

To bring the benefits of ICT to the last mile to ensure transparent, timely, and hassle-free delivery of citizen services, the Government of India has initiated the e-governance program in the late 1990s. After that, India's union government has approved the National e-Governance Plan (NeGP), comprising 27 Mission Mode Projects (MMPs) and 8 components on May 18, 2006, to give a boost to e-governance initiatives in India. The Department of Electronics and Information Technology (DEIT) and Department of Administrative Reforms and Public Grievances (DAR&PG) have formulated the NeGP. The objective is to make all government services available to the citizens of India via electronic media.

E-governance in India has steadily evolved from government departments' computerization to initiatives that encapsulate the finer points of governance, such as citizen centricity, service orientation, and transparency. The NeGP takes a holistic view of e-governance initiatives across the country, integrating them into a collective vision and a shared cause. E-governance in agriculture aims to disseminate useful information about improved technology to the farming community and service providers in rural areas. Indian Government (InDG) will create a platform for different levels in the rural agricultural landscape—farmers, cooperatives and professional bodies, farm machinery vendors, fertilizer and chemical companies, insurance regulators and agronomists, consultants, and farm advisors. The primary focus of the agriculture sector presently in the InDG portal is about agricultural credit, policies and schemes, MGNREGA, market information, agricultural best practices, on and off farm enterprises, and various products and services (https://en.wikipedia.org/wiki/National_e-Governance_Plan: last accessed on 29.01.2020).

Like any other organization (business, services, and manufacturing), the agricultural organizations also operate in an environment characterized by dynamic market forces and customer demands. Deregulation, consolidation, and increased competition are forcing them to re-evaluate their technology and business processes. The changing regulatory requirements, tariff issues, environmental concerns, and health and safety issues continue to pressure these organizations. The need and urgency of developing statistical and analytical skills among executives for data analysis, Six Sigma, and project management implementation have been rapidly increasing in agricultural organizations for efficiency and effectiveness (Muralidharan & Gupta, 2016).

Information governance helps organizations to get a better handle on customers' needs, unearth competitive threats, and uncover new business opportunities. The results—measured in profits, market share, and customer satisfaction—can be an eye-opener, all because organizations have a strategic, dynamic view of information governance that goes beyond compliance and regulations. Since much of the business activities are centered on analytics and big data, the use of IT resources must be done more strategically and cost-effectively. Big data analytics is all about tapping into diverse data sets, finding, and monetizing unknown relationships. Therefore, a completely data-driven process technique like Six Sigma must be integrated into the process of improving the system (Muralidharan, 2008). Since a large volume of data is to be transacted and disseminated in one go, there will be a considerable loss (waste) of quality information and a large number of information hazards. To overcome this and make business process smart, we need analytics and a structured way of a problem-solving approach, which is guaranteed by GS, GSSCM, and Six Sigma initiatives.

### 6.6.3   E-Governance in Education

Although e-governance has percolated to all sectors in India, its impacts on the education sector are appreciable. Education is the foundation on which the development of every citizen and the nation as a whole is built on. India has one of the most significant advanced education frameworks on the planet. Our country has about 1.5 million schools with more than 260 million students enrolled and around 751 universities and 35,539 colleges. In the recent past, India has made massive progress in increasing primary education enrollment, retention, regular attendance rate, and expanding literacy to approximately two-thirds of the population. India's improved education system is often cited as one of the main contributors to the economic development of India. E-governance, in this case, requires several elements of good governance, such as transparency, accountability, participation, social integration, cultural transformation, and development. It includes an extensive range of services for almost all segments of society. The COVID-19 menace has further boosted the importance of e-learning in a massive way. Almost all the academic institutions are now imparting the education electronically, just to avoid the loss of academic year.

Dr. APJ Abdul Kalam, former President of India and a visionary in e-governance, has aptly summarized the fundamental challenge lying before the country in this regard: "*e-Governance has to be citizen-friendly. Delivery of services to citizens is considered a primary function of the government. In a democratic nation of over one billion people like India, e-Governance should enable seamless access to information and seamless flow of information across the state and central government in the federal set up. No country has so far implemented an e-Governance system for one billion people. It is a big challenge before us.*"

In the case of higher education in India, what we need at present is an educational revolution that can help to tap the potential of our human brains. In the last few years, higher education institutes and student strength have also increased in leaps and bounds. Even after having some excellent institutes, universities, and curriculum, we cannot retain the best brains for our public use. Poor infrastructure, research facilities, and insufficient funding have also contributed to this menace. The highly educated people are either moving abroad for greener pastures or are not interested in the prosperity of society and nation because of a lack of confidence, encouragement, motivation, and job opportunities. Many commissions and review committees were appointed to study the situations from time to time. Still, the quality of education has not improved much. The only solace is that the system has accepted the changes happening locally and globally. Therefore, there is an urgent need to build confidence and connect the educated people to the country's administrative network. To recharge the current education system, the government should initiate confidence-building measures and open up sustainable opportunities for all graduates and degree holders. Promoting the e-commerce concept in education can probably streamline the entire educational process for better discharge of duties and responsibilities. To some extent, this may be possible through GS GSSCM or Lean Six Sigma like approaches as these initiatives can capture the benefits in terms of numbers and figures of e-governance adaptability in education.

The Six Sigma philosophy or LSS philosophy, if applied in the service sector (government establishments) departments including educational institutions, can help facilitate in unleashing the immense potential and create synergy between government systems on the one hand and IT on the other. With improved process efficiency and organizational effectiveness, and information networking, a new dimension would be added to the concept of e-governance. The DMAIC philosophy discussed in Sect. 4.12, applied to the service industry, helps identify the people's requirements, measure performance according to specifications, identify and implement best practices, and sustain the improvements as per the requirement, choice, time, and need. It also facilitates to establish standard measures to maintain performance and standardize the curriculum and syllabi. The entire aspect of student–faculty management is best possible through the implementation of lean approaches, as discussed in various chapters. See Table 6.1 for the necessary critical-to-quality parameters of service-related processes and activities, including education.

### 6.6.4  Quality in e-Governance

Quality has acquired a more significant dimension. It is no longer confined to the narrow view of limiting defects or control of nonconformance. Now it encompasses all aspects that go toward making a business (including service management) more effective or achieving what may be called business excellence ("excellence in government," in case of public institutions). The quality is *standard of something when it is compared to set benchmarks or expectations.*

Quality of services, among others, should be the touchstone of judging the success of e-governance. It has to be a key factor. In the context of "quality service," the critical consideration has to be how close to the "expectations of citizens" the services can be provided to meet their aspirations. It would require evolving quality parameters and ensuring them through improved systems. The important parameters need to be looked at in the context of "making services available to people." Based on the urgency and the availability of resources (for providing the services), two levels of services have been suggested. Level-1 of the quality standard is initially the emphasis is to be laid, and then after achieving it, we can move to level-2, where the quality

**Table 6.1** Critical quality parameters for a select list of services

| SN | Sector | Services (level-1) | Important quality parameters (level-2) |
|---|---|---|---|
| 1 | Health | Hospital functioning | Access, cleanliness/tidiness, responsiveness, cost |
| | | Immunization | Reliability, consistency, cost |
| | | Medicine | Access, timeliness, conformance |
| | | Medical checkup for infants and vaccination | Timeliness, safety, availability |
| 2 | Hygiene and sanitation | The functioning of the tube-wells/pipe water system | Reliability, cost, transparency |
| | | Drinking water quality | Access, conformance, consistency |
| | | Servicing of equipment | Serviceability, cost, reliability |
| 3 | Education | Institution functioning | Access, cleanliness/tidiness |
| | | Scholarship | Access, timeliness, transparency |
| | | Posting of teachers | Reliability, consistency, expertise |
| | | Literacy | Access, facilities, friendliness, resources |
| 4 | Land | Records of rights | Timeliness, security, availability |
| | | Mutation of records | Timeliness, security |
| 5 | Agriculture | Agriculture inputs | Timeliness, cost, reliability, conformance |
| | | Commodity prices | Consistency, timeliness |
| | | Technology transfer | Relationship management, partnership, and learning reach of technology |
| 6 | Essential commodities | Availability | Consistency, conformance, timeliness, flexibility (volume, variety) |
| | | Taste and smell | Conformance (of quality) |
| 7 | Rural development | Participatory development | Friendliness, participation, Relationship management. Partnership and learning |
| 8 | Natural disaster | Warning system | Reliability, consistency, frequency |

(continued)

**Table 6.1**  (continued)

| SN | Sector | Services (level-1) | Important quality parameters (level-2) |
|---|---|---|---|
|  |  | Availability of relief | Access, courtesy, friendliness, perceived quality, responsiveness, relationship management, partnership and learning, flexibility (volume, variety, response) |

*Source* Muralidharan and Gupta (2016)

parameters are subject to modifications and refinement. The levels are subjective and sector depended.

It is often argued that improved quality would be associated with loss of efficiency or high cost. Also, there is a trade-off between quality, price, and availability. Such an argument may be valid in the short run. But, while attempting for improved quality, the attention must go on to the people involved in the delivery of services. Since services are usually "labor-intensive," the primary resource inputs determining quality and productivity are personnel, capabilities, training, and motivation. The success of the service department should emphasize selecting the right people for training, motivating, and rewarding them. This way, the service department may become efficient. Motivated employees, if provided opportunities and encouragement, are likely to think of the many forms of improving the processes. The satisfaction that comes from having contributed to the organization, if properly recognized and rewarded, may reinforce the motivating effects of employee participation and encourage further efforts. Service departments that can involve employees in quality and productivity efforts are likely to see the complementary impact of quality and productivity, as the process continuously improves.

Integrating e-governance and sustainable business practices into corporate management can lead to increased business, improved business performance, and further enhancement of the company's credibility with stakeholders. However, the current efforts under e-governance through all efforts are mainly in the area of providing basic amnesties to people of any segment. Their access to government machinery and programs is limited to either utility services or grievances. Hence, it is demanded that the focus should come on providing "quality" services and information to citizens with a perfect mingle of IT and IT-enabled services. According to an estimate of India National Digital Literacy Mission, approximately 40% population is living below the poverty line, the illiteracy rate is more than 25–30%, and digital literacy is almost nonexistent among more than 90% of India's population. To tap the full potential of IT, we need to prioritize improving government systems to enhance the performance standard of the delivery system. E-business is also helping project management a flexible and pleasurable experience for many organizations. E-project management facilitates continuous monitoring and comparing with other ongoing and competing projects a high value-added component of the project.

The resurgence of mobile technology is also helping the promotion of e-commerce activity to an all-time high. 5G mobile technologies have already started showing their presence in every aspect of human life. It is virtually connecting everyone and everything, including machines, objects, and devices. 5G technology is a unified, more capable air interface. It has been designed with an extended capacity to enable next-generation user experiences, empower new deployment models, and deliver new services (https://www.qualcomm.com/invention/5g/what-is-5g last accessed on 10.04.2020). With high speeds, superior reliability, and negligible latency, 5G will expand the mobile ecosystem into new realms. 5G will impact every industry, making safer transportation, remote health care, precision agriculture, digitized logistics, and so on. It is guaranteed to provide higher performance and improved efficiency, which will empower new user experiences and connect new industries. It can be a reason for global growth by improving the economy and progress of the people. High-end machines, enhanced bandwidth, and extreme quality products and services will be universal and available for all. Wait and watch for more.

As mentioned previously, the COVID-19 pandemic has changed the phase of online education in India as well as other countries. As a part of the Digital India project, many colleges, universities, and academic institutions offer online correspondence courses, which are now going to be an everyday affair. The online education market in India was worth $247 million in 2016, which is expected to grow by about $1.96 billion by 2021. The number of users enrolled for various online learning courses is estimated to be 1.6 million in 2016, which is expected to grow by about 9.6 million by the end of 2021. It is estimated that there is a 175% increase in the cost of classroom education; this gives online training more preferred because it is cost-effective (https://eduxpert.in/online-education-india/: last accessed on 20.06.2020). Even otherwise, the demand for online learning is going to increase further. Therefore, e-governance in e-learning is really positing a challenge for the government and administration.

## 6.7　Big Data Analytics

Big data analytics is going to be the next frontier for innovation, productivity, and competitive advantage. "Big data" is often defined as more data made possible by the Internet, when somebody uses a wireless device (Franks, 2012; Dumbill, 2012; Gobble, 2013; Kudyba, 2015). It is all about tapping into diverse data sets, finding and monetizing unknown relationships, and, therefore, ultimately a data-driven process technique. Data itself is a record of an event or a transaction like.

- A purchase of a product
- A response to a marketing initiative
- A text sent to another individual
- A click on a Web link.

Big data analytics (BDA) is characterized by the volume, velocity, and variety of data. Because of the advent of IT and the growth of digitalization, all business processes involve a large volume of available data in many forms (variety). The speed (velocity), at which the terabytes of data (volume) are accessed, recorded, disseminated, and used for further analysis, is now becoming a big challenge for all decision-makers. The problem is further escalated by the uncertainty and variations involved in every process. Simply put, big data can be described as

$$\text{Big data} = \text{Transactions} + \text{Interactions} + \text{Observations}$$

In the future, all organizations are going to face the data explosion problem. Automated data collection tools and mature database technology lead to tremendous amounts of data stored in databases, data warehouses, and other information repositories (cloud-based storage). Hence, there is a need to convert such data into knowledge and information. Any deficiency of proper technology and analytical tools will hamper the entire process of decision making. However, the new information technology is making it possible for companies to capture, store, and analyze reams of data to extract valuable market and customer knowledge.

The most critical challenge for BDA is to ensure the quality of the data, the tools used for analyzing the data, and the interpretation of the analyzed results. Most of the time, the data is unstructured and textual. The ability to access and analyze such a massive flow of data in real time could transform maintenance practices that consume a vast amount of operational resources. To achieve such a vision, the challenge goes beyond gathering and storing vast amounts of data. In many cases, it was labor-intensive to pull together into a useable format for statistical analysis that was never done before. The timely intervention of data mining tools, appropriate designs, related to change management, incentives, and people accountability is essential for simplifying this.

The best way to bring quality in the data analytics is to integrate technically strong quality philosophies like Six Sigma or LSS methodologies in the data analytic process. Fogarty (2015) pointed out that consolidating LSS with advanced analytics is even more effective at taking improvement projects to a new level. Since we are living in an era of the "Internet of everything," the proliferation of big data in the commercial world poses challenges for both methodologies and technicians. The combination of LSS and advanced analytics goes hand in hand due to their process-oriented nature and rigorous search for the truth in data. Moreover, these paradigms combined will yield improved results over their separate application. In particular, it was noted that advanced analytics could be used in the measure and analyze phase of the DMAIC improvement process. During the measure phase, a focus on advanced statistical techniques going beyond the simple control chart could allow businesses to understand how a process works (Muralidharan, 2018a).

Along with challenges, there are many opportunities for improvement as well. Companies can reinvent themselves through the digital transaction of data and make more sense of the data, leading to faster execution of processes and projects. The application areas of BDA are customer profiling and segmentation, customer life-time value analytics, customer sentiment analytics, customer satisfaction analytics, campaign management analytics, forecasting analytics, utility analytics, telecommunication data analytics, supermarket data analytics, transport data analytics, fraud analytics, etc. Any process associated with these business areas generates a considerable amount of raw data and equally generates unutilized and unstructured data, usually termed waste in LSS philosophy. See how the sectors like banking, finance, and insurance are swimming through the data every day. By some estimates, organizations in all industries have at least 100 terabytes of data, many with more than petabytes of data. Even scarier, many predict this number to double every six months of operation. The three key technologies that can help you get a handle on big data—and even more importantly, extract meaningful business value from it—are

- Information management for big data
- High-performance analytics for big data, and
- Flexible deployment options for big data.

These key technologies warrant the use of some quality assessment through quality function deployment (QFD), designed for Six Sigma (DFSS), Kaizen philosophies, and many other techniques used in the Six Sigma project implementation to realize the business values as mentioned above. According to John Sall, co-founder and executive vice president of SAS Institute, USA, "big statistics" is the term that deals with a lot of data and statistics that are important and significant. Taking big data and turning it into "big statistics" is vital for genomics, industry, and business. It is a fact that the industry has recognized the increasing importance of the practice of statistics to solving the complex problems faced by the market, industry, and government entities. "The oft-quoted McKinsey & Company report foretells a shortage of up to 190,000 workers with great analytical skills and 1.5 million managers and analysts to manage data projects (as per the 2015 estimates). A large number of these workers will be holders of bachelor's degrees in statistics." Ultimately, big data and Six Sigma are all statistical inference-based studies, and therefore, statistical research is vital for any quality improvement and decision making.

BDA is transforming how companies are using sophisticated information technologies to gain insight from their data repositories to make informed decisions. This data-driven approach is unprecedented as the data is collected via the Web, and social media is escalating every second. We will see the rapid, widespread implementation and use of big data analytics across the organization and the industry in the future. As it becomes more mainstream, issues such as guaranteeing privacy, safeguarding security, establishing standards and governance, and continually improving the tools and technologies would garner attention.

When speaking of big data, one must consider the source of data and its reliability. This involves the technologies that exist today and the industry applications that are facilitated by them. These industry applications are prevalent across the commerce realm and continue to increase in countless activities, including marketing and advertising; health care; transportation; energy; retail, etc. Analytic methods can range from simple reports, tables, and graphics to more statistically based endeavors to quantitative-based methods. In this connection, Muralidharan (2018b) advocates a six-phase strategic, tactical, and operational framework *Initiate–Measure–Analyze–Validate–Verify–Implement* (IMAVVI) of Six Sigma analytics in the following way:

- *Initiate*: Initiate the need for the BDA project, project attributes, and objectives.
- *Measure*: Identify relevant data variables and measure the performance of the process; what is the problem being addressed; why is it appropriate and exciting?, etc.
- *Analyze*: Identify sources of unacceptable variations, justify the use of advanced multivariate techniques, test–retest the hypothesis, revisit and reinvent the implementable solutions, etc.
- *Validate*: Validate strategic processes based on analysis and outputs; gain insight from productions.
- *Verify*: Verify the operational objectives as per the solutions obtained; if needed, go to the previous steps for validation of the goals and derive policy implications.
- *Implement*: Implement tactical solutions and make informed decisions about the data.

When it comes to problem-solving and process optimization, the statistical tools and Six Sigma tools could be the key to turning big data into manageable inferences. Hence, Six Sigma analytics offers smart methods coupled with focused technology and actionable decisions. Therefore, the above framework assures the right amount of quality in data analytics if it is done with accountable people, sound analysis, efficient processes, and sustainable technology. The growth of 5G mobile technology is an added support for the growth of BDA.

Business analytics, database analytics, data mining, data science, explorative data analysis (EDA), knowledge discovery of data (KDD), statistical analysis, quantitative analytics, Web mining, etc., are the other synonyms used for BDA. Many names are associated with big data now. All big, medium, and small organizations, R&D institutions, and utility service stations are running pillar to post to implement analytics in their business processes. Unfortunately, most businesses have made slow progress in extracting value from big data. And some companies are attempting to use traditional data management practices on big data, only to learn that the old rules no longer apply. Probably that is the reason; BDA intoxicates the CEOs, senior MDs, managers, executives, IT professionals, academicians, consultants, workers, and so on, alike. It adds value to their profile, competitive edge and image, and above all, the integrity of professional skills. For a CEO, if BDA is a tool for business growth, then it is a way of enhancing career path for an executive. It is a new discipline in the reckoning and is here to stay for a long time.

## 6.8  Industry 4.0

With the advent of digital technology, the Industrial Revolution has also changed rapidly upward. The entire working culture and tradition of manufacturing have contributed to high-yield products and services. The relationship between machine and man is improved with the advent of Industry 4.0. Industry 4.0, also known as *Smart Manufacturing*, is an attempt to aid complex manufacturing in the recent era. Industry 4.0 is the representation of the current trend of automation technologies in the manufacturing industry. It mainly includes the cyber-physical production systems (CPPS), Internet of things (IoT), robotics, big data, augmented reality, horizontal and vertical integration, and cloud computing (CC) (Saša Zupan Korže, 2019; Kumar & Kumar, 2019; Dohale & Kumar, 2018). The German government coined Industry 4.0 in the context of its high-tech strategy in 2011. At its early stage, it was related to the "factories of the future" or "smart factories."

It is also interesting to look back at the paradigm shift in the Industrial Revolution over the last six decades. During the First Industrial Revolution, manufacturing production facilities were developed with the help of water and steam powers. The Second Industrial Revolution brought mass production in the manufacturing sector with the help of electrical energy. The Third Industrial Revolution introduced electronic and information technologies that furthered production automation. Over the last few years, Industry 4.0 has emerged as a promising technology framework used for integrating and extending manufacturing processes at both intra-organizational and interorganizational levels (Xu et al., 2018). It is expected that machines will be able to communicate production data with each other using embedded network systems. In Industry 4.0, the use of small, handheld mobile devices to control and monitor the production system will become prevalent.

The increasing information complexity and its influence on production efficiency is a multidisciplinary challenge, where knowledge from cognitive psychology, ergonomics, operations management, communication technology, computer science, industrial design, IoT-based human–computer interaction (HCI), manufacturing technology, instrumentation engineering, and others must be addressed for the optimal design of control and display units in CPPSs (Kumar & Kumar, 2019). Contemporary research literature has shown that the CPPS concept enables an ecosystem of cyber manufacturing in the factory environment. Such CPPS machines are becoming more intelligent and may control and monitor the processes on their own by employing machine learning algorithms (Lee et al., 2015). At this stage, one challenge will be the poor quality of data and its efficient management by the users. Hofmann and Rüsch (2017) point out that there are many challenges, risks, and barriers associated with the implementation of Industry 4.0. They can be defining appropriate infrastructures and standards, ensuring data security and educating employees on the issues that need to be addressed on the road to Industry 4.0. Kamble et al. (2018) propose a sustainable Industry 4.0 framework based on the review's findings with three critical components: Industry 4.0 technologies, process integration, and sustainable outcomes.

Big data is an essential feature of Tourism 4.0. The implementation of big data in the industry means having real-time information about man, machine, materials, methods, management, and the organization's overall operations. Augmented reality (AR) technology changes a person's perception of their physical surroundings when viewed through a particular device (Augment, 2016). The technology has similarities with virtual reality (VR), but AR does not replace the real-world environment but augments it by overlaying digital components. The AR technology makes it possible to layer digital enhancements over an existing reality or a real-life scenario. In years to come, with the emergence of Industry 4.0, machines will be the primary drivers of factory work, with humans serving in the troubleshooting and oversight roles.

## 6.9  Reverse Logistics

Reverse logistics has been defined as *"the process of planning, implementing, and controlling the efficient, cost-effective flow of raw materials, in-process inventory, finished goods, and related information from the point of consumption to the point of origin, to recapture value or proper disposal"* (Rogers & Tibben-Lembke, 1999). Horvath et al. (2005) indicate that RL refers to a set of programs or competencies aimed at moving products in the reverse direction in the supply chain (i.e., from consumer to producer). With the progressive increase in environmental concerns, the efficiency focus, importance of value delivery through co-creation and co-production as well as the need for improving core competencies while strategically positioning in the global competitive market, the understanding of RL shifts toward the "coordinated," "centralized," "consolidated," and "integrated" network value chain (Flygansvaer et al., 2008).

Reverse logistics differs from waste management as the latter is mainly concerned with the efficient and effective collection and processing of waste, that is, products for which there is no longer any reuse potential (De Brito & Dekker, 2003). The definition of "waste" in this context is essential from a legal perspective as the act of "importing waste is often forbidden" is a fact taken for granted by all (Fleischmann, 2001). However, there are similarities between some of the processes used by product recovery networks and waste disposal networks, as often seen in the urban setting (Shakantu et al., 2002). These are most evident in the supply side, where used products are collected from many and need to be consolidated for further processing. However, major differences do exist between these network types on the "demand" side. While a recovered flow of products should be directed toward a reuse market, waste streams eventually end at landfill sites or incineration plants after various treatment processes (Fleischmann et al., 2000).

Ramazan et al. (2014) argue that, given the complexity of RL supply chains and the uncertainty return flows, effective information technology is necessary to support the management of return flows. Efficient information systems are supportive of individually tracking and tracking the gains of the product, linking with the previous sales

(Biehl et al., 2007). Another vital barrier affecting RL is the quality of the end-of-use/end-of-life returned products. Performance metrics form the basis of integrated work management systems. Therefore, successful RL programs will create performance measurement systems that provide data as to whether the designed RL is performing up to the expectations. RL's cost is many times higher than the price of forwarding logistics because the distribution of the new manufactured goods can be consolidated with the involvement of multiple firms and shared resources.

## 6.10   Benchmarking and Reverse Engineering

One of the most cost-effective ways to understand and to predict what your competition is going to do is by benchmarking your competitive products. This type of benchmarking is called *reverse engineering*. It is one of the best sources of competitive reliability and design data that is available. A *benchmark* is a point of reference by which performance is judged or measured for quality. Improving processes and products is often aided by comparing the current state with outstanding processes or products.

In some cases, the comparison will be within or between processes and products. The use of these comparisons is called *benchmarking*. The information for benchmarking may come from various sources, including publications, professional meetings, research, customer feedback, site visits, and analysis of competitors' products. A downside of benchmarking within a particular industry is that it tends to put one behind the best. Benchmarking is useful for driving breakthrough improvement over continuous improvement. According to Kubiac and Benbow (2010), several types of benchmarking exist, and each has its advantages and disadvantages:

- *Internal benchmarking* provides easy access to other departments within the same company. However, the search for exceptional performance is often limited by its culture, norm, and history.
- *Competitive benchmarking* is the continuous process of measuring products, services, and practices against the company's fiercest competitors or renowned industry leaders (Camp, 1994).
- *Functional benchmarking* compares similar functions typically outside the organization's industry and provides ample opportunities to seek out benchmarking partners.
- *Collaborative benchmarking* refers to the cooperation between various functions or organizations to achieve benchmark results. This kind of benchmarking permits access to specific benchmarking partners that may not exist with the other types of benchmarking.

The benchmarking process applies to subjects such as products, customer services, and internal processes, as discussed above. It forces an organization to take an external perspective. However, focusing on industry practices may limit opportunities to achieve high-performance levels, if a given industry is not known for its quality. Hence, the competitive product benchmarking or reverse engineering works as a catalyst for quality and reliability of the process and outcome of any company. The steps involved in reverse engineering are:

1. Understand and obtain customer's products.
2. Analyze order and delivery cycle.
3. Analyze packaging and documentation.
4. Characterize the benchmarking items.
5. Perform life tests.
6. Perform safety factor analysis.
7. Perform environmental tests.
8. Compare performance results.
9. Perform disassembly analysis.
10. Compare product design and production methods.
11. Define the competitor's competitive advantage.

Here, the steps 1–4 are specific to the benchmarking process. Steps 5–7 are purely engineering and science-related, and rest are merely supportive management actions. Life tests and reliability is an essential statistical method to understand the survival of product life cycle, estimation of the complete life of the test, and testing of components and systems. The environmental criteria are designed to define how the product functions under extreme external conditions. Typically, these tests are performed at 10–20% higher stress levels than the actual external environment that the item is required to operate under. Typical environmental tests are temperature, vibration, shock, input voltage fluctuation, humidity, static discharge, etc. ESS and accelerated life testing discussed previously are two such critical environmental tests carried out to observe the quality performance of items subject to life.

The unique step to reverse engineering is the product disassembly analysis shown in step 9. One of the best ways to understand how competitors manufacture their products is by disassembling competitor's products and comparing the product design, assembly methods, and each part to your product. Typical things that reverse engineering can reveal are (Harrington and Anderson, 1999):

• Number of different parts required to accomplish a specific function
• Level of standardization of parts used by the competitor
• The suppliers used by the competitor
• Actual tolerance levels and measurements
• Assembly methods
• Lubrication used
• Materials used
• Ease of repair and maintenance, etc.

The disassembly analysis process can also provide you with additional improvement opportunities and insight into why the competitor's products perform better than your products during the test phase. These days, understanding your competitor is more important for judging your product quality, cost, performance, and reliability (or durability). Copycats are everywhere and are an accepted norm in all the fields. Look at packaged food items, automobile cars, mobile instruments, and so on. Everything is an advanced, modified, and updated version of the existing one. Lots of statistical data analysis go into the investigation, evaluations, and prediction of quality efforts in these cases. Thus, reverse engineering provides sources of ideas for improvement and evidence that better practices exist and must be instituted to be competitive.

Quality improvement in every aspect of design, development, and manufacturing is a requirement of the time and should be ongoing. Toward this, automation of design activities is done on a large scale to escalate the best part of engineering to translate all specifications into manufacturing operations and processes. This is where the recent emphasis on *concurrent engineering* concepts emerged to reduce the time from product concept to market, prevent quality and reliability problems, and reduce costs. The concurrent engineering, also called *simultaneous engineering*, is a team approach to design, with specialists in manufacturing, quality engineering, and other disciplines working together with the product designer at the earliest stage of the product design process (Gryna et al., 2007; Montgomery, 2003). The concept which strikes a balance between performance and cost is *value engineering*, which is a technique for evaluating the design of a product to ensure that the essential functions are provided at the minimal overall cost to the manufacturer or user (Park, 1999). Similarly, completely redesigning or restructuring a whole organization, an organizational component, or a complete process is called *reengineering*. In terms of improvement approaches, reengineering is contrasted with incremental improvement as promoted by Kaizen activities (Kubiac and Benbow, 2010).

## 6.11  Reliability Engineering

Reliability is quality over time. As products and processes become more complex, failures increase with operating time. Traditional efforts of design, although necessary, are often not sufficient to achieve both the functional performance requirements and a low rate of failures with time. To prevent these failures, specialists have created a collection of tools called *reliability engineering* (Gryna et al., 2007). The function of reliability engineering is to develop the reliability requirements for the product, establish an adequate reliability program, and perform appropriate analyses and tasks to ensure the product will meet its obligations. Reliability engineering is closely associated with *maintainability engineering* (engineering principles applied to repairs and compared to requirements) and *logistics engineering* (the engineering principles applied to all movements of logistics and transportation). Many problems from other fields, such as security engineering, can also be approached using reliability engineering techniques. Reliability engineers rely heavily on statistics, probability theory, and reliability theory to estimate, test, and predict the capability and performance of the quality characteristics of a system. Many engineering techniques are used in reliability engineering, such as reliability estimation, reliability prediction, Weibull analysis, reliability testing, accelerated life testing, and so on.

In software engineering and systems engineering, reliability engineering is the sub-discipline of ensuring that a system (or a device in general) will perform its intended function(s) when operated in a specified manner for a specified length of time. Reliability engineering is performed throughout the entire life cycle of a system, including development, test, production, and operation. Design for reliability (DFR) is another emerging discipline that refers to designing reliability into products. This process encompasses several tools and practices and describes the order of their deployment that an organization needs to have in place to drive reliability into their products (Muralidharan & Syamsundar, 2012). The most common reliability program tasks are documented in reliability program standards, such as MIL-STD-785 and IEEE 1332. Failure reporting analysis and corrective action systems are other approaches commonly used for product/process reliability monitoring.

The lifetime data is modeled using reliability distributions. The most frequently used distributions for modeling life data are exponential, gamma, and Weibull distributions. A reliability parameter characterizes them often called failure rate or instantaneous failure rate (or hazard rate). For exponential distribution, the hazard rate is constant. Another critical parameter is the mean time between failures (MTBF). This parameter is handy for systems that frequently operate, such as transport vehicles, machinery, and electronic equipment. Reliability of most of the systems increases as the MTBF increases. The MTBF is usually specified in hours, but can also be used with other units of measurement such as miles or cycles (Leemis, 1995; Meeker & Escobar, 1998). The other parameters or parametric functions used in repairable and maintenance systems are the survival function, mean time to failure (MTTF), and mean time to repair (MTTR). We will discuss all these measures precisely here.

Let $X$ be the lifetime of a unit or component of a system and $f(x)$ be the corresponding probability density function (PDF). If $F(x) = P(X \leq x)$ is the probability distribution function or cumulative density function (CDF) of the random variable $X$, then

(i)   The *survival function* (or *reliability function*) is defined as the probability that a device will perform its intended function beyond a specified period, say $t$, under some stated condition. Mathematically, it is expressed as

$$R(t) = P(X > t) = 1 - F(t) \qquad (6.1)$$

(ii)  The *hazard function* (or instantaneous *failure rate function*) is defined as $h(t) = \frac{f(t)}{1-F(t)} = -\frac{d}{dt}[1 - F(t)]$. Since $1 - F(t) = R(t)$, the relationship between $f(t), R(t)$, and $h(t)$ can be summed up as

$$f(t) = h(t)R(t) \qquad (6.2)$$

If $M(t) = \int_0^t h(x)dx$ is the cumulative intensity function, then the reliability function and failure rate function are related as

$$R(t) = \exp\left\{\int_0^t h(x)dx\right\} \qquad (6.3)$$

(iii)   *Mean time between failures* (MTBF) is defined as the ratio of the total time
        of use to the total number of failures. Here, the total time of use is the sum of
        average time to failure of a repairable product and the average time to repair
        or replace it.
        For instance, if the number of failures reported for a system is 20 and the
        system available for use is 10,000 h, then the MTBF is $10000/20 = 500$ h.

(iv)    The *mean time to failure* (MTTF) is given by the average time the system is
        available in service. Technically, it is provided by

$$\text{MTTF} = E(T) = \int_0^\infty tf(t)dt \qquad (6.4)$$

(v)     The *mean time to repair* (MTTR) is defined as the expected (average) value
        of the time to repair and is given by $\text{MTTR} = \int_0^\infty tf(t)dt$, where $f(t)$ is the
        repair time distribution.
(vi)    (vi) The *mean residual life* (MRL) function is given by $E[X - t|X\rangle t]$.

## 6.11.1  Exponential Distribution

One of the most frequently used lifetime distributions in reliability theory is the *exponential distribution.* It is found to be useful for characterizing uncertainty attached
to machine life, breakdown of machines, the life of an electronic unit, duration of
telephone calls, service time and waiting time of a queue, and so on. An exponential
distribution in its purest form is defined as

$$f(x) = \theta e^{-\theta x}, x \geq 0, \theta > 0 \qquad (6.5)$$

where $\theta$ is the failure rate. The corresponding survival (reliability) function is

$$R(t) = e^{-\theta t} \qquad (6.6)$$

The hazard function of the exponential distribution, according to (6.2), is $h(t) = \theta$
and is constant. The MTTF of the distribution is $\theta^{-1}$. In Fig. 6.3, we present the plot
of $f(x)$, $R(x)$, and $h(x)$, of an exponential distribution.

As mentioned above, the exponential distribution enjoys many new properties.
One such feature is the *memory-less property.* The property says that if a random
variable $X$ follows an exponential distribution, then its conditional property obeys
$P(X > t + s|X > t) = P(X > s)$, for $t < s$. Exponential distribution can also be

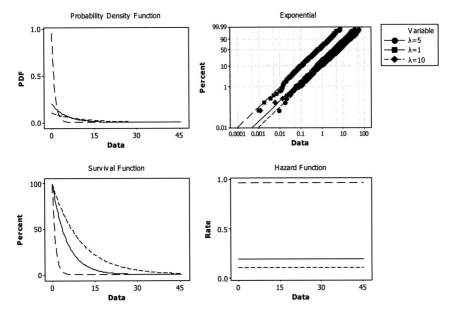

**Fig. 6.3** Plot of $f(x)$, $R(x)$, and $h(x)$ of an exponential distribution

characterized by the mean as its parameter $\theta$; in that case, the probability density function can be expressed as

$$f(x) = \frac{1}{\theta} e^{-\frac{x}{\theta}}, x \geq 0, \theta > 0 \tag{6.7}$$

According to Harrington and Anderson (1999), the exponential distribution is the model that universally quantifies product reliability. This is because everyone requiring product reliability information uses a mutually understood number to plan their future work activity. Both (6.5) and (6.7) facilitate the values of failure rate $(h(t))$, total failures with time $(F(t))$, and failures from the remaining survivors of the initial set or group of products $(f(t))$ in one place. We now consider some examples based on this distribution:

**Example 6.1.** A company has 1000 units of a product operating with a failure rate of 0.0025 per hour. What is the number of failures requiring replacement in the next 300 h?

*Solution.* Given here is failure rate, i.e., $\theta = 0.0025/h$. So, the number of units survived 300 h is

$$R(t) = e^{-(0.0025*300)} = 0.47$$

Hence, the total number of surviving units is $1000*0.47 = 470$. So, 530 units require replacement.

**Example 6.2.** IMOTO Company is producing the back-end switch-enabled panel for its latest mobile instrument to be launched across the Asian continent. A sample of 5,000 units of the product was selected for inspection. After seventy-two hours of electromechanical testing, it was found that 120 panels had some electromechanical problems (failure) with an average estimated operating time of 70 h. Suppose the failure process is according to an exponential distribution, then how many of these units will fail in the first 200 operating hours?

*Solution.* Let $\theta$ be the average failure time, and it is obtained as

$$\theta = \frac{\text{Total failures}}{\text{Total operating time}} = \frac{120}{5000 * 70} = 0.000343 \text{ failure/hour}$$

Since the failure time distribution is according to an exponential distribution, the reliability at 200 h is computed as

$$R(t) = e^{-0.000343 * 200} = 0.933727$$

Thus, the total number of units survived after 200 h is 5000*0.933727 = 4669.

### 6.11.2   Weibull Distribution

Another vital distribution used in reliability studies is the Weibull distribution, where exponential distribution becomes a particular case. Unlike exponential distribution, Weibull distribution has two parameters that can be used to model reliability growth and deterioration. In 1939, Waloddi Weibull introduced the distribution connected with his study on the strength of a material in a series of volumes edited by the Royal Swedish Institute for Engineering Research. The probability density function corresponds to Weibull distribution given by

$$f(x) = \theta \beta x^{\beta-1} e^{-\theta x^{\beta}}, x \geq 0, \theta > 0, \beta > 0 \qquad (6.8)$$

For $\beta = 1$, (6.8) reduces to a constant failure rate distribution, i.e., exponential distribution. For $\beta < 1$, the failure rate decreases, and for $\beta > 1$, the failure rate increases. The survival function and hazard function are, respectively, given as

$$R(x) = e^{-\theta x^{\beta}} \qquad (6.9)$$

And

$$h(x) = \theta \beta x^{\beta-1} \qquad (6.10)$$

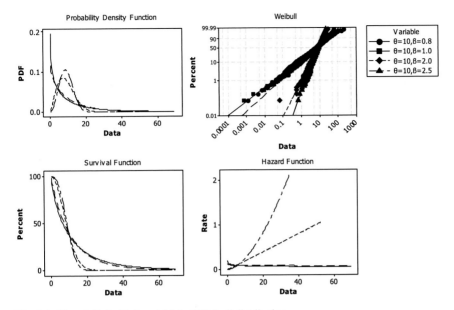

**Fig. 6.4** Plot of $f(x)$, $R(x)$, and $h(x)$ of Weibull distribution

Note that the shape parameter $\beta$ decides the shape of the distribution. Accordingly, Weibull distribution accommodates many other distributions. For example:

- $\beta = 2$, the distribution reduces to Rayleigh distribution.
- $\beta = 2.5$, the distribution reduces to lognormal distribution.
- $\beta = 3.6$, the distribution reduces to normal distribution.
- $\beta = 5$, the distribution reduces to normal peaked distribution.

The graph of $f(x)$, $R(x)$, and $h(x)$ is shown in Fig. 6.4.

Like exponential distribution, Weibull distribution can also be modeled as product reliability if the failure rate varies with time, and all products in a set have the same failure rate. But, because of the shape parameter in the power of the time, the estimation of parameters poses some difficulty. There are plenty of modifications and generalizations available for Weibull distributions now. One may refer to the book by Murthy et al. (2003) and Meeker and Escobar (1998) and references contained therein.

*Example. 6.3.* A company produces electronic devices in which the failure rate distribution has been modeled by a Weibull distribution with $\theta = 0.01$ and $\beta = 0.6$. Obtain the failure rate and the fraction of devices expected to survive 3000 h. What proportion of items will survive between 3000 and 3500 h?

Solution. Given X ~ Weibull ($\theta = 0.01$, $\beta = 0.6$). Then,

$$h(t) = \theta t^{\beta - 1}$$

$$\text{i.e., } h(3000) = 0.01 * 3000^{0.6-1} = 0.00041$$

and

$$R(t) = e^{-\theta t^{\beta}}$$

$$R(3000) = e^{-0.01*3000^{0.6}} = 0.2953$$

The proportion of items will survive between 3000 and 3500 h and is obtained as

$$R(3000) - R(3500) = 0.033 \cong 3.3\%$$

Thus, the theory of reliability studies the failure laws of systems that are operating within the stipulated limits under given operating conditions. A complex system, even if adequately designed, is prone to failure. Hence, it is essential to assure the proper functioning of the system from not only a strictly technological standpoint but simultaneously considering the failure properties of its components (Barlow & Proschan, 1965). Hence, reliability is an intrinsic property of operating systems, and, as such, it has to be specified as part of the system characteristics. Therefore, it becomes imperative to develop a quantitative basis for describing the system's failure process to measure their reliability unambiguously. Among all models, the probability models described above are the best way to describe those system characteristics, enabling one to compare the performance of a similar system on a reliability basis and develop methods for improving the reliability of a particular system.

There are many managerial considerations to improve the reliability of a system. That includes decreasing the system's failure intensity through enhanced technology of the component, introducing redundancy at the component or network, and improving the overall design of the system to achieve maximum reliability. According to Sivazlian and Stanfel (1975), the technological considerations for decreasing the system's failure rate or intensity are:

- Simplification of the system itself
- Selection of the most reliable components
- Leaning the operating conditions of the system
- Rejection of parts of low reliability
- Standardization of components
- Improvement of production technology
- Statistical control of component quality
- Elimination of initial-type failures using break-ins
- Minimization of wear-out-type failures through an effective preventive maintenance program
- Reducing the hazard contribution of human errors by proper automation and increasing the skill of the servicing personal.

Improving the reliability of a system often necessitates the knowledge of failure patterns and the conditions affecting the system's quality planning and components. Most of the time, equipment, instruments, and tools also require maintenance and monitoring of variation. Otherwise, system reliability can diminish over some time. A powerful method to improve the reliability of a system is the reliability-centered maintenance (RCM). RCM is a method of planning and prioritizing resources and actions. The goal of RCM is to ensure process reliability through data collection, analysis, and detailed planning. According to Juran and De Feo (2010), prioritization is the foundation of RCM. The basic premise is to allocate resources as effectively as possible to eliminate unplanned downtime, reduce deteriorating quality, or ensure planned output. Assets are prioritized into reactive, preventive, and predictive maintenance. These assets could include non-critical components, redundant equipment, small, simple items, and all breakdowns. If successfully implemented, RCM can deliver significant business benefits. According to Juran's principles, reactive maintenance costs are two to three times higher than preventive and preventive is two to three times higher than predictive. Measuring improvement in reliability should include several dimensions. The most encompassing is overall equipment effectiveness (OEE) as it measures the cumulative effect of all losses due to equipment condition—machine availability, machine efficiency, and machine quality performance.

## 6.11.3 Software Reliability

*Software reliability* is a unique aspect of reliability engineering. Traditionally, reliability engineering focuses on critical hardware parts of the system. Since the widespread use of digital integrated circuit technology, the software has become an increasingly crucial part of most electronics and, hence, nearly all present-day systems. There are significant differences, however, in how software and hardware behave. Most hardware unreliability is the result of a component or material failure that results in the system not performing its intended function. Repairing or replacing the hardware component restores the system to its original state. However, the software does not fail in the same sense that hardware fails. Instead, software unreliability is the result of unanticipated results of software operations.

A typical reliability metric is the number of software faults, usually expressed as faults per thousand code lines. This metric, along with software execution time, is crucial to most software reliability models and estimates. The theory is that software reliability increases as the number of faults (or fault density) goes down. Establishing a direct connection between fault density and MTBF is challenging; however, because of how software faults are distributed in the code, their severity and the probability of the combination of inputs are necessary to encounter the error. Nevertheless, fault density serves as a useful indicator for the reliability engineer. Other software metrics, such as complexity, are also used for identifying the software faults (Muralidharan & Syamsundar, 2012).

The last two decades have seen tremendous development in software tools, methodologies, and new programming languages for simplifying customer requirements. This has necessitated the need for improving software reliability and quality, as the technology landscape is increasing without any boundary. According to Juran and De Feo (2010), some of the best practices for software quality excellence are:

- Exceed customer and business requirements by creating innovative, breakthrough software products and services that consistently perform as planned, to documented requirements and specifications, without defects.
- Create, document, communicate, and maintain the complete process to be used to generate, test, release, and maintain software products and services, which is sometimes referred to as the software or systems development life cycle (SDLC).
- Establish a software quality leadership role with authority to act as a customer ombudsman and direct the overall planning and implementation of the software quality system.
- Create an environment and process for efficient collaboration, knowledge sharing, code reuse, and open communications among all stakeholders, employees, and customers.
- Manage, prioritize, and organize the software process in close alignment with the business strategy and goals with attention to fitting the organizational culture.

## 6.12 Simulation and Process Model Generation

Simulation is a method of solving decision-making problems by designing, constructing, and manipulating a real system model. For creating models, data may be generated by producing parametric draws from a known model (once or many times) or by repeated resampling with replacement from a specific data set (where the correct data-generating model is unknown). For resampling studies, the actual data-generating mechanism is unknown, and resamples are used to study the sampling distribution. Hence, the entire simulation mechanism duplicates the essence of a system or activity without actually obtaining the reality.

The system here is a collection of entities that act and interact with the accomplishment of some logical end. Hence, it will have a group of variables necessary to describe the status of the system at any time, called states of the system. For the experiment, one can create a physical or mathematical model. The simulation uses a mathematical or statistical model for creating a model. There are various types of simulations. They are Monte Carlo simulation, Markov chain Monte Carlo (MCMC) simulation, operational gaming, system simulation, Bayesian simulation, and so on. The Monte Carlo simulation applies to business problems that exhibit chance or uncertainty. The basis of the Monte Carlo simulation is experimentation on the probabilistic elements or inputs realized through a random sampling method. The steps involved in creating any type of simulation model are:

- Develop a flow diagram or technical statement of the problem or set up the model.

- Characterize all steps of the process.
- Determine input values and probability estimates and cumulative probability distribution.
- Generate random number.
- Determine appropriate measure scheme and simulate trials.
- Validate the model parameters with the existing models, if any.
- Repeat the above steps till a workable mode is ready.

### 6.12.1  Simulation of the Probability Model

We will start with an example of creating a simulated probability model here. We know that $t$, $F$, $x^2$ are obtained as the sampling distribution of normal distribution. For instance, if $x$ is a standard normal random variable and if $y$ is a chi-square ($x^2$) random variable with $n$ degrees of freedom, and if $x$ and $y$ are independent, then

$$t = \frac{x}{\sqrt{y/n}}$$

(6.11)

is distributed as $t$-distribution with $n$ degrees of freedom. Like normal distribution, $t$-distribution is symmetric, symmetric at mean $\mu = 0$, and $\sigma^2 = \frac{n}{n-2}$, $n > 2$. If $n \to \infty$, the $t$-distribution reduces to a standard normal distribution, $N(0,1)$.

Here, we use Minitab software to simulate the $t$-distribution for various degrees of freedom by using the property stated above. The steps involved are: In the software, go to **Cal > Random data,** select normal, generate 10,000 random samples, and assign it to $x$. In the same panel, select chi-square, generate 10,000 random samples with degrees of freedom 3, and assign them to $y1$. Similarly, generate chi-square random variables $y2$, $y3$, and $y4$ with degrees of freedom 5, 10, and 30. Now again, go to the **Cal > Calculator** panel and compute $t(3) = x/\sqrt{y1/3}$. Similarly, compute $t(5) = x/\sqrt{y2/5}$, $t(10) = x/\sqrt{y3/10}$, and $t(30) = x/\sqrt{y4/30}$, respectively, as $t$-variables. The plot of simulated $t$-distributions $t(3)$, $t(5)$, $t(10)$, and $t(30)$ is shown in Fig. 6.5. Note that all the distributions are symmetric at zero with variances 3.435, 1.275, 1.114, and 1.027, respectively.

### 6.12.2  Simulation of an Engineering Model

We now consider a business case concerning a chemical process of mixing to form a particular mix of cleaning powder. Through a cause and effect diagram, several CTQ variables were prioritized for further to know their influence in the process. The effect variable ($y$) here is the variation in weight in an ounce. The predictor variables (input variables) that influence the response variable were finally decided based on multiple regression techniques. They include size ($x_1$), and time is taken

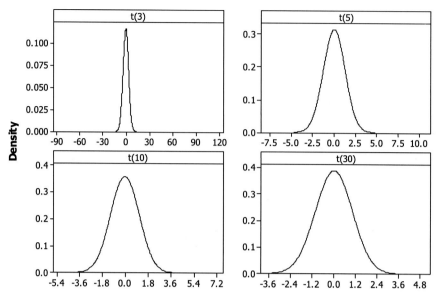

**Fig. 6.5** Simulated t-distribution

to fill the weighing machine ($x_2$), speed of flow of chemical ($x_3$), holding pressure during the flow ($x_4$), and holding pressure at the end of the flow ($x_5$). The graphs of all variables are shown in Fig. 6.6. It is evident that the variables $x_1$, $x_2$, and $x_5$ are

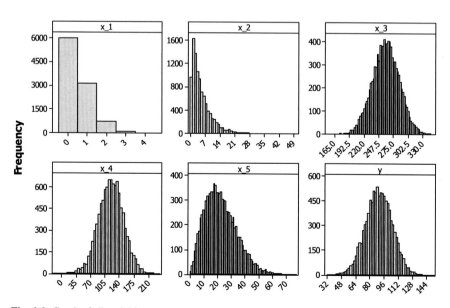

**Fig. 6.6** Graph of all variables

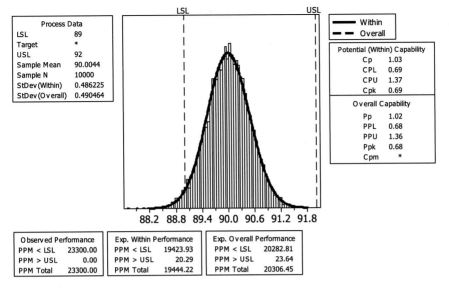

**Fig. 6.7** Process capability of $y$

skewed, and the other variables are normal. Also, the variable $x_1$ was measured in over- or undersized and hence is a binary variable. The process is finally accepted if the proportion of weights less than the lower specification limit does not exceed 2%.

A close investigation of each variable reveals that $x_1$ follows a binomial distribution with parameter $n = 10$ and $p = 0.05$, $x_2$ follows an exponential distribution with mean 5, $x_3$ follows a normal distribution with mean 260 and standard deviation 25, $x_4$ also follows a normal distribution with mean 120 and standard deviation 30, and $x_5$ follows a Weibull distribution with scale parameter 25 and shape parameter 2. This information is necessary for carrying out the Monte Carlo simulation for $y$, where the transfer function is given by $y = 92.5 - 0.046x_1 - 0.036x_2 - 0.00124x_3 - 0.0142x_4 - 0.012x_5$.

The process capability of $y$ is shown in Fig. 6.7. Now $P(y < \text{LSL}) = P(Z < -2) = 0.02275$, which is slightly more than the process entitlement, 2% as decided by the process owner. Since the difference is minimal, the acceptance or rejection of the simulated model lies with the management.

### 6.12.3  Simulation of Reliability Model

A reliability engineer identifies the best survival distribution of the pressure $P$ generated by a pressure balance to customize the future production of new scales used in auto-parts. The quantity $P$ is given implicitly by the measurement model:

$$P = \frac{M * G\left(1 - \frac{M_a}{M_m}\right)}{A(1 + \alpha * T))} \tag{6.12}$$

where the model parameters and their distributions are:

- M: the total mass applied—$N(0.01, \ 0.001)$
- $M_a$: the mass densities of air—$N(0.001, \ 0.001)$
- $M_m$: the mass density of applied mass—$N(0.005, \ 0.0005)$
- G: the local acceleration due to gravity—Weibull $(\alpha = 0.05, \ \beta = 0.25)$
- A: the active cross-sectional area of the balance at zero pressure—$N(0.1, \ 0.01)$
- $\alpha$: the temperature coefficient—gamma $(\alpha = 10, \ \beta = 3)$
- T: the temperature in degree Celsius—$N(40, \ 0.0025)$.

Note that there are seven input variables in the model. As per the above information, we simulated each variable from their respective distributions. Then, the output variable $P$ is computed, and a test of the goodness-of-fit test is carried out in Minitab software. It is found that only Weibull distribution fits well with a $p$-value more significant than 0.25. The shape and scale parameters are, respectively, estimated as $\beta = 0.24843$ and $\alpha = 0.00005$. The distribution overview plot is shown in Fig. 6.8. The six-pack process capability is shown in Fig. 6.9.

There are many advantages of creating models through simulation. They

- Allow one to analyze significant, complex problems for which analytical results are not readily available.

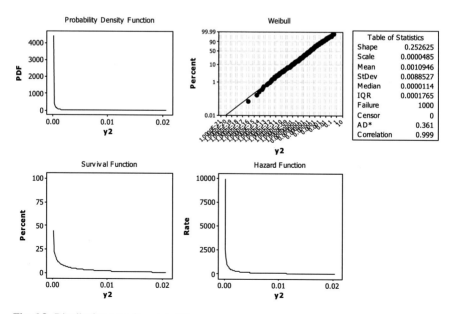

**Fig. 6.8** Distribution overview plot of the pressure

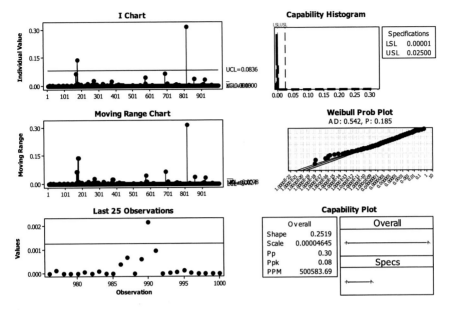

**Fig. 6.9** Six-pack capability of pressure measurement

- Allow the decision-maker to experiment with many different policies and scenarios without actually changing or experimenting with the actual system. This allows "what-if?" questions and answers.
- Allow you to compress time and enable the study of interactions.
- Mitigate process performance risk through experimentation with new processes before building and implementing them.
- Assess process sensitivity by understanding how various activities within processes interact and how changes will affect overall performance.
- Require little or no complex mathematics, and so maybe intuitively more understandable.

Since the probability distribution depends on a particular *seed* (an initial value that determines the sequence of random numbers supplied by the user) of a *pseudo-random number* (one that is created by a random number generator), the numerical results based on the specific set of random numbers can alter the reality of the situation. That is, each model is unique and not typically transferable to other or similar problems. All statistical packages capable of Monte Carlo simulation use a pseudo-random number generator. Each random number is a deterministic function of the current "state" of the random number generator. Therefore, to obtain more accurate results and minimize the likelihood of making a wrong decision, one should use a large number of trials in each simulation and repeat the simulation a large number of times. Naturally, a large number of repetitions can require significant amounts of computer time. However, this is not a big issue now, as computing power's capacity can be shared and minimized these days effectively. Since each simulation requires its distinctive design to mimic the actual scenario under investigation and its associated computer program, it may request significant development effort. This is a technical challenge.

Simulation studies provide us with empirical results for specific scenarios. For this reason, simulation studies will often involve more than one data-generating mechanism to ensure coverage of different situations. For example, it is ubiquitous to vary the sample size of simulated data sets. Performance simulation studies are computer experiments that involve creating data by pseudo-random sampling from known probability distributions. They are invaluable tools for statistical research, particularly for the evaluation of new methods and the comparison of alternative methods (Morris et al., 2017). With the advent of computing technology, the importance of simulation has also increased tremendously, as high-end programming language and compilers are available for any complex tasks. The futuristic technologies like artificial intelligence and machine learning also advocate using simulation in a big way.

The MCMC methods mentioned previously are used to sample from complicated multivariate distributions with normalizing constants that may not be computable, and direct sampling is not feasible. Therefore, the main issue here is the convergence of the sequence. Since the chain is simulated for a long-run infinite time, the method always produces an approximate sample. MCMC is facilitated through two famous samplers, such as the Metropolis–Hastings (MH) algorithm (Hastings, 1970; Metropolis et al., 1953) and Gibbs sampler (Geman & Geman, 1984). For simple distributions, exact samples are obtained by inversion, tabulation, transformation, and composition methods (Ripley, 1987). For practical purposes, rejection sampling is fast enough if we can easily sample a proposal envelope distribution that is reasonably close to the target distribution. Hence, the model preparation and evaluation through simulation are emerging as viable alternatives for future generations' real systems.

## 6.13   Green Statistics and Quality Certifications

To develop a roadmap for policymakers to improve the efficiency and quality, we propose the following guidelines for Green Statistics quality certifications (Muralidharan, 2013). These guidelines will help improve the effectiveness of the environmental regulations and help individual firms adopt the best innovative practices to comply with environmental laws and improve performance. These guidelines are stated based on various categories like environmental regulations, sustainable business practices, voluntary actions, pressures from stakeholders, business innovations, performance (financial and environmental), etc., and are briefly stated as follows:

- Prepare environmental mission and vision statements for every process in the organization.
- Set a defined standard of environmental regulations through economic parameters.

- Involve pollution control and waste reduction in every production process.
- Check for harmful emissions and unnecessary energy consumption frequently.
- Specify emission targets, greenhouse gases, and carbon budgets. If necessary, specify their management methods.
- Implement economically viable waste resource for collection and disposal.
- Implement environmentally friendly practices like recycling, remanufacturing, etc.
- Record all public environmental statement and make available for public concern.
- Record and act upon all environmental compliance violations.
- Develop innovative processes and IT savvy technical know-how's.
- Promote our own sources of electricity generations.
- Practice sustainable business practices to save energy consumption.
- Establish a transparent environmental management system to collect data and responsibilities.
- Promote eco-friendly vehicles and transportations for goods and services movement.
- Assess the achievements of the company in terms of energy conservation and cost saving on sustainable business practices.
- Assess the performance of the company in terms of the growth and market share on sustainable practices.
- Assess the product reach and geographical area covered due to sustainable practices.
- Publish all CSR-related activities and report to confederation and association of industries.
- Report all other specific methods adopted for energy recovery and saving.

The above-stated guidelines may be applied as per the size, need, and objective of the organization. Depending on the organizational environment, one may revise or modify the instructions to suit their purpose. Studies have shown that environmental performance, financial performance, and business innovations are strongly correlated (Mohd Zamri et al., 2013). Instead, business innovations always contribute to environmental performance and financial performances. Thus, the regulations stated above are significant for the business to sustain and improve the organization's quality. Some of the above guidelines have a strong base of quantitative and qualitative characteristics and, therefore, can be evaluated for assessment anytime in the organization. And it should be made mandatory for any quality certification. Thus, the GS described in this chapter can act as a core ingredient for lean, green, and clean quality certifications and evaluations.

We believe that the corporate objectives should be linked with GS for any improvement and setting a particular target. The interrelationship between GS, GSSCM, and other supply chain dynamics and various green performance measures facilitates the management to follow compliance measures. The importance of GS in the supply chain management system necessary to identify the uncertainty in the green movement of inventory and dynamics directs the management to have a flexible evaluation system in the organization. Therefore, we strongly advocate that, like ISO certification and carbon auditing, Green Statistics certification should also be made mandatory as documentation and auditing process of company activities. At least for those organizations that are into the service of humanity, it should be made compulsory. Organizations whose products and services address environmental compliance issues and climate change concerns should be promoted for full acceptance and demand.

## 6.14  Exercises

6.1.   Define Green Statistics. State all the characteristics of GS.
6.2.   Compare GS with supply chain dynamics pointing out the similarities between the two.
6.3.   State and describe all the components of Green Statistics Supply Chain Management.
6.4.   Discuss briefly recursive partitioning and contingent valuation method.
6.5.   What is the significance of e-commerce and e-governance? Discuss the varieties of e-commerce activities.
6.6.   State all benefits of e-governance.
6.7.   Discuss how e-governance help is improving information and communication technology?
6.8.   State the quality components associated with e-governance in education.
6.9.   What is big data analytics? What are its applications?
6.10.  Discuss the DMAIC phases of big data analysis.
6.11.  What is reverse logistics? How does it compare with Lean Six Sigma?
6.12.  What is benchmarking?
6.13.  What is reverse engineering? State all the steps involved in it.
6.14.  Distinguish between reliability engineering and maintainability engineering.
6.15.  Define the following terms:

    (i)     Survival function
    (ii)    Instantaneous failure rate
    (iii)   Mean time to failure
    (iv)    Mean time to repair
    (v)     Mean time between failures.

6.16.  State all the applications of exponential distribution.
6.17.  Suppose that the amount of time one spends in a queue for railway reservation at window of a railway station is approximately exponentially distributed with mean 40 min.

    (i)     What is the probability that a customer spends more than one hour?
    (ii)    What is the probability that a customer will spend more than 1 h given that he is still in the queue after 40 min?

6.18.  What are the applications of Weibull distribution?

6.19. State all the Green Statistics quality guidelines.
6.20. What is the importance of simulation? State some applications of simulation.

# References

Aldred, J. (1994). Existence value, welfare and altruism. *Environmental Values, 34*, 381–402.

Augment. (2016). Augmented reality applications in the tourism industry, 6 Jan 2016. Available at: https://www.augment.com/blog/augmented-reality-in-tourism/(13.3.2020).

Barlow, R. E., & Proschan, F. (1965). *Mathematical theory of reliability*. Wiley.

Biehl, M. U., Marion, R., & McKelvy, B. (2007). Complexity leadership theory: Shifting leadership from the industrial age to the knowledge era. *The Leadership Quarterly, 18*(4), 298–318.

Breiman, L. (1984). *Classification and regression trees*. Chapman & Hall/CRC.

Daniel, E. M., Hoxmeier, J., White, A., & Smart, A. (2004). A framework for the sustainability of e-marketplaces. *Business Process Management Journal, 10*, 277–290.

Davis, T. (1993). Effective supply chain management. In *Sloan Management Review* (pp. 35–45). Summer.

Dohale, V., & Kumar, S. (2018). *A review of literature on industry 4.0, conference paper*. https://www.researchgate.net/publication/328345685.

Dumbill, E. (2012). What is big data? An introduction to the big data landscape. *O'Reilly Strata*. [Blog post, January 2011.]. http://strata.oreilly.com/2012/01/whatis-big-data.html.

De Brito, M. P., & Dekker, R. (2003). *A framework for reverse logistics*. Institute of Management Report Series Research in Management, Erasmus University.

Edwards, J., Wang, Y., Potter, A., & Cullinane, S. (2010). E-business, e-logistics, and the environment. In Mckinnon, A., Cullinane, S., Browne, M., Whiteing, A. (eds.) *Green logistics*. Kogan Page.

Franks, B. (2012). *Taming the big data tidal wave: Finding opportunities in huge data streams with advanced analytics*. Wiley.

Fogarty, D. (2015). Lean Six Sigma and Data Analytics: Integrating complementary activities. *Global Journal of Advanced Research, 2*(2), 472–480.

Fleischmann, M., Krikke, H. R., Dekker, R., & Flapper, S. D. P. (2000). A characterization of logistics networks for product recovery omega. *International Journal of Management Science, 28*, 653–666.

Fleischmann, M. (2001). *Reverse logistics network structures and design*. Erasmus Research, Institute of Management Report Series Research in Management, ERS-2001-52-LIS.

Flygansvaer, B. (2008). Coordinated action in reverse distribution systems. *International Journal of Physical Distribution & Logistics Management, 38*(1), 5–20.

Geman, S., & Geman, D. (1984). Stochastic relaxation, Gibbs distributions, and Bayesian restoration of images. *IEEE Transactions on Pattern Analysis and Machine Intelligence, 6*, 721–741.

Gobble, M. A. (2013). Big data: The next big thing in innovation. *Research and Technology Management, 56*(1), 64–66.

Gryna, F. M., Chua, R. C. H., & Defeo, J. (2007). *Juran's quality planning & analysis for enterprise quality*. Tata McGraw-Hill.

Geary, S. C., & Pand, T. D. (2002). Uncertainty and the seamless supply chain. *Supply Chain Management Review, 6*(4), 52–61.

Gupta, D. N. (2008). *e-Governance: Comprehensive framework*. New Century Publication.

Hans, J. S., Barzilai-Nahon, K., Jin-Hyuk, A., Olga H. P., & Barbara Re. (2009). E-Commerce and e-Government: How do they compare? What can they learn from each other? In *Proceedings of the 42nd Hawaii international conference on system sciences*, (pp. 1–10).

Hanemann, M., Loomis, J., & Kanninen, B. (1991). Statistical efficiency of double-bounded dichotomous choice contingent valuation. *American Journal of Agricultural Economics, 73*(4), 1255–1263.

Harrington, H. J., & Andeerson, L. C. (1999). *Reliability simplified: Going beyond quality to keep customers for life*. McGraw-Hill.

Hastings, W. K. (1970). Monte Carlo sampling methods using Markov chains and their applications. *Biometrika, 57*, 97–109.

Hervani, A. A., Helms, M. M., & Sarkis, J. (2005). Performance measurement for green supply chain management. *Benchmarking: an International Journal, 12*(4), 330–353.

Hofmann, E., & Rüsch, M. (2017). Industry 4.0 and the current status as well as future prospects on logistics. *Computers in Industry, 89*, 23–34.

Hodge, G. A., & Greve, C. (2016). On public-private partnership performance: A contemporary review. *Public Works Management & Policy*, 1–24.

Horvath, L. (2001). Collaboration: The key to value creation in supply chain management. *Supply Chain Management: An International Journal, 6*, 205–207.

Horvath, M., Martinez-Cruz, B., Negro, J., J., & Gody, J. A. (2005). An overlooked DNA for non-invasive genetic analysis. *Journal of Avian Biology, 36*(1), 84–88.

Horowitz, J. K., & Mcconnell, K. (2005). Willingness to accept, willingness to pay and the income effect. *Journal of Economic Behavior & Organization, 51*(4), 537–545.

http://www.supplychainmetric.com/

https://businessjargons.com/e-governance.html

https://en.wikipedia.org/wiki/National_e-Governance_Plan

https://www.qualcomm.com/invention/5g/what-is-5g

Jones, M. R., & Towill, D. (1999). Using the information decoupling point to improve supply chain performance. *International Journal of Logistics Management, 10*(2), 13–26.

Juran, J. M., & De Feo, J. A. (2010). *The Juran's quality handbook: A complete guide to performance excellence*. The McGraw-Hill Companies, Inc.

Kamble, S. S., Gunasekaran, A., & Gawankar, S. A. (2018). Sustainable Industry 4.0 framework: A systematic literature review identifying current trends and future perspectives. *Process Safety and Environmental Protection, 117*, 408–425.

Kubiak, T. M., & Benbow, D. W. (2010). *The certified six sigma black belt handbook* (2nd ed.). Dorling Kindersley (India) Pvt. Ltd.

Kudyba, S. (2015). *Big data, mining, and analytics*. CRC.

Lee, J., Bagheri, B., & Kao, H. A. (2015). A cyber-physical systems architecture for Industry 4.0 based manufacturing systems. *Manufacturing Letters, 3*, 18–23.

Leemis, L. M. (1995). *Reliability: Probabilistic models and statistical methods*. Prentice-Hall.

Marche, S., & McNiven, J. D. (2009). E-Government and E-Governance: The future isn't what it used to be. *Canadian Journal of Administrative Sciences/revue Canadienne Des Sciences De L'administration., 20*, 74. https://doi.org/10.1111/j.1936-4490.2003.tb00306.x

Metropolis, N., Rosenbluth, A. W., Rosenbluth, M. N., Teller, A. H., & Teller, E. (1953). Equations of state calculations by a fast computing machine. *The Journal of Chemical Physics, 21*, 1087–1091.

Meeker, W. Q., & Escobar, L. A. (1998). *Statistical methods for reliability data*. Wiley.

Mohd Zamri, F. I., Hibadullah, S. N., Mogd Fuzi, N., & Habidin, N. F. (2013). Green lean six sigma and managerial innovation in malaysian automotive industry. *International Journal of Innovation and Applied Studies, 4*(2), 366–374.

Montgomery, W. D. (1972). Markets in licenses and efficient pollution control programs. *Journal of Economic Theory, 5*, 395–418.

Montgomery, D. C. (2003). *Introduction to statistical quality control*. Wiley-India.

Morris, T. P., White, I. R., & Crowther, M. J. (2017). Using simulation studies to evaluate statistical methods. *Tutorial in Biostatistics, Statistics in Medicine,*. https://doi.org/10.1002/sim.8086

Muralidharan, K. (2008). Six sigma: A success story of quality management. *University Today, XXVIII*(22). (New Delhi. ISSN 1051-9580).

Muralidharan, K. (2018a). Lean, Green and clean: Moving towards sustainable life. *Communiqué, Indian Association for Productivity, Quality and Reliability, 40*(4), 3–4.

Muralidharan, K. (2018b). Quality assurance in big data analytics. *Communique, Indian Association for Productivity, Quality, and Reliability, 40*(1), 2–4.

Muralidharan, K., & Gupta, D. N. (2016). Six Sigma approach for e-Governance: Indian perspectives. *Communications on Dependability and Quality Management, an International Journal, 19*(1), 55–65.

Muralidharan, K. (2013). Green statistics: Some quality guidelines. *International Journal of Industrial Engineering and Technology, 3*(1), 7–20.

Muralidharan, K. (2015). *Six sigma for organizational excellence: A statistical approach.* Springer Nature.

Muralidharan, K., & Ramanathan, R. (2013). Green statistics: Management perspectives. *Journal of Industrial Engineering, VI*(9), 15–20.

Muralidharan, K, and Syamsundar, A. (2012). *Statistical methods for quality, reliability and maintainability.* PHI Learning in PVT.

Murthy, D. N. P., Xie, M., & Jina, R. (2003). *Weibull models.* Wiley.

Kumar, N., & Kumar, J. (2019). Efficiency 4.0 FOR Industry 4.0, *15*(1), 55–78.

Ngai, E. W. T., and Wat, F. K. T. (2002). A literature review and classification of electronic commerce research. *Information & Management*, 39, 415–429.

Nielsen. (2008). *Trends in online shopping: A global Nielsen consumer report.* February 2008, Haarlem.

Park, R. (1999). *Value engineering.* Lucie Press.

Ramazan, K., Ipek, K., & Ali, E. A. (2014). The role of reverse logistics in the concept of logistics centers, Procedia—Social and behavioral sciences. In *2nd World conference on business, economics, and management-WCBEM 2013* (vol. 109, pp. 438–442)

Ripley, B. (1987). *Stochastic simulation.* Wiley.

Rogers, D. S., & Tibben-Lembke, R. S. (1999). *Going backward: Reverse logistics trends and practices.* Reverse Logistics Executives Council

Rosen, G., Harvey, S., & Ted (2008). *Public finance* (pp. 71–103). McGraw-Hill Irwin.

Saša Zupan Korže. (2019). From Industry 4.0 to tourism 4.0. *Innovative Issues and Approaches in Social Sciences, 12*(3), 29–52.

Saugata, B., & Masud, R. R. (2007). *Implementing E-governance using OECD model (modified) and Gartner model (modified) upon agriculture of Bangladesh.* IEEE, 1-4244-1551-9/07.

Scholl, H. J., Barzilai-Nahon, K., Jin-Hyuk, Ahn, Popova, O. H., & Barbara Re. (2009). E-commerce and e-Government: How do they compare? What can they learn from each other? In *Proceedings of the 42nd Hawaii international conference on system sciences 2009*, (pp. 1–10).

Shakantu, M., Tookey, J. E., & Bowen, P. A. (2002). Defining the role of reverse logistics in attaining sustainable integration of materials delivery with construction and demolition waste management. In *Proceedings of creating a sustainable construction industry, Proceedings of the CIB W107 International conference on creating a sustainable the construction industry in developing countries* (pp. 97–103). 11–13 November.

Sivazlian, B. D., & Stanfel, L. E. (1975). *Analysis of systems in operations research.* Prentice Hall Inc.

Singh, R. K., Murty, H. R., Gupta, S. K., & Dikshit, A. K. (2009). An overview of sustainability assessment methodologies. *Ecological Indicators, 9*(2), 189–212.

Stevens, G. (1989). Integrating the supply chain. *International Journal of Physical Distribution and Materials Management, 19*(8), 3–8.

Towill, D. R. (1991). Supply chain dynamics. *International Journal of Computer Integrated Manufacturing, 4*(4), 127–208.

Towill, D. R. (1999). Simplicity wins twelve rules for designing effective supply chain control. *The Institute of Operations Management, 25*(2), 9–13.

Turban, E., Leidner, D., McLean, E., & Wetherbe, J. (2007). *Information Technology for Management: Transforming business in the digital economy* (6th ed.). Wiley.

Teddy (2010). *TERI energy data directory and yearbook*.

Wang, Y., Li, J., Liu, P., & Yang, F. (2007). Electronic commerce research review: Classification and analysis. In 3rd International conference on wireless communications, networking, and mobile computing (WiCom 2007) (pp. 3505–3508). IEEE.

Xu, L. D., Xu, E. L., & Li, L. (2018). Industry 4.0: State of the art and future trends. *International Journal of Production Research, 56*(8), 2941–2962.

Zhiyuan, F. (2002). E-government in the digital era: Concept, development, and practice. *International Journal of the Computer, 10*(2), 1–22.

# Chapter 7
# Achieving and Promoting Lean, Green, and Clean Quality Objectives

*Let us sacrifice our today so that our children can have a better tomorrow.*
—A. P. J. Abdul Kalam

## 7.1 Introduction

We know that industrial and commercial organizations are relevant if they are committed to providing lean (fast, flexible, waste-free, and defect-free) products and satisfaction to the consumer. For them, competitiveness in quality is central to profitability and crucial to business survival. In today's tough and challenging business environment, the comprehensive quality policy development and implementation are not merely desirable; it is essential for survival as well (Gryna et al., 2007; Oakland, 2005). In this context, the various tenets of LGC principles discussed so far can play a vital role in creating environmentally sustainable business practices in an organization with guaranteed results. These principles may be made mandatory for those organizations, wherever GHG and climatic conditions are worst. The least organizations bothered about the environmental, economic, and social aspects of their product and services will have slow growth and suffer economically. Besides, product quality depends on some specific green methods of ingredients practiced in the organization. They can be passive or reactive type and proactive or preventive type. Setting acceptable quality levels for the product, inspecting to measure compliance concerning the process, etc., are active type approaches. The dynamic or preventive nature includes ensuring design quality in products and processes, identifying sources of variation in processes and materials, and finally, monitoring the process performance.

When an unreliable product component breaks down and requires replacement, it creates social loss (inconvenience), environmental damage (carbon emissions embodied in the replacement component, which would not have been released if the element had been reliable) as well as the economic loss (cost of replacement). A business that is carbon positive and wastes nothing is regenerative to both environments and climate and hence will have a positive effect on the lives of people. It is estimated that automobiles, worldwide, emit around 10% of global emissions from fossil fuels and 6% of all global warming potential sources (DeCicco & Fung, 2006). Thus, it is not surprising that there are legislative, social, and political pressures on manufacturers to reduce their global warming contribution during production and use. Achieving green objectives is, therefore, mandatory for sustaining international competition and stakeholder pressures.

## 7.2  Achieving Lean Quality

All business process improvement demands a strategy where the leadership is ready for implementing a change. The author of this book believes that any process improvements can be achieved successfully through lean culture. Lean culture is often defined as a mansion of prosperity, having a new way of thinking at its foundation supported by two main pillars, respect for people and continuous improvement. A better way to appreciate lean culture is through *Shingo Process*. In his book "*Good to Great*" (Collins Business, 2001), Jim Collins talks about the flywheel effect. Great organizations do not become that way overnight. He writes, "They become that way because continuous small pushes create a breakthrough velocity. This velocity sustains growth. At the point where the momentum of change reaches breakthrough velocity (the tipping point), the organization moves forward along its Lean journey." The organization typically contains a majority of the "anchor draggers," few "early adaptors," and the rest "fence-sitters" (Patel, 2016a). While traditional managers spend their efforts focusing on the anchor draggers, they should be spending time with the early adopters, providing cover and support. The organization's focus should be on positive reinforcement, which promotes a forward shift in the fence-sitters. When we talk about the best roadmap to implement "lean," we should use Fig. 7.1, which describes the guiding principles and how to support them:

During the last quarter of the twentieth century, nearly everyone in senior management thought that business processes had been perfected. Henry Ford's great innovation, the moving assembly line, had been refined over the five decades after 1910, had served as the magic bullet during World War II, and by the mid-1960s was operating efficiently, at mass-production scale, in a wide range of industries around the world. Silently, in Japan, Taiichi Ohno and his engineering colleagues at Toyota were perfecting what they named the Toyota Production System, which we now can call as lean business processes. In more recent years, this new understanding of lean has evolved into a richer appreciation of the power of its underlying management disciplines: putting customers first by truly understanding what they need and then delivering it efficiently; enabling workers to contribute to their fullest potential; continually searching for better ways of working; and giving meaning to work by connecting a company's business strategy and goals in a transparent, coherent approach across the organization.

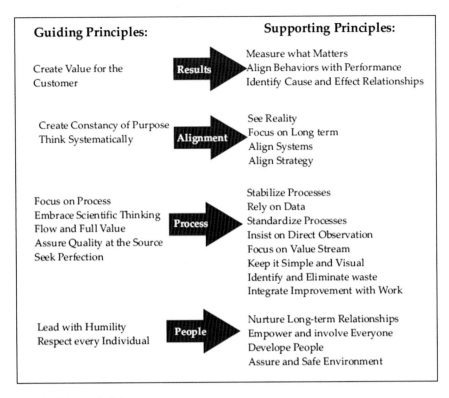

**Fig. 7.1**   Shingo principles

Toyota itself is pushing the boundaries of lean, rethinking the art of the possible in production-line changeovers, and for example, bringing customer input more directly into factories, flexible working hours as per employee choice, and so on. A leading service-based company like *Amazon.com* is extending the value of lean further into areas beyond manufacturing like fulfillment centers and the latest initiative called "Amazon Go," a new way of shopping. The main idea is to use an app on your phone, create a customer ID on this app, and then you can scan yourself in the Amazon Go store utilizing this ID. Once you are in the store, you just pick up the items you want, and then, you can leave straight away without queuing or paying as the payment is made online through the mobile app. Many companies are now following this kind of user-friendly applications to woo their customers.

There was a belief that lean works best for manufacturing and automobile sectors only. This is incorrect. Below are the examples of how lean can improve various business service processes (Patel, 2016b; Patel & Muralidharan, 2019):

- *Aviation industry*: Quality service is their prime concern, as there is stiff competition between airline services. Some of the airlines are working with Six Sigma capability. Aircrafts worth millions or more routinely sit idle at gates. Turnaround times between flights typically vary by upward of 30 percent generally. Lean techniques can cut hours to minutes with an improved changeover system.

- *Retail banking*: Lean has transformed the entire flow of modern-day banking activities to a new look and glamor. Retail banking involves many physical processes like; handling of paper and application checks, tendering and receipt of money and credit card slips lending and clearing, etc., where not only quality service but also speedy execution of service demand is necessary. With the excellent management of lean techniques, banks are now recovering a huge layover of time and increased transactions by adding value to all aspects of banking services. Whether they are nationalized or private, several banks in India are conducting Lean Six Sigma projects to improve their quality of service.

- *Health management*: Health or hospital management is a big area for quality improvement, as the service is offered round the clock where people and patients move quite often. The health personnel work on shifts, and they get transit to various departments and sub-departments. There are many chances for variation as man, machine, materials, and methods are involved in the whole process and are unpredictable. Uniform quality of service is possible only with a sound database management system in place. Nevertheless, hospitals can often reduce their variability a good deal in operations like prescription distribution, use of consumables and patient registration, and out-patient processes with lean principles.

- *Hotel chains and foodservice*: Lean techniques seek to improve product and service quality while simultaneously reducing waste and labor costs. The cancelation of hotel bookings cost a lot for the operators. Every removal is a waste for both the operator and the customer. Both qualities of food and service matter to this chain of an industry. Waste is universal, but disposal is even more severe. With a well-designed lean principle, the problems associated with food wastes can be reduced to a great extent. Many hotels have introduced their waste disposal methods for reducing the menace from food- and other dormitory-related wastes. For foodservice operators, the additional trick is to link such improvements to customer loyalty.

- *Designing products for value*: Data is now available through telemetric machine devices having embedded sensors. These small data sensors monitor installed equipment in the field and give companies insights into how and where products are used, how they perform, the conditions they experience, and how and why they break down. The next step is to link this information back to product design and marketing—for example, by tailoring variations in products to the precise environmental conditions in which customers use them.

Industries like electricity, telecommunication, mobile, and other utility service stations can also reap benefits with the efficient deployment of lean principles. Savvy companies these days use the data to show customers evidence of unmet needs they may not even be aware of and to eliminate product or service capabilities that are not useful. Applying lean techniques to all these new insights arising at the interface of marketing, product development, and operations should enable companies to make further strides in delighting their customers and boosting productivity. Therefore, it is necessary to understand the necessity of quality policies and audits from this perspective. In a recent article, Muralidharan and Raval (2019) opine that the focal point of past marketing efforts is on managing pleasurable total customer experience (TCE), as marketing activity has progressed from traditional promotion-based profit-making business function to relationship-based customer-oriented business function. In this connection, authors have introduced the Six Sigma Marketing (SSM) concept for overall quality management of the marketing component of a business organization. See also the contributions by Creveling et al. (2007), Webb and Gorman (2006), Reidenbach (2009), Muralidharan (2015), and references contained therein. SSM is actually *"a fact-based process approach executed through coordinated marketing efforts, managed through measurement-based Key Process Indicators (KPI) and root cause analysis."*

A recent paper by Muralidharan and Raval (2019) recommends that SSM enables high performance and sustains growth in various transactional activities of an organization. They argue that SSM can

- Influence customers in such a way that it would motivate them to opt for the offered products or services.
- Seek to remove the randomness from marketing and make it systematic and predictable about the transactions.
- Allow an organization to concentrate its limited resources on the most exceptional opportunities to increase sales and achieve sustainable competitive advantage.
- Ensure that organizations are continuing to meet the needs of their customer and are getting appropriate value in return.

Thus, SSM enables companies to improve marketing's strategic, tactical, operational, and personal aspects of processes to enhance the top line to drive revenue. By applying Six Sigma to marketing, organizations can identify leading indicators of growth and become proactive about performance improvement throughout a business cycle (Muralidharan, 2015). To make the marketing process meaningful, one must play with management by facts rather than management by objectives. SSM is, therefore, strategic in identifying the problem, tactical in understanding the customer requirements, and operational in implementing the strategy. Managing with Six Sigma may not be as exciting for some people as traditional management practices. However, governing data, facts, measurement, analysis, and experiments minimize the politics, power struggles, personality clashes, and drama of organizational life.

The DMAIC philosophy of SSM, according to Muralidharan and Neha (2017), is summarized as follows:

- *Define.* This phase involves defining and clarifying customer requirements (both internal and external). Those requirements are then get translated and parsed into supplier (or technical) specifications.
- *Measure.* This phase facilitates the measurement of customer purchase activity, marketing program results and conversion rates, actual costs for programs and people, lead quality data and lead cost, win/loss ratios defections that occur in the

buying process, etc. The more it is quantified, the better the decision be. Three of the significant marketing quality parameters that need special attention are critical to quality (CTQ), critical to cost (CTC), and critical to process (CTP). A Six Sigma strategy guarantees to incorporate the customer requirements through these critical parameters and enable the metric for quality assessment performance evaluation.

- *Analyze.* Performance improvement through data analysis helps the professionals to validate gaps in requirements versus current metrics with vital causes. Therefore, this phase helps marketing professionals to prioritize root causes affecting the marketing quality, establish a proper process model in terms of deterministic or probability model, suggest possible relationships and associations between control variables, test the significance of various input/output variables, and conduct a revised root cause analysis to identify the vital causes and judge the current performance with the customer requirements.
- *Improve.* This phase enables developing solutions targeted at confirmed causes of variation and waste generated through various marketing activities. The objective also extends to verify that the established reasons are statistically and practically significant and to optimize the process or product/service with the improvements. Further, it facilitates solutions with a short-term and long-term approach. Select and implement the improved marketing process and metric, measure results, conduct a designed experiment as per the marketing requirements, evaluate whether improvements meet targets, and evaluate for marketing risk.
- *Control.* This phase does the entire monitoring and surveillance of the marketing process. It helps to ensure whether market requirements as working according to customer's perception, the efficiency, and effectiveness of the marketing planning process are consistent, transparency in the marketing processes are intact, and the collaboration between marketing and other groups within the business is improved or not.

To make the marketing process more meaningful, one must play with management by facts. Make sure that these facts are measurable and subject to statistical treatment. Such information facilitates transparency of communication and confidence and excites customers for long-term loyalty and sustenance. Submitting to the rule of data demands a maturity of mind, respect for reality, and dedication to standards (Muralidharan, 2015). For the marketing process to succeed, all these are required.

## 7.3   Green Quality Policy

Quality policy shall be consistent with the company's strategy. The top management must share all the necessary measures to ensure that its corporate quality policy is clearly understood, implemented, and maintained. This will ensure the extent of the commitment of top management to quality. Quality managers shall assist in its formulation and implementation. It must exhibit quality intentions and directions of an organization, as formally expressed by senior management. Quality improvement and strategic planning should cover the areas of design, conformance to model, field service, and marketing availability of products and spare parts in the market (Juran & Gryna, Jr., 1980). Further, the quality policy shall include the following aspects:

- It must clearly state the key elements of quality, such as fitness for use, performance, safety, and reliability.
- Product design shall fully meet the customer's quality requirements, including functional, safety, and aesthetic and life characteristics, with adequate attention to the economy of manufacture.
- There shall be wholehearted attempts to strictly adhere to specifications during manufacture, dispatch, installation, and operational instructions, commissioning with emphasis on defect prevention.
- Use scientific guidelines for vendor selection, development, evaluation, vigilant supervision, and the vendor shall be assisted in achieving the standards of the purchaser.
- It should carry planned and a systematic program for acquainting the customer adequately on the proper usage of products.
- The policy shall emphasize continuous product review and improvement based on market survey, feedback of field performance data, and significant development of facilities shall be made available.
- The policy may link with the proposed increase in productivity, standardization, inventory control, factory automation, reduction in scrap, targeted delivery schedules, and reduction in cost and expenses.
- It should include the flexible provision of motivating the employees for quality, incentives for excellence.

Thus, green quality policy guidelines work as a catalyst for organizational health and intended expectations, as it covers all the organization's divisions. The quality policy can also influence the organization's culture and character by promoting user-friendly approaches and attitudes.

## 7.4  Green Quality Audit

Quality audit is an appraisal of the quality system of an entire organization. This is just like an accounting audit, and it evaluates the product, the process, and the system for achieving product quality. It inquires about the adequacy of the entire system of handling quality function. This involves bringing together the data from the following investigations (Muralidharan & Syamsundar, 2012):

To examine whether

- the organization adheres to environmental compliances
- the organization subscribes to environmental taxations
- the design meets the functional requirements completely

- the design specifications are clear cut without ambiguity
- the design fulfills the customer's requirement
- the manufacturing specifications are complete and transparent
- the customer's quality complaints are resolved in time or not
- the process and quality control charts are in place or not
- the inspection policies and its implementation, etc., are done following the customer complaints or not
- The program's scope and organizations ensure that all personnel understand the actions they must take to achieve the quality needs of the customers, etc.

Depending upon the nature of the company, the quality audit may be conducted as per the need arises. The frequency of inspection, revision of control, and modification of audit may also be done as per the requirements. Usually, a quality audit covers the following subject areas: policy on quality, the economics of quality, customer relations, specification of quality, organization of excellence, vendor relations, quality goals and results, quality control personnel, etc. The executive report of an audited quality survey will have the information like measures of customer satisfaction or dissatisfaction with the product, i.e., complaint rate, return rate, gain or loss of customers due to quality, etc.; service calls per some specified number of units sold; the cost of service calls due to quality faults; damage due to predominant defects; analysis of shipping damage; the cost of acceptance; the cost of prevention; summary of defects on purchased materials; list of assembly defects; measures of how well the overall system of quality is being carried out; report on action to correct factory troubles, etc. (Muralidharan & Syamsundar, 2012).

A green quality audit must be practiced at the organizational level for maintaining a certain level of quality throughout an intended period of evaluation. Among various components included for audit, the process quality audit should be given maximum importance as process conformance to standards and specifications are always valued by the stakeholders. A process quality audit includes any activity that can affect final product quality. This on-site audit is usually done on a specific process by one or more persons and uses the process operating procedures. Adherence to existing procedures is emphasized, but audits often uncover situations of inadequate or nonexistent systems. According to Gryna et al. (2007), audits must be based on a foundation of hard facts that are presented in the audit report in a way that will help those responsible for determining and executing the required corrective action. Thus, a quality audit followed by the quality policy can facilitate a strong base for sustainable business practices in the organization.

## 7.5  Promoting Lean, Green, and Clean Thinking

Here are some simple ways by which one can promote lean, green, and clean thinking in our day to day activities that will help reduce the overall environmental impact and the imbalance of our biodiversity. Many of them help in multiple ways and can be experimented with full potential. Some of them are as follows:

- Identify the use, necessity, and source of energy in each activity.
- Measure and monitor the quantity of energy used. Minimum the best.
- Incorporate renewable energy and energy efficiency into homes and organizations.
- Have your sources of waste disposal methods.
- Use a holistic method of manufacturing methods for sound carbon emissions.
- Monitor and practice measurement-based evaluations and interpretations.
- Monitor carbon management and GHG emissions.
- Safeguard flora and fauna of your ecological system.
- Use environmentally friendly products for safe use, consumption, and practice.
- Promote multi-user transportation methods like public transport and vehicle-pooling.
- Encourage the use of bicycles.
- Encourage walking for short duration destinations.
- Be content and learn to compromise with the availability of natural resources.

The above promotional activities can be done at the individual level as well as at the institutional level. Most of them are flexible and straightforward for practice and may be incorporated daily for inclusive excellence. Make sure that everyone is involved in some sustainable practices to a better living and better tomorrow. Accounting for quality is the best way to improve the quality of sustainable life. Also, ensure that the liability and assets are in tandem with each other. Otherwise, the quality aspect will be compromised. Let there be minimum use of resources and space. Use everything wisely so that the future generation will have no hardships and problems.

## 7.6 Benefits of Being Lean, Green, and Clean

Organizations that attain superior results by designing and continuously improving their products and services are often called *world-class* or *performance excellence*. Such world-class goods and services exceed customer's expectations in terms of quality and sustainable characteristics. This is realized through the lean and green thinking process. They require and rely on the support of organizational values and true empowerment, which is instrumental to the sustainability of lean. The absence of lean thinking harms morale, leading to increasing workers' unhappiness and withdrawal, ultimately leading to a failure (Jones, 2014). Thus, the LGC thinking in an organizational setup can

- Significantly reduce process cycle time
- Eliminate costly non-value-added activities
- Enhance value creation and team focus
- Reduce non-recurring and indirect costs

- Improve the utilization of scarce resources
- Maximize synergy among all related processes
- Ensure the participation of all stakeholders in the activities
- Transform "I" to "We"
- Create a coexistence between all living beings
- Create a win-win situation for all
- Lead to a less corrupt system.

Note that the successful implementation of any lean or Green quality projects depends entirely on the appropriate organizational tools and techniques. This should also involve a strong leadership at the top who can percolate the leadership qualities to the people at the bottom. Hence, every institution should encourage the best practices to improve the service communication between all the people involved with the processes. This can help build a strong, healthy, and responsible atmosphere for realizing sustainable practices, goods, and services as required.

Many case studies have shown that there is a strong relationship between Green Lean Six Sigma and financial performance in the production and automotive industries (Muralidharan, 2015; Mohd Zamri et al., 2013). According to Hart and Ahuja (1996), waste reduction can help save costs as the cost of raw material and waste disposal costs lower, while it is also able to prevent contamination. This is echoed by McWilliams and Siegel (2001) and Ghassemi (2002). Organizations are looking for more innovative and cost-effective methods that are always a step ahead of reducing waste and inefficiency. Right systems and processes in the application of environmental management systems are critical that the opportunity to reduce and eliminate waste and inefficiency would be higher. However, according to Porter and Van der Linde (1995a), this paradigm has been challenged by some analysts during the last decade. It was argued by Ambec and Lanoie (2008), Iwata, and Okada (2011) that a company's environmental performance could better lead to economic or financial independence as well. Besides, pollution commonly associated with the waste of resources (material and energy) can offset the cost of compliance with these policies. It is also essential to look at both sides of the balance sheet: increasing revenues and reduces costs to improving the organization's environmental performance. This can guide to better economic or financial performance and not necessarily to an increase in cost (Porter and Van de Linde, 1995b).

Studies have also shown that proper use of LGC concepts can lead to increased organizational efficiencies and growth sizably. In modern-day business, "*outsourcing*" is the order of the day. This is an excellent example of sustainable business practices. One can outsource human resources, technology, machines, processes, etc., to a remote party, who is not directly connected to your organization. The quality standards have also gone up with the growing promotion of outsourcing and should be promoted. We strongly believe that all the organizational and corporate objectives should be linked with LGC philosophies for any visible improvement and setting particular standards. For any goals to sustain, one needs to focus more on improving customer value, close firm–customer linkage, and stable firm–customer communication channel for a big transparent and informed decision (Muralidharan

& Raval, 2019). Toward this, various sustainable aspects presented in this chapter can be directed toward achieving long-term significance in modern-day organizational activities. A confident institutional leadership will be automatically in place if there is proper lean, green, and clean thinking.

## 7.7 Exercises

7.1. Discuss various features of Shingo principles.
7.2. What is Six Sigma Marketing? How does it improve the efficiency of marketing activity?
7.3. Discuss the DMAIC philosophy of Six Sigma Marketing.
7.4. What are the main ingredients of quality policy?
7.5. What are the steps involved in a quality audit?
7.6. State all the benefits of LGC activities.

## References

Ambec, S., & Lanoie, P. (2008). Does it pay to be green? A systematic overview. *Academy of Management Perspective, 22*(4), 45–62.

Creveling, C. M., Hambleton, L., & McCarthy, B. (2007). *Six Sigma for marketing processes: An overview for marketing executives, leaders, and managers.* Prentice-Hall.

Collins, J. (2001). *Good to Great: why some companies make the leap and others don't.* Random House, Business Books.

DeCicco, J., and Fung, F. (2006). *Global warming on the road, environmental defense.* http://www.edf.org/sites/default/files/5301_Globalwarmingontheroad_0.pdf

Ghassemi, A. (2002). *Handbook of pollution control and waste minimization* (p. 3).

Gryna, F. M., Chua, R. C. H., & Defeo, J. (2007). *Juran's quality planning & analysis for enterprise quality.* Tata McGraw-Hill.

Hart, S. L., & Ahuja, G. (1996). Does it pay to be green? An empirical examination of the relationship between emission reduction and firm performance. *Business Strategy and the Environment, 5,* 30–37.

Iwata, H., & Okada, K. (2011). How does environmental performance affect financial performance? Evidence from Japanese manufacturing firms. *Ecological Economics,* 1691–1700. https://doi.org/10.1016/j.ecolecon.2011.05.010

Jones, B. F. (2014). The human capital stock: A generalized approach. *American Economic Review, 104*(11), 3752–3777.

Juran, J. N., & Gryna, F. M. (1980). *Quality planning and analysis: From product development through use.* Tata McGraw-Hill.

McWilliams, A., & Siegel, D. (2001). Corporate social responsibility: A theory of the firm perspective. *Academy of Management, 26*(1), 117–127.

Mohd Zamri, F. I., Hibadullah, S. N., Mohd Fuzi, N., Auni Chiek Desa, F. N., and Habidin, N. F. (2013). *Business Management and Strategy, 4*(1), 97–106.

Muralidharan, K. (2015). *Six sigma for organizational excellence: A statistical approach.* Springer India.

Muralidharan, K., & Neha, R. (2017). Six Sigma marketing and productivity improvement. *Productivity: A Quarterly Journal of the National Productivity Council, 58*(1), 107–114.

Muralidharan, K., & Syamsundar, A. (2012). *Statistical methods for quality, reliability and maintainability.* PHI Learning Private LTD.

Muralidharan, K., & Raval, N. (2019). Realizing total customer experience through six sigma marketing: An empirical approach. *Productivity: A Quarterly Journal of the National Productivity Council, 60*(3), 303–315.

Oakland, J. S. (2005). *Statistical process control.* Elsevier.

Patel, S. (2016a). *Business excellence.* CRC.

Patel, S. (2016b). *Lean Transformation.* CRC.

Patel, S., & Muralidharan, K. (2019). *Implementation of lean management approach utilizing lean principles, lean systems, and tools to facilitate business process improvements* (Unpublished research article).

Porter, M. E., & Van der Linde, C. (1995a). Green and Competitive. *Harvard Business Review, 73*(5), 120–134.

Porter, M., & Van der Linde, C. (1995b). Toward a new conception of the environment-competitiveness relationship. *Journal of Economic Perspective, 9*(4), 97–118.

Reidenbach, E. (2009). *Six sigma marketing: From cutting costs to growing market share.* American Society for Quality.

Webb, M. J., & Gorman, T. (2006). *Sales and marketing: The six sigma way.* Kaplan.

# Chapter 8
# Control, Monitoring, and Deployment of Lean, Green, and Clean Activities

*"Yesterday I was clever, so I wanted to change the world. Today I am wise, so I am changing myself"*: Jalaluddin Rumi.

## 8.1 Introduction

Quality has become one of the most crucial consumer decision factors in the selection among competing products and services. Quality can be defined in many ways, ranging from "satisfying customers' requirements" to "fitness for use" to "conformance to requirements." The phenomenon is widespread, regardless of whether the consumer is an individual, an industrial organization, a retail store, or a military program. Consequently, understanding and improving quality is a crucial factor in business success, growth, and an enhanced competitive position (Montgomery, 2003). Businesses achieving higher quality in their products enjoy a significant advantage over their competitors; hence, the personnel responsible for the design, development, and manufacture of products must understand adequately the concepts and techniques used to improve the quality of products (Chandra, 2016). According to Garvin (1987), the eight dimensions of quality and their purposes are:

- *Performance*—Will the product do the intended job?
- *Reliability*—How often does the product fail?
- *Durability*—How long does the product last?
- *Serviceability*—How easy is it to repair the product?
- *Aesthetics*—What does the product look like?
- *Features*—What does the product do?
- *Perceived Quality*—What is the reputation of the company or its product?
- *Conformance to Quality*—Is the product made precisely as the designer intended?

The other quality dimensions closely related to the above are size (large, medium, small), color, texture, flexibility, and appearance, taste, intensity, and so on. Any deviations or variations in any of these dimensions can cause severe implications for the quality of the product or services. As mentioned in various chapters of this book, excessive variability in process performance often results in waste. Hence, quality improvement is synonymous to waste reduction.

Although quality concepts are perceptional and notional, monitoring and control of quality are equally essential as we do it in the case of processes and products for understanding assignable and chance causes of variations. The quality observed at one point may not be the same at some other time point and can change as per context and objectives. Similarly, quality perceptions may also fluctuate as per the variations in quality dimensions described above. Therefore, this chapter will contribute some technical tools to define, analyze, and perceive quality aspects of LGC sciences.

Note that variation is the cause of all defects and is inherent in all processes in varying amounts. It is natural to minimize the variation of a process to make a product defect-free. Hence quantifying the amount of variation in a process is the first and the critical step toward improvement. This will further enhance the understanding of the type of causes and decide the course of action to reduce the variation. This action will have a lasting impression on the quality of the product. Since statistical terms can only describe variability, statistical methods play a central role in quality improvement efforts. The analytical tools and techniques offer a wide range of quality control tools to detect variations in the process (Muralidharan, 2015).

There is a certain amount of variability in every product; consequently, no two things are exactly alike. For instance, a process is bound to vary from product to product, materials to materials, operation to operation, service to service, machine to machine, man to man, shift to shift, day to day, etc. However, if the variation is substantial, the customer may perceive the unit as undesirable and unacceptable. As mentioned earlier, quantifying the information is the first step toward the improvement and control of quality. The quantification is done based on the characteristics under study. The characteristics can be either qualitative (or attributes) or quantitative (or variables) in nature. Variables data is further divided into two categories: *discrete* and *continuous*. Measurements such as length, radius, temperature, thickness, weight, viscosity, pressure, etc. are continuous measurements. Attributes data or qualitative data, on the other hand, are either *nominal* or *ordinal*. They are generally described in the categorical form, often taking the form of counts. Analysis and interpretation of both variable data and attribute data are different and should be done with proper tools and techniques. Statistical principles allow extracting useful and relevant information from both types of data, and hence it is popular.

Let us now understand the concepts of quality assurance and quality control from monitoring, tracking, and deployment perspectives. After that, we will discuss various statistical tools of monitoring variation and control of the quality of a process. The stress is given to visual presentations for better communication and appeal.

## 8.2 Quality Assurance

According to the American Society for Quality (2000), *quality assurance* (QA) is defined in ISO 9000–2000 as "Part of quality management focused on providing confidence that quality requirements are fulfilled." *Quality control* (QC) refers to the activities or tools that are used to provide this assurance. In other words, a system of activities whose purpose is to provide assurance to customers by the company and show evidence that the overall quality control is effective is nothing but quality assurance. An assurance system involves verification, audits, and evaluation of quality factors, which influence the specifications, manufacturing process, and the use of products or services by customers. The basic structure of QA includes various components of each subsystem and their interrelationships. Considering an integrated view of all aspects involved in QA, it is found that an attempt should be made to build quality consciousness at all stages of life cycle of a product (conception, design, manufacturing, marketing, operation of the product, maintenance and replacement, etc.).

As per international standards organization, quality assurance is those planned and systematic actions necessary to provide adequate confidence that a product or service will satisfy the given requirements for quality. Quality assurance provides confidence to its management that the intended quality is being achieved and sustained. A well-organized quality assurance system facilitates:

- The trade-off between quality and quantity of product
- Interaction between quality and cost
- Length of service desired from the product
- Define clearly the service offered
- Complete elimination of waste
- Improve customer satisfaction
- Enhance profitability.

The modern concept of quality assurance is normally used to achieve a balance between three aspects, namely quality, reliability, and maintainability concerning cost. If the *quality* of the product is taken as relative goodness in fulfilling its purpose, then *reliability* is the probability of the product performing its intended role adequately for some time, intended under standard operating conditions, and *maintainability* is to ensure timely replacement and speedy repair from the stage of product design to on operating stage (Muralidharan & Syamsundar, 2012). These activities are necessary to evaluate the *cost of poor quality* (COPQ), which is the total monetary loss of products and processes that do not achieve their quality objectives. Knowledge of the COPQ leads to the development of a strategic plan consistent with the overall organization of goals (Gryna et al., 2007). Thus, LGC activities can play a crucial role in coordinating the three core areas quality, reliability, and maintainability of the organizational process to sustain quality assurance and future competitions.

Quality circle and the quality council can play a vital role in monitoring the quality assurance of the organization. *Quality circle* or *quality control circle* is a small team of people usually coming from the same work area who voluntarily meet regularly to identify, investigate, analyze, and solve work-related problems. This idea was first popularized in Japan by Dr. K. Ishikawa in 1962, and later, it was adopted by many other Asian and European counties. Quality circles take a democratic process and introduce a participative management culture in the organization. The circle presents the solutions to the management and implements them after approval. The circle is also responsible for the review and follow-up of the implementation of the solutions. These groups are some time called the leadership group or *quality council* as per Six Sigma project philosophy. Most of the developing countries are now stressing the need of quality circles in their organizations. Many such quality circles are currently working in India either centered at the organizational level itself or working as non-governmental organizations.

Various names have been used for programs similar to QC circles. Some of these are industrial democracy, work place democracy, employee participation groups, participative quality control, people implementing procedures and saving (PIPS), and success through everybody's participation (STEP). QC circles, as a management tool, are mainly based on the following basic principles (Wardsworth et al. 2004):

- People want to do a good job.
- People want to be recognized as intelligent, interested employees and to participate in decisions affecting their work.
- People want the information to understand the goals of their organization better and to make informed decisions.
- Employees want recognition and responsibility and a feeling of self-esteem.

New programs of total strategic quality recognize the importance of employee contribution and provide programs of empowerment, teamwork, and recognition. Creating a vision with values, guiding principles, and goals is the beginning of all improvement, including quality, productivity, profits, customer satisfaction, and employee satisfaction. Unfortunately, the new generation management is volatile to all these traditional tools, because they are interested in promoting instant results and fast delivery with automated technology and human resources. But to sustain the business competition, they will be bound to depend entirely on the integrated approach of the old and new philosophies in a new bottle with a positive outlook for quality.

Quality assurance can well be ascertained through the implementation of quality control techniques. Quality control is operational techniques to make inspection more efficient and to reduce the costs of quality through continuous monitoring of variation in the process. The methods like statistical quality control (SQC) and statistical process control (SPC) are the best-known tools for monitoring variation in a process. Both these techniques involve the measurement and evaluation of variation in a process and the efforts made to limit or control such variation. In its most common applications, SPC helps an organization or process owner to identify possible problems or unusual incidents so that action can be taken promptly to resolve them (Oakland, 2005; Muralidharan, 2015). As discussed at various places in various chapters, we reiterate that statistical tools and techniques are the underlying principles of any quality improvement program.

## 8.3 Quality Control

The operational techniques to make inspection more efficient and to reduce the costs of quality are done through quality control. Much of the earliest documented work in quality control centered on the Bell Telephone System. Realizing the necessity of continuous flow of transmission, AT&T, in the year 1892, recorded the importance of managing quality of all products intended for the rapidly growing network. Since quality control involves the inspection of the items for analyzing quality problems and improving the production process's performance, it is sometimes termed as *quality inspection*. Inspection means checking the material, components, or components itself at various stages in manufacturing. It often involves a decision on product acceptance, regulating manufacturing process, rating overall product quality, measuring inspection accuracy, and sorting out the faulty or defective items. The control and inspection act includes the interpretation of a specification, measurement of the product, comparison of specification and measurement, etc. The objectives of the inspection are:

- *Receiving inspection*: Inspection of incoming materials and purchased parts to ensure that they are according to the required specifications.
- *In-process inspection*: Inspection of raw materials as it undergoes processing form one operation to another.
- *Finished goods inspection:* To inspect the final finished product to detect the defects and its sources.
- *Gauge maintenance:* Control and maintenance of measuring instruments and inspection gauges.
- *The decision of salvage:* It is necessary to decide on the defective parts. Some of these parts may be acceptable after minor repairs.

Hence before embarking for complete control of the process, we need to achieve a certain quality through inspection. At this stage, the people associated with the quality will be able to bring the process to operate within a predictable range of variation. That is to bring a process to a stable and consistent level. It was in the year 1925, Dr. Walter Shewhart published his classic paper on a control chart in the *Journal of American Statistical Association*, which became a primary manufacturing tool for quality control used around the world. The use of statistical principles was in use in the development of *Acceptance sampling* techniques developed by Dr. H. F. Dodge and H. G. Romig in 1928. The purpose of the *design of experiments* developed by Dr. R. A. Fisher in the UK began in the early 1930s. The end of World War II saw increased interest in quality through statistical principles, primarily among the industries in Japan, developed by Dr. W. E. Deming. Since the early 1980 s, US industries have strived to improve quality with Dr. Deming's assistance, Philip Crosby, Dr. Joseph M. Juran, and Dr. Genichi Taguchi.

Statistical quality control (SQC) should be viewed as a kit of statistical tools that may influence decisions related to the functions of specifications, productions, or inspections. They increase the knowledge of the process by efficiently controlling the variation of the process. And hence, it is called statistical process control (SPC). There are many quality control tools practiced by engineers and scientists. Among them, the Shewhart control chart is a powerful tool for monitoring quality and variations in a process. As we know, the variation in a process tells us what that process is capable of achieving (tolerance). In contrast, specifications tell us what we want a process to be able to make. The first step in understanding variation should always be to plot a process data in time order. The time order can be hourly, daily, weekly, monthly, quarterly, yearly, or any suitable period.

Generally, two types of variations are encountered in a process. They are the special cause or assignable cause of variation and the random or irregular kind of variations. The goal of any process then will be to identify the causes and eliminate the causes to make an unstable process to a stable process. A process with only common cause variation is called stable. A stable process has controlled variation and is described through a normal distribution, as shown in Fig. 8.1.

- *Special Cause variation*: The variations which are relatively abundant in magnitude and viable to identify are the special causes of variation or assignable causes. They may come and go sporadically. Quite often, any specific evidence of the lack of statistical control gives a signal that a special cause is likely to have occurred. Hence such variations are local to the process and are unstable.
- *Common Cause variations*: This type of variation is the sum of the multitude of effects of a complex interaction of random or common causes. It is common to all occasions and places and always present to some degree. Variation due to common

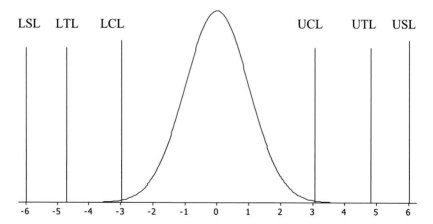

**Fig. 8.1** A normal (controlled) process

causes will almost always give results that are in statistical control. These kinds of variations are generally stable and can be controlled with proper treatment.

## 8.3.1 Control Chart

The best way of analyzing the special cause variation and common cause variation is the control chart techniques. A control chart is a device intended to be used at the point of operation, where the process is carried out and by the operators of that process. The control charts are frequently used to monitor the process variation and to signal the maintenance and revisit the ongoing process. People use control charts to separate common cause and special cause variation in a process. A control chart has three zones: a zone in which no action should be taken (presence of common causes); another zone which suggests more information should be obtained and the third zone in which one requires some effort to be made (presence of special causes).

A statistically controlled chart will be called a normal process whose process means (a measure of accuracy, say $\mu$) and standard deviation (a measure of precision, say $\sigma$) will work according to the tolerance level suggested by the customer. Control limits are statistical limits set at $\pm 3$ standard deviations from the mean (i.e., $\mu \pm 3\sigma$). Thus, lower control limit (LCL) is corresponding to $\mu - 3\sigma$, and upper control limit (UCL) is corresponding to $\mu + 3\sigma$. For practical problems, samples of sizes (say $n$) are selected for a fixed number of units (say $k$), and then sample characteristics are computed. There are two types of control charts: control charts for variables (quality characteristics which have absolute measurements) and control charts for attributes (quality characteristics which have no absolute measurements). They are described below.

### 8.3.1.1 Control Chart for Variables

Variable charts are the most sensitive charts for tracking and identifying assignable causes of variation. Here the changes in process variability can be distinguished from changes in process average, and small shifts in process average can be detected very quickly. To control a process using variable data, it is necessary to check the current state of the accuracy (average) and precision (spread) of the distribution of the data. The $\bar{x}$-chart and $R$-charts (or range chart) are the most frequently used control chart for variables. Although they are used in both administrative and manufacturing applications, they are the tools of the first choice in many manufacturing applications. If $\bar{x}$-chart is a plot of the subgroup means, then $R$-chart is a plot of the subgroup ranges. Instead of plotting the subgroup ranges, if we plot the subgroup standard deviation along with the $\bar{x}$-chart, we get $\bar{x}$-$s$ chart. Both $\bar{x}$-$R$-charts and $\bar{x}$-$s$ charts are two separate charts of the same subgroup data. They are considered to be the most sensitive charts for tracking and identifying assignable causes of variation based on control chart factors that assume a normal distribution within subgroups. Table 8.1

**Table 8.1** Control limit for variable control charts

| Chart | Standard values are given | | Standard values are not given (Estimated) | |
|---|---|---|---|---|
| | Centerlines | 3 $\sigma$ control limits (LCL, UCL) | Centerlines | 3 $\sigma$ control limits (LCL, UCL) |
| $\bar{x}$ | $\mu_0$ | $\mu_0 - A\sigma_0, \mu_0 + A\sigma_0$ | $\bar{\bar{x}}$ | $\bar{\bar{x}} - A_2\bar{R}, \bar{\bar{x}} + A_2\bar{R}$ or $\bar{\bar{x}} - A_3\bar{s}, \bar{\bar{x}} + A_3\bar{s}$ |
| $R$ | $d_2\sigma_0$ | $D_1\sigma_0, D_2\sigma_0$ | $\bar{R}$ | $D_3\bar{R}, D_4\bar{R}$ |
| $s$ | $c_4\sigma_0$ | $B_5\sigma_0, B_6\sigma_0$ | $\bar{s}$ | $B_4\bar{s}, B_3\bar{s}$ |

presents the control limits for variable charts. The constants $\mu_0$ and $\sigma_0$ are prespecified values of $\mu$ and $\sigma$, respectively. The quantities $d_2$, $A_2$, $A_3$, $B_3$ $B_4$, $B_5$, $B_6$, $D_1$, and $D_2$ are constants associated with the control chart and are available in Appendix Table 8 for various values of $n$ and subgroup sizes $k$.

As a practical significance, we first draw $R$-chart to see whether the variation is in control. If $R$-chart is in control, only then it is meant to draw the $\bar{x}$-chart. When $R$-chart is out of control, then the out-of-control points are removed from the analysis and a new value of $\bar{R}$ is computed. With this new value of $\bar{R}$, the revised limits are calculated for $\bar{x}$-chart and $R$-chart, and the chart is drawn. If the sample size is very small, then it is not advisable to remove those out-of-control points and revise the limits. For large samples, this kind of adjustment will tighten the limits on both charts, making them consistent with a process standard deviation $\sigma$. This estimate of $\sigma (= \bar{R}/d_2)$ could be used as the basis of a preliminary analysis of process capability. The effective use of any control chart will require periodic revision of the control limits and center lines. While revising control limits, it is highly recommended to use at least 25 samples or subgroups in computing control limits. We now consider an example to illustrate the use of a variable control chart.

***Example 8.1*** The data given in Table 8.2 corresponds to an industrial process, where the measurement of piston diameter for 20 ($= k$) samples of size $n = 5$ is collected for assessing the presence of variation to judge the quality of the process. Decide whether the process is statistically in control.

*Solution.* We first prepare a Box plot (Whisker plot) to understand the presence of any inconsistent observations or outliers/inliers in the data. Outliers/inliers in the data signify the presence of assignable causes of variation. Sample values across the process units can be compared with the use of the Box plot. The Box plot of the data given in the above example is shown in Fig. 8.2. As such, there is no influence of outliers or inliers in this data set. If they are present, then they generally lie away from the whisker on either side. Note that the line in the middle of each box is the median ($Q_2$) of the data. The bottom of the box represents the lower quartile ($Q_1$) of the data value, and the upper portion of the region corresponds to the third quartile ($Q_3$). Since there are no disturbances in the data, one can prepare the control chart.

From Appendix Table 8, we obtain, the values $A_2$, $D_3$, and $D_4$ for subgroup size 5 are 0.577, 0, and 2.115, respectively. Hence, the control limits are:

**Table 8.2** Measurement of piston diameter

| Sample number | $x_1$ | $x_2$ | $x_3$ | $x_4$ | $x_5$ | Range | $\bar{x}$ |
|---|---|---|---|---|---|---|---|
| 1 | 15.8 | 16.3 | 16.2 | 16.1 | 16.6 | 0.8 | 16.2 |
| 2 | 16.3 | 15.9 | 15.9 | 16.2 | 16.4 | 0.5 | 16.14 |
| 3 | 16.1 | 16.2 | 16.5 | 16.4 | 16.3 | 0.4 | 16.3 |
| 4 | 16.3 | 16.2 | 15.9 | 16.4 | 16.2 | 0.5 | 16.2 |
| 5 | 16.1 | 16.1 | 16.4 | 16.5 | 16.0 | 0.5 | 16.22 |
| 6 | 16.1 | 15.8 | 16.7 | 16.6 | 16.4 | 0.9 | 16.32 |
| 7 | 16.1 | 16.3 | 16.5 | 16.1 | 16.5 | 0.4 | 16.3 |
| 8 | 16.2 | 16.1 | 16.2 | 16.1 | 16.3 | 0.2 | 16.18 |
| 9 | 16.3 | 16.2 | 16.4 | 16.3 | 16.5 | 0.3 | 16.34 |
| 10 | 16.6 | 16.3 | 16.4 | 16.1 | 16.5 | 0.5 | 16.38 |
| 11 | 16.2 | 16.4 | 15.9 | 16.3 | 16.4 | 0.5 | 16.24 |
| 12 | 15.9 | 16.6 | 16.7 | 16.2 | 16.5 | 0.8 | 16.38 |
| 13 | 16.4 | 16.1 | 16.6 | 16.4 | 16.1 | 0.5 | 16.32 |
| 14 | 16.5 | 16.3 | 16.2 | 16.3 | 16.4 | 0.3 | 16.34 |
| 15 | 16.4 | 16.1 | 16.3 | 16.2 | 16.2 | 0.3 | 16.24 |
| 16 | 16.0 | 16.2 | 16.3 | 16.3 | 16.2 | 0.3 | 16.2 |
| 17 | 16.4 | 16.2 | 16.4 | 16.3 | 16.2 | 0.2 | 16.3 |
| 18 | 16.0 | 16.2 | 16.4 | 16.5 | 16.1 | 0.5 | 16.24 |
| 19 | 16.4 | 16.0 | 16.3 | 16.4 | 16.4 | 0.4 | 16.3 |
| 20 | 16.4 | 16.4 | 16.5 | 16.0 | 15.8 | 0.7 | 16.22 |
| | | | | | | $\bar{R} = 0.475$ | $\bar{\bar{x}} =$ 16.268 |

$$\text{LCL} = \bar{\bar{x}} - A_2 \bar{R}$$
$$= 16.268 - 0.577 * 0.475$$
$$= 15.9939$$

and

$$\text{UCL} = \bar{\bar{x}} + A_2 \bar{R}$$
$$= 16.268 - 0.577 * 0.475$$
$$= 16.5421$$

Similarly, according to LCL $= D_3 \bar{R}$ and UCL $= D_4 \bar{R}$, the control limits for R-chart are obtained as 0 and 1.021, respectively. The control chart prepared in MINITAB software is shown in Fig. 8.3. Both the R-chart and $\bar{x}$-the chart does not show any kind of abnormal behavior. The points are randomly distributed, and no

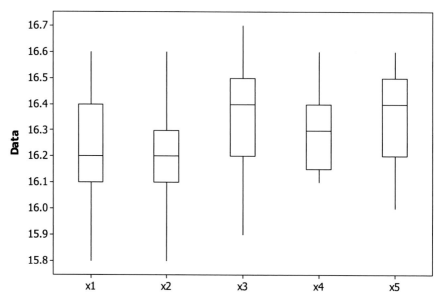

**Fig. 8.2**  Box plot of the sample values

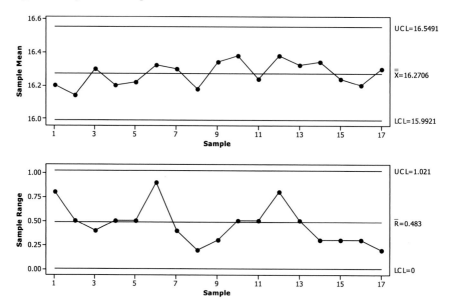

**Fig. 8.3**  Control chart corresponds to piston diameter

points are falling outside the control limits, and hence the process is statistically in control.

Control charts are used to establish and maintain statistical control of a process. They are also used for estimating process parameters, particularly in process capability studies. The use of a control chart requires that the engineer or analyst select a sample size, a sampling frequency or interval between samples, and the chart's control limits. The selection of these three parameters is usually called the design of the control chart (Montgomery, 2003). They typically involve selecting the sample size and control limits. The average run length of the chart to detect a particular shift in the quality characteristic and the average run length of the procedure when the process is in control are equal to specified values. The frequency of sampling is rarely treated analytically. Usually, the practitioner is advised to consider such factors as the production rate, the expected frequency of shifts to an out-of-control state, and the possible consequences of such process shifts in detecting the sampling interval (Muralidharan, 2015).

If $p$ is the probability that any point exceeds the control limits, then the *average run length* (ARL) of any control chart is calculated as

$$ARL = \frac{1}{p} \tag{8.1}$$

For the piston diameter example discussed above, $\hat{\sigma} = \frac{\bar{R}}{d_2} = \frac{0.475}{2.326} = 0.204$. Therefore, the ARL is obtained as

$$
\begin{aligned}
p &= P(x < LSL) + P(x > USL) \\
&= P(x < 15.99) + P(x > 16.55)\backslash \\
&= P\left(z < \frac{15.99 - 16.27}{0.204}\right) + P(z > \frac{16.55 - 16.27}{0.204}) \\
&= P(z < -1.3726) + P(z > 1.3726) \\
&= 0.1706
\end{aligned}
$$

Therefore, the ARL of $\bar{x}$-chart, when the process is in control, is $\frac{1}{p} = 5.8616 \cong 6$. That is, an out-of-control signal will be generated at every 6th samples, on average. It is also convenient to express the control chart's performance in terms of its *average time to signal* (ATS). If samples are taken at fixed intervals of time that are $k$ hours apart, then ATS $= k * ARL$, in this case, the ATS is 120. The process capability of the measurement is $C_p = \frac{USL - LSL}{6\sigma} = 0.457$.

### 8.3.1.2  Control Chart for Attributes

As mentioned above, attributes are quality characteristics that do not have any measurements, but they can be counted. They are generally classified under acceptable or unacceptable outcomes. An unacceptable item can be called a defect or defectives. A *defect* (non-conformities) is any variation of a required characteristic of the product (or its parts) or services or process output which is far enough from its target value to prevent the product from fulfilling the physical and functional requirements of the customer/business, as viewed through the eyes of the customer/business manager. A *defective* (non-conforming) item may have one or more non-conformities called defects. Hence, the inspection of an item mainly concentrates on identifying theses defects. There are two types of attribute charts. They are defective charts (*p*-chart and *np*-chart) and defect charts (*c*-chart and *u*-chart).

p-chart plots the percentage defective per subgroup and *np*-chart plots the number of defectives per lot. Since *p*–chart and *np*-charts represent the proportion of nonconforming; it is expected that the sampling is done through a proper sample size and distribution to get the required capability. The charting based on both these proportions generally requires the inspection of a much higher fraction of the total units to achieve system evaluation and monitoring goals effectively. For analysis, it is suggested to have 20 or more subgroups. The probability law associated with these charting is done using Binomial distribution. *c*-chart plots the number of defects per subgroup, and *u*-chart plots the number of defects per unit. This kind of chart is useful in situations where the average number of non-conformities per unit is a more convenient basis for process control. The probability law associated with these charting is done using Poisson distribution. Table 8.3 presents the control limits for attribute characteristics: The constants $p_0$, $c_0$, and $u_0$ are prespecified values.

We now consider an example where the quality characteristic is measured in terms of imperfections (defects) in the manufacturing of rolls of paper. As seen in Chap. 4, Sect. 4.2, the imperfections (or the number of defects) follow a Poisson distribution with an average number of imperfections. For SPC/SQC applications, Poisson distribution is frequently used for modeling rare events of occurrence in large volume of sample information.

**Table 8.3**  Control limit for attributes control charts

| Chart | Standard values are given | | Standard values are not given | |
|---|---|---|---|---|
| | Centerlines | 3 $\sigma$ control limits | Centerlines | 3 $\sigma$ control limits |
| $p$ | $p_0$ | $p_0 \pm 3\sqrt{\frac{p_0(1-p_0)}{n}}$ | $\bar{p}$ | $\bar{p} \pm 3\sqrt{\frac{\bar{p}(1-\bar{p})}{n}}$ |
| $np$ | $np_0$ | $np_0 \pm 3\sqrt{np_0(1-p_0)}$ | $n\bar{p}$ | $n\bar{p} \pm 3\sqrt{n\bar{p}(1-\bar{p})}$ |
| $c$ | $c_0$ | $c_0 \pm 3\sqrt{c_0}$ | $\bar{c}$ | $\bar{c} \pm 3\sqrt{\bar{c}}$ |
| $u$ | $u_0$ | $u_0 \pm 3\sqrt{\frac{u_0}{n}}$ | $\bar{u}$ | $\bar{u} \pm 3\sqrt{\frac{\bar{u}}{n}}$ |

**Table 8.4** Data on imperfections in rolls production

| Day | Number of rolls produced ($n_i$) | Total number of imperfections ($x_i$) | $u = \frac{x_i}{n_i}$ |
|---|---|---|---|
| 1 | 18 | 12 | 0.67 |
| 2 | 18 | 14 | 0.78 |
| 3 | 24 | 20 | 0.83 |
| 4 | 22 | 18 | 0.82 |
| 5 | 22 | 15 | 0.68 |
| 6 | 22 | 12 | 0.55 |
| 7 | 20 | 11 | 0.55 |
| 8 | 20 | 15 | 0.75 |
| 9 | 20 | 12 | 0.60 |
| 10 | 20 | 10 | 0.50 |
| 11 | 18 | 18 | 1.00 |
| 12 | 18 | 14 | 0.78 |
| 13 | 18 | 9 | 0.50 |
| 14 | 20 | 10 | 0.50 |
| 15 | 20 | 14 | 0.70 |
| 16 | 20 | 13 | 0.65 |
| 17 | 24 | 16 | 0.67 |
| 18 | 24 | 18 | 0.75 |
| 19 | 22 | 20 | 0.91 |
| 20 | 21 | 17 | 0.81 |

*Example 8.2* The concerned paper mill uses a control chart to monitor the imperfection in finished rolls of paper. Production output is inspected for 20 days, and the imperfection in rolls production on each day is recorded. The complete data is shown in Table 8.4. Discuss whether the process is statistically in control.

*Solution.* Since the production of rolls, every day is not constant; we use a $u$-chart with varying sample sizes to prepare the chart. Also, since $X_i$ is the number of defects in the $i$-th sample, then $u$ is the defects per unit follows a Poisson distribution with mean $\bar{u} = \frac{1}{20} \sum_{i=1}^{20} \frac{x_i}{n_i} = 0.70$.

The control limits are computed as

$$\text{LCL} = \bar{u} - 3\sqrt{\frac{\bar{u}}{\bar{n}}}$$

$$= 0.7 - 3\sqrt{\frac{0.7}{20.55}}$$

$$= 0.1456$$

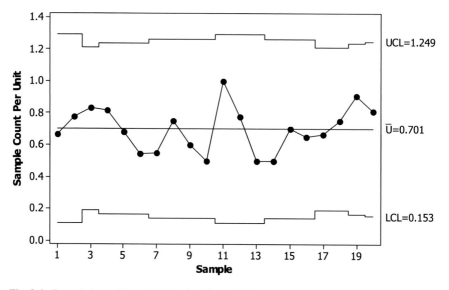

**Fig. 8.4** Control chart of imperfections in rolls production

and

$$LCL = \bar{u} + 3\sqrt{\frac{\bar{u}}{\bar{n}}}$$
$$= 0.7 + 3\sqrt{\frac{0.7}{20.55}}$$
$$= 1.253$$

The control chart for the data is presented in Fig. 8.4. The slight difference in the control limit value is due to variable sample sizes and approximations. The control limits are supposed to be different for different units. The control chart prepared in MINITAB takes care of this fact (see Fig. 8.4 for details). The graph clearly shows that the imperfections are perfectly in statistical control. That is, the day-to-day production of rolls is producing controlled defectives.

***Example 8.3*** The data shown in Table 8.5 is collected for monitoring variations in fraction non-conforming with variable sample sizes. Discuss whether the non-conforming units are statistically in control.

*Solution.* Since the sample sizes for each unit is varying in size, the control limits also will be different for different situations. The calculation of control limits is also shown in the table. We have $\bar{p} = \frac{1}{25}\sum_{i=1}^{25} p_i = 0.0952$. Since the sample sizes are different, the control limits are obtained through the formula: $\bar{p} \pm 3\sqrt{\frac{\bar{p}(1-\bar{p})}{n_i}}$. The computations are shown in Table 8.5. The negative value of LCL is set at zero.

**Table 8.5** Data of non-conforming units

| Sample number ($i$) | Sample size ($n_i$) | Number of non-conforming units ($D_i$) | $p_i = \frac{D_i}{n_i}$ | LCL | UCL |
|---|---|---|---|---|---|
| 1 | 100 | 12 | 0.12 | 0.007153 | 0.183247 |
| 2 | 80 | 8 | 0.10 | −0.00324 | 0.19364 |
| 3 | 80 | 6 | 0.08 | −0.00324 | 0.19364 |
| 4 | 100 | 9 | 0.09 | 0.007153 | 0.183247 |
| 5 | 110 | 10 | 0.09 | 0.01125 | 0.17915 |
| 6 | 110 | 12 | 0.11 | 0.01125 | 0.17915 |
| 7 | 100 | 11 | 0.11 | 0.007153 | 0.183247 |
| 8 | 100 | 16 | 0.16 | 0.007153 | 0.183247 |
| 9 | 90 | 10 | 0.11 | 0.00239 | 0.18801 |
| 10 | 90 | 6 | 0.07 | 0.00239 | 0.18801 |
| 11 | 110 | 20 | 0.18 | 0.01125 | 0.17915 |
| 12 | 120 | 15 | 0.13 | 0.014824 | 0.175576 |
| 13 | 120 | 9 | 0.08 | 0.014824 | 0.175576 |
| 14 | 120 | 8 | 0.07 | 0.014824 | 0.175576 |
| 15 | 110 | 6 | 0.05 | 0.01125 | 0.17915 |
| 16 | 80 | 8 | 0.10 | −0.00324 | 0.19364 |
| 17 | 80 | 10 | 0.13 | −0.00324 | 0.19364 |
| 18 | 80 | 7 | 0.09 | −0.00324 | 0.19364 |
| 19 | 90 | 5 | 0.06 | 0.00239 | 0.18801 |
| 20 | 100 | 8 | 0.08 | 0.007153 | 0.183247 |
| 21 | 100 | 5 | 0.05 | 0.007153 | 0.183247 |
| 22 | 100 | 8 | 0.08 | 0.007153 | 0.183247 |
| 23 | 100 | 10 | 0.10 | 0.007153 | 0.183247 |
| 24 | 90 | 6 | 0.07 | 0.00239 | 0.18801 |
| 25 | 90 | 9 | 0.10 | 0.00239 | 0.18801 |

The MINITAB output of the chart is presented in Fig. 8.5. Since there is one point that falls on the upper control line, the process is not statistically in control.

The charts discussed so far are highly useful to detect, sharp, intermittent changes to a process. However, if one is interested in a small, sustained shift in process, other types of control charts may be preferred, for example, the *Cumulative Sum* (CUSUM) chart and *moving average* chart are some of them. A CUSUM chart originally developed by Page (1954) is a chronological plot of the cumulative sum of deviations of a sample statistic from a reference value (or nominal or target value). The moving average chart is a chronological plot of the moving average. It is calculated as the average value updated by dropping the oldest individual measurement and adding the newest individual measurement. Thus a new average is calculated with

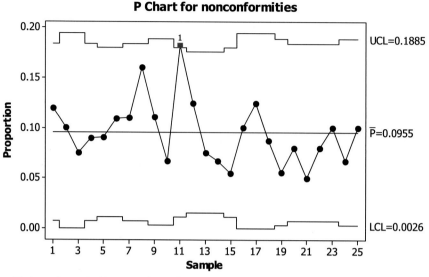

**Fig. 8.5** p-chart for non-conformities

each measurement. Another widely used chart is the *Exponentially Weighted Moving Average* (EWMA) chart for quality characteristics following an exponential increase (Bower (2000)). See also Champ and Woodall (1987) and Montgomery (2003). These charts are also called time-weighted control charts.

Recently many authors have proposed control charts for non-normal data as well. They are called nonparametric control charts. Chakraborti et al. (2001) are credited with giving an overview of such studies and have reviewed nonparametric control charts up to 2000, for the first time. Chakraborti and Graham (2007), and Chakraborti et al. (2011) have further reviewed the literature through 2010. There is now a vast collection of nonparametric control charts in the literature. These include nonparametric control charts equivalent to Shewhart-type chart, CUSUM chart, EWMA chart, and other variants of these charts. These methods have also been shown to perform well compared to their parametric counterparts.

Whatever may be the type of characteristics, the control charts attempt to differentiate assignable (or special) causes of variation from common (or random) causes. Common causes of variations are expected part of the process, and hence much less concern to the manufacturer than assignable sources of variation. Therefore, the distinction between common causes and assignable causes is context-dependent. A common cause today can be an assignable cause tomorrow. The terminology may change with a change in the sampling scheme. However, one wants to react only when a cause has a sufficient impact that it is practical and economical to remove it to improve quality. This is ultimately achieved by the SQC and SPC techniques. Both these techniques involve the measurement and evaluation of variation in a process, and the efforts made to limit or control such variation. In its most common applications, SPC helps an organization or process owner to identify possible problems or unusual incidents so that action can be taken promptly to resolve them.

Note that, the control limits have no relationship with *specification limits* (a limit specified by customers usually); however, both the limits can be the same or near to it. As a compromise, quality control personal's compute *tolerance interval* (an interval estimator determined from a random sample to provide a level of confidence that the interval covers at least a specified proportion of the sampled population), which generally falls between the control limits and specification limits. However, any points falling outside statistical limits signal a special cause and are a matter of concern. For various issues related to charting techniques, one may refer to Woodall (1997), and Montgomery (2003) and details contained therein.

### 8.3.2 Time Series Plot

Another quality tool similar to the control chart is the time series chart or run chart: A *run chart* shows the degree and pattern of variation in a process, product, or other factors. It helps to detect change timing patterns from moment to moment, day to day, etc. The plot (see Fig. 8.6 for the variable $x_1$ of Table 8.2) is done by considering the time sequence of occurrences on the $X$-axis and any continuous or count measure, including percentage, number of defects and temperature, etc. on $Y$-axis. The main

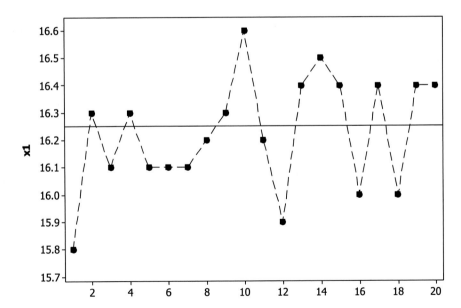

**Fig. 8.6** Time series chart of $x_1$

**Table 8.6**  Accidental data due to natural causes

| Year | 2005 | 2006 | 2007 | 2008 | 2009 | 2010 | 2011 | 2012 | 2013 | 2014 | 2015 |
|------|------|------|------|------|------|------|------|------|------|------|------|
| Total deaths | 22415 | 21502 | 25153 | 23993 | 22255 | 25066 | 23690 | 22960 | 22759 | 20201 | 10491 |

objective of drawing Time series plot is (i) to see how a process or critical factor in responding to change and (ii) to see how process improvements are impacting performance.

One should use a run chart:

- To study observed data for trends or patterns over a specified period.
- To focus attention on significant changes in the process.
- To track useful information for predicting trends.
- To understand variation in the process.
- To compare a performance measure before and after implementation of a solution to assess the solution's impact.
- To detect trends, shifts, and cycles in the process.

A diagram similar to the Time series chart is the Time series plot, where $y$ (usually a continuous variable) is plotted against the $x$ variable (often the time) to understand the trend of the constant variable over some time. This plot can visually depict variations like a trend, seasonal, cyclical, and random variation present in a time series. The analysis of time series has been useful to economists and business peoples, sociologists, scientists, etc. It has also found its utility in meteorology, seismology, oceanography, geomorphology, etc. It has several applications in earth sciences, medical sciences, space technology, and so on. Time series analysis helps in understanding the following phenomena: It helps in (i) knowing the real behavior of the past, (ii) predicting the future action, and (iii) planning the future operations. Many policy decisions about business, government, and administration can be designed and executed as per the behavior of time series data.

As an illustration, consider the number of accidental deaths due to natural causes that happened in India for the period 2005–2015 (the data is accessed through http:// mospi.nic.in/statistical-year-book-india/2018/207, Source: Accidental Deaths and Suicides in India' Report 2016, National Crime Records Bureau, Ministry of Home Affairs). The year-wise total number of deaths is reported in Table 8.6.

The time series plot of the above data is shown in Fig. 8.7. The number of deaths over a period is showing declining growth.

## 8.3.3  Pareto Chart

A Pareto chart is a quality tool used for understanding the pattern of variations in attribute or categorical type data sets. The Pareto chart organizes data to show which

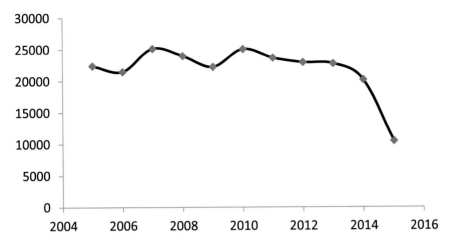

**Fig. 8.7** Time series plot of accidental deaths in India

items or issues have the most significant impact on the process, system, or service. It is a pictorial representation of the data stratified according to the largest to the lowest number of items according to the frequency of occurrences in each group. This works on the principle of the Pareto principle and is also called 80:20 rules. That is, 80% of the effect(s) or a problem is the result of 20% of the causes. The objective of this is to develop a plan to work on issues that will give us the most significant return for process improvement efforts.

Thus dividing data into categories, Pareto chart judges the relative impact of various parts of a problem (quantifying the problem) and track down the most prominent contributor(s) to a problem. A Pareto chart helps to decide where your improvement efforts will have the biggest payoff. You can use a Pareto chart only when the problem under study can be broken down into categories, and the number of occurrences can be counted for each type. Once a Pareto chart has been prepared, the office worker and team can start developing plans and actions for improving the process, beginning with the most frequently occurring issue or problem. As that problem is improved, the team can go to the next most frequently recurring problem and work on that, and so on.

As an illustration, suppose a bottling company prepares a breakdown of causes of lost time of production to study the characteristic of improvement. The compiled data is presented in Table 8.7. Generally, the data may not be available in the compiled form, as shown in the table. However, statistical software like SPSS, MINITAB, and R-programs automatically collect the same in an abridged form and prepare the chart for raw data. The Pareto chart corresponds to the same is presented in Fig. 8.8. It is evident from both the table and chart that the primary cause of the breakdown is caused due to downtime and repair, followed by cleaning and maintenance. One can also investigate further the reasons behind the breakdown by drawing separate Pareto chart for downtime and repair and other causes.

**Table 8.7** Data on causes of lost time of production

| Item | Causes in % |
|---|---|
| Downtime and repair | 38.7 |
| cleaning and maintenance | 20.8 |
| Electricity | 2.0 |
| Water | 3.0 |
| Lack of workforce | 1.5 |
| Lack of bottles | 1.0 |
| Defective caps | 0.4 |
| Worker transportation service | 0.4 |
| Lack of forklift | 0.2 |
| Lack of CO2 | 0.1 |
| Others | 31.0 |

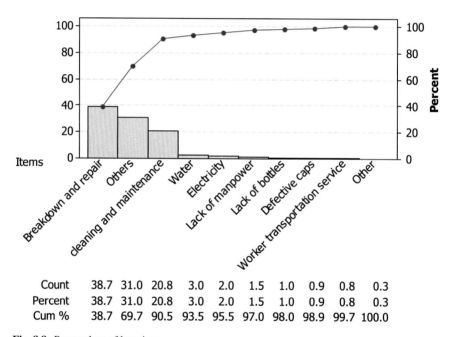

| | | | | | | | | | | |
|---|---|---|---|---|---|---|---|---|---|---|
| Count | 38.7 | 31.0 | 20.8 | 3.0 | 2.0 | 1.5 | 1.0 | 0.9 | 0.8 | 0.3 |
| Percent | 38.7 | 31.0 | 20.8 | 3.0 | 2.0 | 1.5 | 1.0 | 0.9 | 0.8 | 0.3 |
| Cum % | 38.7 | 69.7 | 90.5 | 93.5 | 95.5 | 97.0 | 98.0 | 98.9 | 99.7 | 100.0 |

**Fig. 8.8**  Pareto chart of lost time

Interestingly, the Pareto chart is frequently used in grouping the variables when there are many process input variables (PIVs) impacting the quality of process output variables (POVs). This is the reason it is used widely in the design of experiments (DOE), root cause analysis (RCA), failure mode effect analysis (FMEA), quality function deployment (QFD), and many other technical and management tools. For Six Sigma and lean projects, Pareto charts are frequently prepared to identify the problem and understand the primary sources of variation. They are also used for setting the goal based on the initial quality level (in terms of sigma level) and capability of the process. Since the Pareto chart is flexible for illustration, LSS professionals often use the Pareto chart to monitor quality in their process activity.

### 8.3.4 Scatter Plots

A scatter plot is a graph that helps to visualize the relationship between two variables. It can be used to check whether one variable is related to another variable and is an effective way to communicate the link you find. The Scatter plots may be used:

- To study and identify possible relationships between the changes observed in two different sets of variables.
- To understand the extent of relationships between variables.
- To discover whether two variables are related linearly or nonlinearly.
- To discover an unusual pair of observations.
- To find out if changes in one variable are associated with changes in the other.

A scatter plot is generated by plotting a dependant variable (say, $y$) against the independent variable (say, $x$). The resulting plot is called a scatter plot. The configuration of each pair of points will decide the strength of the relationship between the two variables. The relationship can be either linear or nonlinear. Identifying $x$ (process input variable) and y (process output variable) is crucial for determining the relationship. Depends on the subject area, one should appropriately define the variables $x$ and y. See Table 8.8 for the various perception of $x$ and y:

Consider the data on the distribution of students and regular players, according to age groups (Table 8.9). It is quite natural to ask what kind of association (or

**Table 8.8** $x$ versus $y$

| $x$-variable | $y$-variable |
| --- | --- |
| Independent | Dependant |
| Input | Output |
| Predictor | Response |
| Problem | Symptom |
| Cause | Effect |
| Control | Monitor |

**Table 8.9** Data on players and age group

| Age groups(yrs.) | 15–16 | 16–17 | 17–18 | 18–19 | 19–20 | 20–21 |
| --- | --- | --- | --- | --- | --- | --- |
| Number of students | 200 | 270 | 340 | 360 | 400 | 300 |
| Regular players | 150 | 162 | 170 | 180 | 180 | 120 |

**Table 8.10** Anxiety scores and marks

| Anxiety | 12 | 16 | 6 | 22 | 17 | 14 | 8 | 27 | 18 | 19 |
|---------|----|----|----|----|----|----|----|----|----|----|
| Marks | 75 | 56 | 91 | 48 | 69 | 73 | 88 | 65 | 72 | 65 |
| Anxiety | 11 | 9 | 21 | 3 | 15 | 17 | 22 | 19 | 13 | 5 |
| Marks | 88 | 94 | 48 | 99 | 78 | 65 | 58 | 52 | 70 | 90 |

correlation) exists between age group and playing habits. Is the number of students and regular players linearly related?.

Let us consider another example: The data on anxiety scores and the marks obtained in an IQ test of twenty students of a group of SSC students are collected in Table 8.10. The experiment is conducted to understand the extent of the effect of variables on each other. Once that is established, then the modality to estimate one based on the other variable can be facilitated.

If one wants to understand the effect of marks scored in IQ given the anxiety score, then the variable $y$ will be marks obtained in IQ, and variable $x$ will be anxiety score. One can investigate the relationship the other way as well. Accordingly, the terminology of the variable interpretation will change. In this example, both situations are unusual. Figure 8.9 presents the scatter plot of the data of marks obtained in the IQ test against anxiety and vice versa.

There are two scatter plots in the same panel. In the first case, anxiety is plotted against marks. Looking at the downward trend of points, one can easily conclude that there is a negative association or correlation between the two. If marks in the test increase, then there is a chance that the anxiety will decrease. In the second panel, the marks scored in the IQ test are plotted against the anxiety score, giving

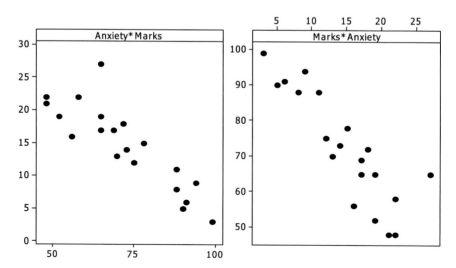

**Fig. 8.9** Scatter plot of anxiety versus marks and vice versa

an identical conclusion. However, the outcome here will be interpreted differently: That as anxiety increases, marks will decrease. So in both the way the meanings are the same. From the graph, one can also observe that one pair of observations (27, 65) seems to be an outlier or unusual. The reason for this is to be investigated further.

As far as the relationship is concerned, there is a linear relationship with a negative slope. The amount of degree of relationship is numerically evaluated using the *correlation* concept, and the actual relationship can be ascertained using the *regression* concept. The correlation (usually denoted by $r$) between the two variables is $-0.864$, and the regression (estimating) equation here is Marks $= 103 - 2.12*$ Anxiety. Suppose anxiety is 12, then according to the estimating equation, the marks obtained will be 77.56, which is very close to the actual value as per the actual data available. The fitted line plot is given in Fig. 8.10. For the exact linear relationship, the correlation $(-1 \le r \le 1)$ will be as high as one, and in that case, the predicted and actual values will coincide. One can also use multiple variables to construct regression equations. The technical aspects of these two concepts will also be discussed subsequently in this chapter.

In the above example, correlation analysis helps to measure the closeness of the linear relationship between two variables in the population. The correlation coefficient indicates how closely the data fit a linear pattern. The correlation coefficient $r$ can be used as a test statistic to test whether the linear relationship between $X$ and $Y$ is statistically significant. Mathematically, the correlation coefficient is defined as the ratio of covariance (common variance) between $X$ and $Y$ to the square root of the product of variance of $X$ and variance of $Y$. That is:

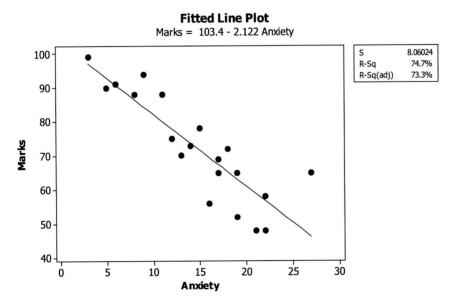

**Fig. 8.10** Fitted line plot

$$r = \frac{\text{Cov}(X, Y)}{\sqrt{\text{Var}(X)\text{Var}(Y)}}$$

$$= \frac{n \Sigma x_i y_i - \Sigma x_i \Sigma y_i}{\sqrt{n \Sigma x_i^2 - (\Sigma x_i)^2} \sqrt{n \Sigma y_i^2 - (\Sigma y_i)^2}} \tag{8.2}$$

The measure of correlation coefficient shown in (8.1) was introduced by Karl Pearson in the year 1895 and hence is known in the same name. The value of $r = 0$ indicates that there is no linear relationship between $X$ and $Y$, while $r = 1$ shows a perfect correlation or linear relationship between two variables. We consider an example to illustrate this formula.

**Example. 8.4** Suppose the systolic blood pressures ($Y$) of randomly selected 12 hypertensive patients of different ages ($X$) were measured and recorded. What is the correlation coefficient between the two variables? Are they linearly related? If so, fit a regression line.

| Observations | $X$ | $Y$ | $X^2$ | $Y^2$ | $XY$ |
|---|---|---|---|---|---|
| 1 | 42 | 153 | 1764 | 23,409 | 6426 |
| 2 | 48 | 156 | 2304 | 24,336 | 7488 |
| 3 | 53 | 157 | 2809 | 24,649 | 8321 |
| 4 | 57 | 155 | 3249 | 24,025 | 8835 |
| 5 | 59 | 161 | 3481 | 25,921 | 9499 |
| 6 | 63 | 164 | 3969 | 26,896 | 10,332 |
| 7 | 65 | 162 | 4225 | 26,244 | 10,530 |
| 8 | 68 | 170 | 4624 | 28,900 | 11,560 |
| 9 | 70 | 166 | 4900 | 27,556 | 11,620 |
| 10 | 72 | 164 | 5184 | 26,896 | 11,808 |
| 11 | 73 | 168 | 5329 | 28,224 | 12,264 |
| 12 | 75 | 167 | 5625 | 27,889 | 12,525 |
| Total | 745 | 1943 | 47,463 | 314,945 | 121,208 |

*Solution.* We have $n = 12$, $\sum_{i=1}^{12} x_i = 745$, $\sum_{i=1}^{12} y_i = 1943$, $\sum_{i=1}^{12} x_i^2 = 47463$, $\sum_{i=1}^{12} y_i^2 = 314945$ and $\sum_{i=1}^{12} x_i y_i = 121208$. Therefore,

$$r = \frac{n \Sigma x_i y_i - \Sigma x_i \Sigma y_i}{\sqrt{n \Sigma x_i^2 - (\Sigma x_i)^2} \sqrt{n \Sigma y_i^2 - (\Sigma y_i)^2}}$$

$$= 0.9028$$

The high positive correlation value of $r$ shows that there is a robust linear relationship between the variables. The systolic blood pressure indeed depends on the age of the people. Note that, the correlation coefficient was established here to assume that the data follow a normal distribution. Suppose this assumption is not valid, then we obtain a nonparametric measure of correlation called rank correlation, proposed by Spearman in 1904 as a measure of the strength and direction of the associations between two variables.

The computational procedure is as follows: Let $(x_i, y_i)$, $i = 1, 2, ..., n$ be the paired repeated measures on $n$ subjects. Let $R_i$ and $S_i$ be the ranks to observed values $x_i$ and $y_i$ respectively, and $d_i = R_i - S_i$, $i = 1, 2, ..., n$, then the rank correlation is obtained as

$$\rho = 1 - \frac{6 \sum d_i^2}{n(n^2 - 1)} \tag{8.3}$$

If there are ties in the observations, then the average of the ranks may be considered. Like Karl Pearson's correlation coefficient, Spearman's rank correlation also ranges from $(-1, 1)$. The interpretations of $\rho$ are similar to $r$.

**Example 8.5** The following table shows the $X$ and $Y$ coordinates of the knee-pain centroid location of the front view of the right leg of a patient. Compute the Spearman's rank correlation coefficient $\rho$.

| Patient | X | Y | $R_i$ | $S_i$ | $d_i$ | $d_i^2$ |
|---------|-----|-----|-------|-------|-------|---------|
| 1 | 38 | 187 | 1 | 7 | -6 | 36 |
| 2 | 49 | 156 | 8 | 6 | 2 | 4 |
| 3 | 47 | 128 | 6 | 1 | 5 | 25 |
| 4 | 50 | 194 | 9 | 8 | 1 | 1 |
| 5 | 46 | 230 | 5 | 10 | -5 | 25 |
| 6 | 43 | 201 | 3 | 9 | -6 | 36 |
| 7 | 48 | 144 | 7 | 3 | 4 | 16 |
| 8 | 39 | 148 | 2 | 4 | -2 | 4 |
| 9 | 45 | 133 | 4 | 2 | 2 | 4 |
| 10 | 52 | 153 | 10 | 5 | 5 | 25 |
| Total | | | | | | 176 |

According to the formula (8.2), we have the rank correlation coefficient as

$$\rho = 1 - \frac{6 \sum d_i^2}{n(n^2 - 1)}$$

$$= 1 - \frac{6 * 176}{10(100 - 1)}$$

$$= -0.0667$$

Since the $\rho$ value is very close to zero, it may be concluded that there is no association between the coordinates.

As seen in example 8.1, there is a strong association between age and blood pressure, which is generally true. What exactly is the relationship is what people are interested in so that the relationship can be used for predicting the response variable ($y$)? Here it is the blood pressure of a person given the predictor ($x$), that is age. Similarly, a relationship can also be established with many predictors ($x_i's$). In continuation to root cause analysis (RCA), it is quite common for the LSS professionals to determine the cause and effect relationship in terms of $y = f(x_1, x_2, \ldots, x_n)$. This is where we use multiple regression analysis. Regression analysis is a powerful statistical tool to investigate the relationships between process variables. Through regression analysis, decision-makers can screen out CTQ variables quickly. A relationship can be linear or nonlinear, logarithmic, exponential, and so on. They can be used to predict/forecast and quantify the strength of the relationship between variables.

Regression analysis is a standalone tool for many managerial decision-making problems on quality. The tool facilitates a decent treatment of uncertainties about a business process. Let us understand the power of a simple linear regression analysis first. A simple linear regression model is

$$Y = \beta_0 o + \beta_1 X + \varepsilon \tag{8.4}$$

where

$\beta_0$ = intercept parameter, showing the $Y$ value when $X = 0$
$\beta_1$ = slope parameter, showing the effect of an additional $X$ value on $Y$, and
$\varepsilon$ = the random "error" term (or uncertainty term) which reflects other explanatory variables that influence dependent variable.

In literature, there are many names associated with the term $\varepsilon$. They are generally called the unaccounted factors, also called "error" or "disturbance" or "noise" factors, which influence the dependent variable. This factor is a random variable, and we take it into account in studying the relationship. To get rid of its effect, we usually make some assumptions (Gauss–Markov assumptions) for the error term. They are: (i) the error terms are statistically independent of each other, (ii) the error term has a constant variance, and (iii) the independent variables are non-random. Put together, we say $\varepsilon_i$ are $N(0, \sigma^2)$ random variables. Under this assumption, we can write the model (8.4) as

$$E(Y) = \beta_0 + \beta_1 X$$

or

$$\hat{Y} = \beta_0 + \beta_1 X \tag{8.5}$$

Thus, Eq. (8.5) is called the regression line of $Y$ on $X$. The unknown constants $\beta_0$ and $\beta_1$ are found by minimizing the error sum of squares (ESS) say $\Sigma e_i^2 = \Sigma(y_i - \hat{y})^2 = \Sigma(y_i - \beta_0) = \Sigma(y_i - \hat{y})^2$, concerning the parameters $\beta_0$ and $\beta_1$. This will lead to two standard equations as follows:

$$\Sigma y_i = n\beta_0 + \beta_1 \sum x_i \tag{8.6}$$

and

$$\Sigma x_i y_i = \beta_0 \Sigma x_i + \beta_1 \Sigma x_i^2 \tag{8.7}$$

Solving (8.6) and (8.7) simultaneously, we get

$$\hat{\beta_1} = \frac{n\Sigma x_i y_i - \Sigma x_i \Sigma y_i}{n\Sigma x_i^2 - (\Sigma x_i)^2}$$

and

$$\hat{\beta_0} = \frac{\Sigma y_i}{n} - \hat{\beta_1}\frac{\Sigma x_i}{n} = \bar{y} - \hat{\beta_1}\bar{x} \tag{8.8}$$

The constants $\beta_0$ and $\beta_1$ are called the intercept and slope of the regression line. A regression line is said to be significant if the slope of the line is significant. The significance of slope (i.e., $H_0 : \beta_1 = 0$ against $H_0 : \beta_1 \neq 0$) can be tested using a one-way analysis of variance (ANOVA), where the total variation will be expressed as a sum of two variations, namely the variation due to regression and variation due to error. That is, if the total sample variance of all responses is given by $s^2 = \frac{1}{n-1}\sum_{i=1}^{n}(y_i - \bar{y})^2 = \frac{TSS}{n-1}$. Then

$$\text{TSS} = \sum_{i=1}^{n}(y_i - \bar{y})^2$$

$$= \sum_{i=1}^{n}(y_i - \hat{y} + \hat{y} - \bar{y})^2$$

$$= \sum_{i=1}^{n}(y_i - \hat{y})^2 + \sum_{i=1}^{n}(\hat{y} - \bar{y})^2 + 2\sum_{i=1}^{n}(y_i - \hat{y})(\hat{y} - \bar{y})$$

$$= \text{ESS} + \text{sum of squares due to regression}$$

If we call the sum of squares due to regression as RSS, for simplicity, then, we have TSS $=$ RSS $+$ ESS. The corresponding degrees of freedom for TSS, RSS, and ESS will be $n-1$, $1$, and $n$-2. Then the mean sum of squares (MSS) correspond to RSS is $\sum_{i=1}^{n}(\hat{y} - \bar{y})^2/df$ and $\sum_{i=1}^{n}(y_i - \hat{y})^2/df$. Hence, the test statistics for testing $H_0 : \beta_1 = 0$ against $H_0 : \beta_1 \neq 0$ is computed as

**Table 8.11** ANOVA for linear regression

| Source | DF | Sum of Squares | Mean SS | F-computed | F-critical |
|---|---|---|---|---|---|
| Regression | 1 | $\sum(\hat{Y} - \bar{Y})^2$ | $\dfrac{\Sigma\left(\hat{y}-\bar{y}\right)^2}{1}$ | $F = MRSS/MESS$ | $F(\alpha, 1, n-2)$ |
| Residual or error | $n-2$ | $\sum(Y - \hat{Y})^2$ | $\dfrac{\Sigma\left(Y-\hat{y}\right)^2}{n-2}$ | | |
| Total | $n-1$ | $\sum(Y - \bar{Y})^2$ | | | |

$$F = \frac{\sum_{i=1}^{n}\left(\hat{y} - \bar{y}\right)^2/df}{\sum_{i=1}^{n}\left(y_i - \hat{y}\right)^2/df}$$

$$= \frac{\sum_{i=1}^{n}\left(\hat{y} - \bar{y}\right)^2/1}{\sum_{i=1}^{n}\left(y_i - \hat{y}\right)^2/n - 2},$$

which follows an $F$-distribution with degrees of freedom ($\upsilon_1 = 1$, $\upsilon_2 = n - 2$) at $\alpha$-level of significance. For various combinations of ($\upsilon_1$, $\upsilon_2$) and $\alpha$, the $F$-values are tabulated and are presented in Appendix Table 7. The ANOVA table corresponds to testing $H_0 : \beta_1 = 0$ against $H_0 : \beta_1 \neq 0$ is presented in Table 8.11.

**Example 8.6** The raw material used in synthetic fiber production is stored in a place with no humidity control. Measurements of the relative humidity and the moisture content of samples of the raw material (both in percentages) on 12 days yielded the results, as shown in Table 8.6. Draw a scatter plot and decide the association between the variables. Fit a regression line that will enable us to predict the moisture content in terms of the relative humidity. Using the regression line, estimate the moisture content when the relative humidity is (i) 38 percent and (ii) 50 percent. Also, test the significance of the regression coefficient.

*Solution.* As per the terminology, the response (output) variable ($y$) is the moisture content, and the predictor (input) variable ($x$) is the relative humidity. The computations involved are shown in Table 8.12.

The estimates are

$$\hat{\beta}_1 = \frac{n\sum x_i y_i - \sum x_i \sum y_i}{n\sum x_i^2 - (\sum x_i)^2}$$

$$= \frac{12 * 6314 - 507 * 144}{12 * 22265 - (507^2)}$$

$$= 0.272431$$

and

$$\hat{\beta}_0 = \bar{y} - \hat{\beta}_1 \bar{x}$$

**Table 8.12**  Measurements on moisture content and relative humidity

| Samples | $x$ | $y$ | $x^2$ | $y^2$ | $xy$ | $\hat{y}$ | $(y-\hat{y})^2$ | $(\hat{y}-\bar{y})^2$ | $(y-\bar{y})^2$ |
|---|---|---|---|---|---|---|---|---|---|
| 1 | 46 | 12 | 2116 | 144 | 552 | 13.02161 | 1.043687 | 1.043687 | 0 |
| 2 | 53 | 14 | 2809 | 196 | 742 | 14.92863 | 0.862348 | 8.576856 | 4 |
| 3 | 37 | 11 | 1369 | 121 | 407 | 10.56973 | 0.185131 | 2.045669 | 1 |
| 4 | 42 | 13 | 1764 | 169 | 546 | 11.93189 | 1.140868 | 0.00464 | 1 |
| 5 | 34 | 10 | 1156 | 100 | 340 | 9.752438 | 0.061287 | 5.051535 | 4 |
| 6 | 29 | 8 | 841 | 64 | 232 | 8.390283 | 0.152321 | 13.03006 | 16 |
| 7 | 60 | 17 | 3600 | 289 | 1020 | 16.83564 | 0.027013 | 23.38345 | 25 |
| 8 | 44 | 12 | 1936 | 144 | 528 | 12.47675 | 0.227289 | 0.227289 | 0 |
| 9 | 41 | 10 | 1681 | 100 | 410 | 11.65946 | 2.753791 | 0.115971 | 4 |
| 10 | 48 | 15 | 2304 | 225 | 720 | 13.56647 | 2.055003 | 2.453835 | 9 |
| 11 | 33 | 9 | 1089 | 81 | 297 | 9.480007 | 0.230407 | 6.350365 | 9 |
| 12 | 40 | 13 | 1600 | 169 | 520 | 11.38702 | 2.601692 | 0.37574 | 1 |
| **Total** | **507** | **144** | **22265** | **1802** | **6314** | **143.9999** | **11.34084** | **62.6591** | **74** |

$$= 12 - 0.272431 * 42.25$$
$$= 0.489784$$

Therefore, the regression line is

$$\hat{y} = 0.489784 + 0.272431x$$

So, when $x = 38$, $\hat{y} = 10.8421$ and when $x = 50$, $\hat{y} = 14.1113$. The sums of squares due to various sources of variation are:

$$\text{TSS} = \sum_{i=1}^{12}(y_i - \bar{y})^2 = 74$$

$$\text{RSS} = \sum_{i=1}^{12}(\hat{y} - \bar{y})^2 = 62.6591$$

and

$$\text{ESS} = \sum_{i=1}^{12}(y - \hat{y})^2 = 11.34084$$

The complete analysis of variance is shown in Table 8.13. The correlation coefficient, according to the formula (8.1) is 0.92. The *Significance F* shown in Table 8.13 is the $p$-value corresponding to the null hypothesis. As per the value, the minimum

**Table 8.13** ANOVA of linear regression

| Source of variation | df | SS | MS | F | Significance F |
|---|---|---|---|---|---|
| Regression | 1 | 62.65916 | 62.65916 | 55.25093 | 2.23E-05 |
| Residual | 10 | 11.34084 | 1.134084 | | |
| Total | 11 | 74 | | | |

level at which the hypothesis $H_0 : \beta_1 = 0$ is rejected is 2.23E-05, which is very small, and therefore the regression coefficient is highly significant.

Quality problems sometimes require a study of the relationship between two or more variables. This is provided by multiple regression analysis. As seen in the simple linear relationship discussed above, the relationships obtained through multiple regression analyses are also used for forecasting and prediction. The strength of the relationship determines the power of the projection. Multiple linear regressions are an extension of bivariate regression analysis, where more than one predictor is used for forecasting the response variable. Let $X_1$, $X_2$, ..., $X_k$ be $k$ independent or explanatory variables, and $Y$ be the dependent or response variable. Then general multiple linear regression models are written as

$$Y = \beta_0 + \beta_1 X_1 + \beta_2 X_2 + \ldots + \beta_k X_k + \varepsilon \tag{8.9}$$

where $\beta_0$, $\beta_1$, ...., $\beta_k$ are parameters and $\varepsilon$, the error terms, are independent $N(0, \sigma^2)$. As we did in a simple regression model, the parameters $\beta_0, \beta_1$, ..., $\beta_k$ are to be estimated using the ordinary least square method by solving $(k + 1)$ normal equations. Since there are many explanatory variables used in multiple regression models, it may possible that the explanatory variables may also be linearly related to themselves. This problem is generally called the problem of multicollinearity. The presence of multicollinearity can seriously affect the standard errors of the estimates of $\beta$ coefficients and hence influences all other inferences. So it is essential to address a severe diagnosis to the variables and its factors before going for model construction. Let us illustrate the use of multiple regressions in an experimental study.

***Example. 8.7*** The data shown in Table 8.14 is the scores and measurement of 10 randomly selected alcohol addicted patients. The variables measured include blood alcohol level ($Y$), insecurity level ($X_1$), anxiety level ($X_2$), depression level ($X_3$), self-esteem level ($X_4$), and age of the patient ($X_5$). The variables $Y$, $X_1$, $X_2$, $X_3$, $X_4$ have compiled scores based on some psychological questions posted to them, and the variable $X_5$ is the actual measurement of age in years. Fit a multiple regression model and see how the explanatory variables are associated with the response variable.

*Solution.* We present the EXCEL output of multiple regression below: Note that the multiple correlation coefficient is 0.99434 (with adjusted $R^2 = 0.9746$), which is very high, showing the variation explained by the input variables is adequate for predicting the response variable. The little value of $F$-significance substantiates this in the ANOVA table (Table 8.15).

**Table 8.14** Data on psychological measurements

| Patient | Y | $X_1$ | $X_2$ | $X_3$ | $X_4$ | $X_5$ |
|---------|------|------|------|------|------|------|
| 1 | 52.8 | 51.9 | 50.2 | 58.7 | 53.8 | 28 |
| 2 | 45.8 | 42.6 | 46.8 | 51.3 | 56.8 | 55 |
| 3 | 50.9 | 53.8 | 59.3 | 61.7 | 55.1 | 68 |
| 4 | 35.7 | 39 | 31.5 | 41.6 | 58.5 | 33 |
| 5 | 58.6 | 60.5 | 50.8 | 67.9 | 50.4 | 42 |
| 6 | 42.7 | 36.8 | 35.7 | 49.4 | 58 | 45 |
| 7 | 34.6 | 20.5 | 18.9 | 25.8 | 58.6 | 48 |
| 8 | 55.3 | 55.9 | 54.3 | 62.6 | 51.5 | 52 |
| 9 | 60.4 | 64 | 62.6 | 65.3 | 47.2 | 22 |
| 10 | 44.6 | 45.2 | 48.4 | 52.6 | 57.9 | 69 |

**Table 8.15** ANOVA table

| Source | df | SS | MS | F | Significance F |
|--------|-----|---------|----------|----------|---------------|
| Regression | 5 | 717.8084 | 143.5617 | 70.06793 | 0.000551 |
| Residual | 4 | 8.195572 | 2.048893 | | |
| Total | 9 | 726.004 | | | |

As far as individual regressor variables are concerned, the variables: depression $(X_3)$ and self-esteem $(X_4)$ are very highly significant, as their $p$-value is less than 0.05. Although the variables insecurity level $(X_1)$, anxiety level $(X_2)$, and age $(X_5)$ is not significant, their presence is required in the model for predicting y. The prediction model is

$$y = 118 - 0.489X_1 + 0.133X_2 + 0.626X_3 - 1.59X_4 + 0.0085X_5$$

There are practical situations where linear regression may not be suitable for predicting the response variable. In those situations, we use nonlinear regression models to explain the process equations. For instance, suppose we wanted to fit a process equation (regression) model, say, $y = ab^x$, which is an exponential model. By a logarithmic transformation, one can convert the given equation into a linear form as $z = \beta_0 + \beta_1 x$, where $z = \log(y)$, $\beta_0 = \log(a)$, $\beta_1 = \log(b)$. The equation $z = \beta_0 + \beta_1 x$ is now a simple linear equation and can be used for predicting $y$ (Table 8.16).

There are plenty of real-life situations where correlation and regressions techniques are used for quality decision making. Many industrial and organizational applications, process input/output variables naturally appear in pairs, for instance, height and weight; pressure and temperature; precision and efficiency; blood pressure and cholesterol; radius and volume; sales and expenditure; thickness and strength;

**Table 8.16** Coefficients' table and their summary statistics

| Variables | Coefficients | Standard error | t Stat | P-value | Lower 95% | Upper 95% | Lower 95.0% | Upper 95.0% |
|---|---|---|---|---|---|---|---|---|
| Intercept | 118.0582 | 18.42278 | 6.408272 | 0.003046 | 66.90833 | 169.208 | 66.90833 | 169.208 |
| $X_1$ | −0.4894 | 0.24376 | −2.00772 | 0.115097 | −1.16619 | 0.187383 | −1.16619 | 0.187383 |
| $X_2$ | 0.133161 | 0.128097 | 1.039527 | 0.357264 | −0.22249 | 0.488816 | −0.22249 | 0.488816 |
| $X_3$ | 0.626468 | 0.181961 | 3.44287 | 0.026225 | 0.121263 | 1.131673 | 0.121263 | 1.131673 |
| $X_4$ | −1.58846 | 0.29072 | −5.46389 | 0.005456 | −2.39563 | −0.78129 | −2.39563 | −0.78129 |
| $X_5$ | 0.007989 | 0.046816 | 0.170638 | 0.872792 | −0.12199 | 0.137971 | −0.12199 | 0.137971 |

length and width, and so on. In these situations, the uses of correlation and regression techniques are always warranted for establishing a relationship between them. Identifying the input and output variable itself can pose lots of challenges when the process is complex, interconnected, and problem-solving become tedious. Since the causes are used as input variables and effect is treated as output, their identification can be realized through the cause and effect analysis, described in detail in the next subsection.

## 8.3.5  Cause and Effect Analysis

A cause and effect diagram (C&E) or fishbone diagram is a graphic representation of the relationship between a given effect and its potential causes. The diagram was introduced by Ishikawa in 1985 and hence called the Ishikawa diagram as well. The possible causes are divided into categories and subcategories so that the display resembles a fish's skeleton. The diagram has the best use at the time of brainstorming, process evaluation, and planning activities.

The first step toward preparing a C&E diagram is identifying the problem were quality issues, and management attention is required. Then start with the definition of effect ($y$ variable) and later determine the potential causes ($x$ variables) associated with the impact. The second step requires lots of exercises and homework if large numbers of issues are necessary to address the problem. The best way to identify the potential causes is through a brainstorming session attended by the think tank of the process or organization. The next step is to group the causes according to the source of variation: whether it is through man, machine, materials, methods, or mother earth (or environment). Then identify the importance of each cause on the effect; for this, one may need a detailed statistical analysis or management expertise or both to identify the most significant variables impacting the quality of the process.

Some frequently identified causes for a production process are material failure, assembly error, incorrect speeds/feeds, incorrect tooling, inadequate venting, misalignment, improper torque, inadequate gauging, overload capacity, out of tolerance, tool damage/burrs, clamping, packaging damage, handling injury, inappropriate tool setup, improper surface preparation, inadequate control system, damaged part, deficient gating, and inadequate holding, etc. (Muralidharan, 2015). A typical C&E diagram for the effect of "high petrol consumption" is shown in Fig. 8.11.

Suppose we want to analyze the above C&E diagram to identify the critical-to-quality variables ($x$-variables) affecting the high petrol consumption ($y$-variable). So we first prepare an excel sheet (spreadsheet) to record the data. One can plan the data collection for three months to six months to understand the impact of each source of variation on the effect. Here, procedure, driver, vehicle, road condition, maintenance, and materials are the primary sources of variation and are termed as $X$'s as per the lean terminology. The source of variation identified in each primary cause may be notated as $x_{11}, x_{12}, x_{13}, \ldots,$ for $x_1$ (say, procedure), $x_{21}, x_{22}, x_{23}, \ldots$ for $x_2$ (say, driver), and so on. Data for each $x_{ij}$ will be recorded for effect ($y$) for each day, and thus, we

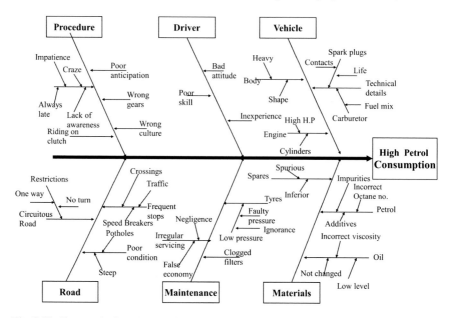

**Fig. 8.11** Cause and effect diagram for high petrol consumption. ( *Source* Muralidharan, 2015)

have massive data. The data collection plan in this particular context is subject to customer-specific notions and ideas and cannot be uniform. However, the tools to identify the CTQs will be most often standard, and the conclusions are drawn from the analysis also will be uniform for the objectives.

For the above example, one may use a Pareto chart to identify the primary source of variation, and then use another Pareto chart within the variable to identify the most significant contributing variables for high petrol consumption. A typical Pareto chart based on the material causes of variation is shown in Fig. 8.12.

It is now clear that impurities followed by additive, viscosity, and frequency of oil change are the primary sources of variation affecting the quality of the materials. The next step will be to see how these causes are associated with the other reasons identified in other main variables. Using a detailed statistical analysis of means, variances, etc., the people working with the process can prioritize and classify the most important causes. This comprehensive analysis facilitates one to identify the *vital few* from the *critical many* causes. DOE is another powerful tool for identifying vital few from many essential parameters of quality, which will be discussed later in this chapter.

Two of the best reason for carrying out C&E analysis is (i) to understand relationships between potential causes and (ii) to track which possible causes have been investigated, and which proved to contribute significantly to the problem. By and large, the possible effect(s) of failure is anything like:

• Poor coordination
• Abrupt use of materials

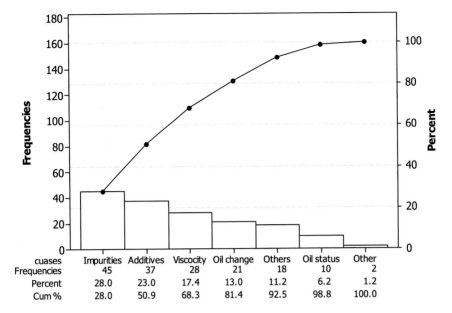

| cuases | Impurities | Additives | Viscocity | Oil change | Others | Oil status | Other |
|---|---|---|---|---|---|---|---|
| Frequencies | 45 | 37 | 28 | 21 | 18 | 10 | 2 |
| Percent | 28.0 | 23.0 | 17.4 | 13.0 | 11.2 | 6.2 | 1.2 |
| Cum % | 28.0 | 50.9 | 68.3 | 81.4 | 92.5 | 98.8 | 100.0 |

**Fig. 8.12**  Pareto chart of material causes

- Air/water leakage
- Brake failure
- Contamination
- Customer dissatisfaction
- Delay in delivery/Payment
- Excess interest
- Poor brake efficiency
- Poor valve performance
- Oil carryover
- Inadequate service life
- Unable to assemble
- Wrong material
- Wrong delivery-short or excess.

A well-developed cause and effect diagram can be a useful tool to identify possible maintainability-related problems. A repeated exercise of the C&E diagram conducted through brainstorming can even lead to the identification of future breakdowns and repairs and the causes leading to them.

### 8.3.6  Testing and Confirmatory Analysis

People, organizations, companies, business establishments, etc. make all sorts of claims and promises. Making claims or hypotheses is common for increased attention, business growth, financial gains, and profits. For instance, consider the following situations:

1.  A coaching center may claim that if a student undergoes their training program, then the expected grade in the national level test will increase by more than 95%.
2.  A sociologist claims that the virus's infection rate will be less in Maharashtra than Gujarat if the lockdown is continued for another two weeks.
3.  A tour operator claims that the actual proportion of tourists' books for a foreign tour for next year will be more than 90%.
4.  An economist claims that community ABC incomes are more spread out than the income in community XYZ.
5.  An educationist may claim that the grades obtained by students in a skill development course follow a normal distribution.
6.  A forest officer claims that finding oversized lions in the Gir forest is more likely than any other place.

If we carefully examine the above situations, a common thing can be found, that they are all statements of uncertain events, but verifiable and testable claims or hypothesis by using quantitative information. In the situations above, it is preferred to make decisions based on experimental data representing a suitable population, as they are all assumptions about a population parameter. The method of making decisions using experimental data is nothing but a *Statistical test of hypothesis*. The statistical tests of the hypothesis make use of a random sample from the population to test the claims or hypotheses. Out of many statistical assumptions or claims described above, some of them are dealing with parameters of well-defined probability distributions, some are about the randomness of phenomena, and some are about specific patterns, etc. If a hypothesis is about the parameter(s) of well-defined probability distributions, then we call them *parametric hypothesis*. Otherwise, it is called a *nonparametric hypothesis*.

Because of the nature of the problem, the statistical test of the hypothesis is often called confirmatory data analysis, in contrast to exploratory data analysis. Statistical hypothesis testing is a key statistical inference technique and is widely used for decision making under uncertainty. The other part of inference is the point estimation and interval estimation. A point estimate proposes a single value as an estimate for the population parameter. The interval estimate offers a range of values for the population parameter with some level of confidence. The confidence interval method of suggesting estimates is a powerful method of estimation and is widely accepted. The general idea of hypothesis testing begins with some sample data from a population, then uses some test criteria and makes a decision either to reject or not to reject the hypothesis. This is done based on some population characteristics called parameters like $\theta$, $\mu$, $\sigma$, $P$, etc.

There are two types of statistical hypotheses. A hypothesis that is being tested is called the *null hypothesis* and is usually denoted by $H_0$. An $H_0$ describes an unbiased statement about the situation. The hypothesis that is tested against $H_0$ is called the *alternative hypothesis*, denoted by $H_1$ or $H_a$. For example, suppose we wanted to determine whether a given coin is fair and balanced (or unbiased). In this case, a null hypothesis might be that half the flips would result in heads and half, in tails. The alternative hypothesis might be that the number of heads and tails would be very different. Symbolically, these statements would be expressed as $H_0 : p = 0.5$ versus $H_1 : p \neq 0.5$. To test this claim, we may toss the coin a good number of times and see how frequently the head and tail appear. If the occurrence is equally likely with probability 0.5, we can conclude that the coin is unbiased. If we flipped the coin 50 times, resulting in 22–28 heads, we might find that the coin was probably fair and balanced. Table 8.17 presents the hypothesis that corresponds to the various situations described above.

A hypothesis can be simple or composite. A hypothesis that specifies all the population parameters under study is called a *simple hypothesis*; otherwise, it is called a *composite hypothesis*. As mentioned previously, the hypothesis tests are a confirmatory analysis, and hence is a postdata concept. That is when a decision is taken, after testing $H_0$, whether to reject or not to reject $H_0$, we still have other possibilities. The hypothesis itself may be correct or not correct. That is, our decision may be right or incorrect. If a testing procedure gave a decision not to reject the hypothesis that does not mean that the hypothesis is correct. The hypothesis may

**Table 8.17** Formulation of hypothesis

| Situation | Hypothesis formulation |
|---|---|
| 1 | Suppose $\mu$ is the expected grade in the national level test, then the claim is to test $H_0 : \mu = 95$ against $H_1 : \mu > 95$ <br> If $H_0$ is rejected, then their claim is correct |
| 2 | Suppose $\mu_1$ and $\mu_2$ are the expected rate of COVID infection in Maharashtra and Gujarat, respectively, then the claim is to test $H_0 : \mu_1 = \mu_2$ against <br> $H_1 : \mu_1 < \mu_2$ <br> If $H_0$ is rejected, then the claim is correct |
| 3 | Suppose the probability of tourists opt for the foreign tour is $p$, then the claim is to test $H_0 : p = 0.9$ against $H_1 : p > 0.9$ <br> If $H_0$ is rejected, then the claim is correct |
| 4 | Suppose $\sigma_1$ and $\sigma_2$ are the spread of income in community ABC and XYZ, respectively, then the claim is to test $H_0 : \sigma_1 = \sigma_2$ against $H_1 : \sigma_1 \neq \sigma_2$ <br> If $H_0$ is rejected, then the claim is correct |
| 5 | Suppose a typical grade obtained by students is x, then the claim is to test $H_0 : x \sim$ Normal distribution against $H_1 : x$ do not follow a normal distribution <br> If $H_0$ is accepted, then the claim is correct |
| 6 | Suppose $p_1$ and $p_2$ are the probability of lions found in Gir and Sundarban forests, respectively, then the claim is to test $H_0 : p_1 = p_2$ against $H_1 : p_1 > p_2$ <br> If $H_0$ is rejected, then the claim is correct |

| Decision | Truth | |
|----------|-------|-|
|  | $H_0$ is true | $H_0$ is not true |
| Reject $H_0$ | Type-I error | Correct decision |
| Not to reject $H_0$ | Correct decision | Type-II error |

**Fig. 8.13** Types of statistical testing errors

be different from the reality of the situation. So we have one of the following four possibilities, as shown in Fig. 8.13.

The figure shows two situations of correct decision and two situations of a wrong decision. The error committed to rejecting the null hypothesis $H_0$ when it is true is called Type-I error, and the error of not rejecting $H_0$ when $H_0$ itself is not true is called Type-II error. The probability of committing a Type-I error is called the *significance level* and is often denoted by α. Similarly, the probability of committing a Type-II error is called *Beta* error and is often denoted by β. The probability of not committing a Type-II error is called the *Power* of the test. Thus,

$$\alpha = P(Reject\ H_0|H_0\ is\ true) \qquad (8.10)$$

$$\beta = P(Not\ reject\ H_0|H_0\ is\ not\ true) \qquad (8.11)$$

and

$$Power = 1 - \beta = P(Reject\ H_0|H_0\ is\ not\ true) \qquad (8.12)$$

The next issue with the testing procedure is to get a criterion for rejecting the null hypothesis. Once that is done, the computation of α and β is routine. In practice, statisticians describe the decision rules in two ways—concerning a $p$-value or concerning a region of rejection or critical region. Let S be the sample space, and $C$ and $\bar{C}$ are such that $C \cup \bar{C} = S$, then region $C$, where the null hypothesis $H_0$ is rejected, is called a *critical region*. In such cases, we say that the hypothesis has been rejected at α level of significance.

The $p$-value measures the strength of evidence in support of a null hypothesis. Suppose the test statistic is equal to say $t(x)$. Then the $p$-value is the probability of observing a test statistic as extreme as $t(x)$, assuming the null hypothesis is true. If the $p$-value is less than the significance level, we reject the null hypothesis. Most of the statistical software uses the $p$-value approach for the rejection of test statistics. A $p$-value is a measure of how much evidence you have against the null hypothesis. The smaller the $p$-value, the more evidence you have against $H_0$. One may combine the $p$-value with the significance level to decide on a given test of the hypothesis. In such a case, if the $p$-value is less than 0.05 (a default value), then you reject the null hypothesis. Most of the statistical software's use 0.05 as a default level of significance.

As an illustration, consider the problem of testing $H_0 : \mu = 10$ against $H_1 : \mu = 20$ in a normal population, $N(\mu, \sigma^2)$. Suppose $\sigma^2$ is known to be 25, and the test criterion is that $H_0$ is rejected if the observation is larger than 16, otherwise, the test is accepted based on a single observation from the normal population. The values of $\alpha$ and $\beta$ are respectively obtained as

$$\alpha = P(Reject\ H_0 | H_0\ is\ true\}$$
$$= P(x > 16 | \mu = 10)$$
$$= P\left(\frac{x - \mu}{\sigma} > \frac{16 - \mu}{\sigma} | \mu = 10\right)$$
$$= P\left(\frac{x - 10}{5} > \frac{16 - 10}{5}\right)$$
$$= P(z > 1.2)$$
$$= 0.12,$$

using standard normal distribution table (see Appendix T-3)
and

$$\beta = P(Not\ reject\ H_0 | H_0\ is\ not\ true)$$
$$= P(x \langle 16 | \mu = 20)$$
$$= P\left(\frac{x - \mu}{\sigma} \langle \frac{16 - \mu}{\sigma} | \mu = 20\right)$$
$$= P\left(\frac{x - 20}{5} < \frac{16 - 20}{5}\right)$$
$$= P(z < -0.8)$$
$$= 0.21,$$

using standard normal distribution table (see Appendix T-3)

Note that both $\alpha$ and $\beta$ are errors and should be minimized for better decision making. Unfortunately, both cannot be reduced simultaneously. Since Type-II error is more severe than Type-I error, we usually prefix $\alpha$ and minimize $\beta$ or maximize $1 - \beta$, i.e., power of the test. A plot of the probability of Type-II errors under various hypotheses gives the *Operating Characteristics* curves (OC curves). An OC curve provides indications of how well a given test will enable us to minimize Type-II errors. In SQC/SPC applications, the value of $\alpha$ and $\beta$ is called the *Producers' risk* and *Consumers' risk*, respectively.

The general perception of statistical tests of a hypothesis as per the LSS philosophy based on a typical problem is shown in Table 8.18.

The detailed steps involved in hypothesis tests are stated below:

**Table 8.18** Process of testing of hypothesis

| Process of hypothesis testing | Situation and its equivalent notion |
| --- | --- |
| Practical problem | Is there any significant difference between the IQ of boys of Science stream and Management stream students? |
| Statistical problem | $H_0 : \mu_S = \mu_M$ against $H_1 : \mu_S \neq \mu_M$ |
| Statistical solution | Suppose we use $t$-test of difference to test the hypothesis: So $t$-value $= -0.88$ $p$-value $= 0.390$ DF $= 17$ Fail to reject the null hypothesis |
| Practical solution | There is no evidence of a significant difference between the IQ of Science and Management students |

1. State the null and alternative hypotheses. The null hypothesis should be an unbiased statement of the situation.
2. Specify a suitable level of significance $\alpha$.
3. Compute the test statistic as per the given hypothesis.
4. Analyze sample data. Find the value of the test statistic (mean score, proportion, t-score, z-score, chi-square statistics, etc.) described in the analysis plan.
5. Interpret results. Apply the decision rule described in the analysis plan. If the test statistic's value is unlikely, based on the null hypothesis, reject the null hypothesis.
6. Write the appropriate conclusions and suggest suitable recommendations.

As mentioned previously, there are many parametric and nonparametric types of statistical tests. The test that makes use of normal distribution assumptions are parametric tests. The test based on normal (or z-test), t-test, F-test is parametric tests. For nonparametric tests, there is no assumption of normality, and hence they are also called *distribution-free tests*. A statistical test can be again one-tailed or two-tailed. *One-tailed tests* have a critical region (or rejection region) entirely on one side of the area of sample space. It can be either on the left side (hence called left-tailed tests) or right side (thus called right-sided tests). If the rejection region is on both the side, then the test will be *two-tailed tests*. Table 8.19 gives critical values of $z$ for both a one-tailed and two-tailed test at various levels of significance. For other levels of significance, one may use standard normal curve areas (Appendix Table 3).

The graphical presentations of two-tailed and one-tailed tests are presented in Fig. 8.14a–c.

**Table 8.19** Critical values of $z$

| Critical values of z | Level of significance, $\alpha$ | | | | |
| --- | --- | --- | --- | --- | --- |
| | 0.10 | 0.05 | 0.01 | 0.005 | 0.002 |
| Left-tailed | −1.280 | −1.645 | −2.33 | −2.580 | −2.880 |
| Right-tailed | 1.280 | 1.645 | 2.330 | 2.580 | 2.880 |
| Two-tailed, |z| | 1.645 | 1.960 | 2.580 | 2.810 | 3.080 |

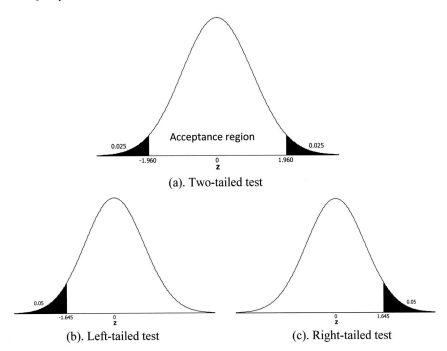

**Fig. 8.14** Rejection region of various types of test

There are statistical tests for mean, variance, proportion, equality of means, equality of several means, etc. The test statistics depend on the type of hypothesis under consideration, and accordingly, one can prepare the most powerful test, unbiased test, similar test, likelihood ratio test, and so on. One may consult books like Lehman (1998), Kale, and Muralidharan (2015) for the theoretical development of test statistics and their distributional properties. Below we briefly discuss some examples to illustrate the tests of the hypothesis.

***Example 8.8*** A research team is willing to assume that systolic blood pressure in a specific population of males is approximately normally distributed with a standard deviation of 16. A simple random sample of 64 male's forms the population had a mean systolic blood pressure reading of 133. At the 0.05 level of significance, do these data provide sufficient evidence to conclude that the population means is greater than 130? Also, propose a 95% two-sided confidence interval for $\mu$. Compute the $p$-value of the test.

*Solution.* Let $X$ be the measure of blood pressure, then $X \sim N(\mu, 256)$. Also given $n = 64$, $\bar{x} = 133$, and $\alpha = 0.05$. It is proposed to test $H_0 : \mu = 130$ against $H_1 : \mu > 130$. Since the sample size is large, and the population standard deviation is known, we use a z-test (normal test) to test the hypothesis. The procedure is as follows:

1.   $H_0 : \mu = 130$ versus $H_1 : \mu > 130$
2.   Significance level, $\alpha = 0.05$
3.   Computation of test statistics

$$z_{cal} = \frac{\sqrt{n}\left(\bar{x} - \mu_{H_0}\right)}{\sigma}$$

$$= \frac{\sqrt{64}(133 - 130)}{16} = 1.5$$

4.   For $\alpha = 0.05, z_\alpha = 1.645$. Since $z_{cal} < z_\alpha$, we do not reject the hypothesis.
5.   Since the hypothesis is not rejected, we conclude that there is no significant evidence that the blood pressure level has increased for the given sample.

The $p$-value of the test is computed as $p$-value $= P(z_{cal} > 1.5) = 0.07$. The graphical presentation of the critical region and the test value is shown in Fig. 8.15. It is evident that the $z_{cal}$ value falls under the acceptance region. The 95% two-sided confidence interval for $\mu$ is obtained as $\bar{x} \pm 1.96\frac{\sigma}{\sqrt{n}} = (129.08, 136.92)$.

***Example 8.9*** Consider the case of a pharmaceutical manufacturing company testing two new compounds intended to reduce blood pressure levels. The compounds are administered to two different sets of laboratory animals. In group one, 71 of 100 animals tested to respond to drug-1 with lower blood pressure levels. In group two, 58 of 90 animals tested to respond to drug-2 with lower blood pressure levels. The company wants to test at the 0.05 levels, whether there is a difference between the effectiveness of these two drugs.

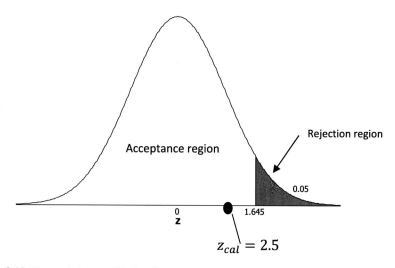

**Fig. 8.15**  Test statistic and critical region

*Solution.* Let $X$ and $Y$ be the effectiveness of drug-1 and drug-2, respectively. Assume that both the population follows a normal distribution. Also given $n_1 = 100, n_2 = 90$, $p_1 = \frac{71}{100} = 0.71$, $p_2 = \frac{58}{90} = 0.64$, and $\alpha = 0.05$. It is desired to test $H_0 : P_1 = P_2$ against $H_1 : P_1 \neq P_2$. This is a test of equality of two proportions, and we use a z-test for equality of proportions to test the hypothesis. The procedure is as follows:

1. $H_0 : P_1 = P_2$ against $H_1 : P_1 \neq P_2$
2. Significance level, $\alpha = 0.05$
3. Computation of test statistics

$$z_{cal} = \frac{(p_1 - p_2)}{\sqrt{p(1-p)\left(\frac{1}{n_1} + \frac{1}{n_2}\right)}},$$

where $p = \frac{n_1 p_1 + n_2 p_2}{n_1 + n_2} = 0.68$. Thus $z_{cal} = 1.04$.

4. For $\alpha = 0.05$, $z_{\alpha/2} = 1.96$. Since $|z_{cal}| < z_{\alpha/2}$, we do not reject the hypothesis.
5. Since the hypothesis is not rejected, we conclude that there is no significant evidence that the effectiveness of the two drugs is different.

The p-value of the test is computed as p-value $= P(|z_{cal}| > 1.04) = 2P(z_{cal} > 1.04) = 0.30$. The graphical presentation of the critical region and the test value is shown in Fig. 8.16. It is evident that the $z_{cal}$ value falls under the acceptance region.

**Example 8.10** Over six consecutive days, the opening prices (in rupees) of two well-known stocks were observed and recorded as follows:

| Stock-1 | 390.9 | 397.0 | 417.7 | 389.6 | 414.2 | 422.6 |
|---------|-------|-------|-------|-------|-------|-------|
| Stock-2 | 423.3 | 391.6 | 421.0 | 409.2 | 464.7 | 450.2 |

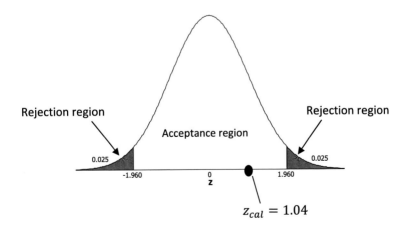

**Fig. 8.16** Z-test statistic and critical region

Test whether the variabilities in the two stocks' opening prices are the same at the 10% level.

*Solution.* Let $X$ and $Y$ be the opening price value of stock-1 and stock-2, respectively. Assume that both the population follows a normal distribution. Also given $n_1 = n_2 = 6$, and $\alpha = 0.10$. It is desired to test $H_0 : \sigma_1^2 = \sigma_2^2$ against $H_1 : \sigma_1^2 \neq \sigma_2^2$. This is a test of equality of two variances, and hence, we use $F$-test for equality of variances to test the hypothesis. The table values of F are given in Appendix Table 7. The procedure is as follows:

1.   $H_0 : H_0 : \sigma_1^2 = \sigma_2^2$ against $H_1 : \sigma_1^2 \neq \sigma_2^2$
2.   Significance level, $\alpha = 0.10$
3.   Computation of test statistics

$$F_{cal} = \frac{s_1^2}{s_2^2} \sim F_{(v_1, v_2), \alpha/2}$$

where $s_1^2$ and $s_2^2$ are the sample variances and are respectively obtained as 210.9987 and 715.7107. Also, $v_1 = n_1 - 1$ and $v_2 = n_2 - 1$ are the degrees of freedom correspond to samples 1 and 2. Since $s_1^2 < s_2^2$, the $F_{cal}$ is computed as

$$F_{cal} = \frac{s_2^2}{s_1^2} = 3.39.$$

4.   For $\alpha = 0.05$, $F_{(v_2, v_1), \alpha/2} = 5.05$. Since $F_{cal} < F_{(v_2, v_1), \alpha/2}$, we do not reject the hypothesis.
5.   Since the hypothesis is not rejected, we conclude that there is no significant evidence to say that the stock prices are different.

The $p$-value of the test is computed as $p$-value $= 2P(F_{cal} > 3.39) = 0.206$. The graphical presentation of the critical region and the test value is shown in Fig. 8.17. It is obvious that the $F_{cal}$ value falls under the acceptance region.

**Example 8.11** A researcher examined the admissions to a mental health clinic's emergency room on days when the moon was full. For the 12 days with full moons from April 2020 through June 2020, the number of people admitted was as follows:

| 5 | 13 | 14 | 12 | 6 | 9 | 13 | 16 | 25 | 13 | 14 | 20 |
|---|----|----|----|---|---|----|----|----|----|----|----|

Assume that the above data is normally distributed. It was observed that the average no. of admissions on other days is 14.2. Test the hypothesis that on a full moon, average number of admissions are higher than on other days at a significance level, $\alpha = 0.05$.

*Solution:* Let $X$ be the number of admissions. Also given $n = 12$, $\bar{x} = 13.33$, and $\alpha = 0.05$. It is proposed to test $H_0 : \mu = 14.2$ against $H_1 : \mu > 14.2$. Since

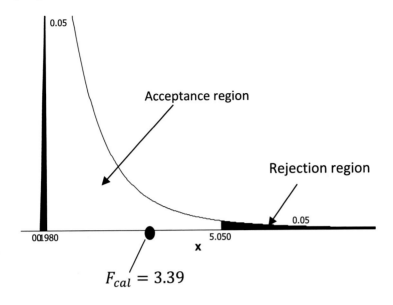

0.01980          5.050
              x

$F_{cal} = 3.39$

**Fig. 8.17**  *F*-test statistic and critical region

the sample size is small, and the population standard deviation is estimated from Appendix Table 5. The procedure is as follows:

1.   $H_0 : \mu = 14.2$ versus $H_1 : \mu > 14.2$
2.   Significance level, $\alpha = 0.05$
3.   Computation of test statistics

$$t_{cal} = \frac{\sqrt{n}\left(\bar{x} - \mu_{H_0}\right)}{s}$$
$$= \frac{\sqrt{12}(14.2 - 13.33)}{5.49} = 0.5489$$

4.   For $\alpha = 0.05$, $t_{0.05,11} = 1.796$. Since $t_{cal} < t_{0.05,11}$, we do not reject the hypothesis.
5.   Since the hypothesis is not rejected, we conclude that there is no significant evidence to say that the blood pressure level has increased for the given sample.

The *p*-value of the test is computed as *p*-value $= P(t_{cal} > 0.5489) = 0.29$. The graphical presentation of the critical region and the test value is shown in Fig. 8.18. It is obvious that the $z_{cal}$ value falls under the acceptance region.

***Example 8.12*** An experiment is conducted to compare the precision of two brands of mercury detectors in measuring mercury concentration in the air. During the noon hour of one day in a downtown city area, seven measurements of the mercury

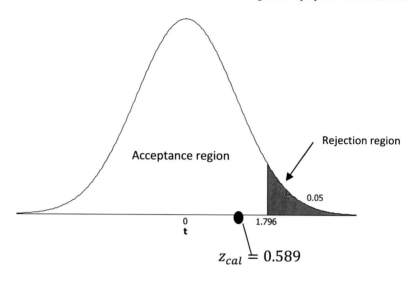

**Fig. 8.18**  $Z$ test statistic and critical region

concentration are made with Brand-A instrument, and six measurements are made with the Brand-B instrument. The measurements per cubic meter of air are:

| Brand-A | 0.95 | 0.82 | 0.88 | 0.96 | 0.91 | 0.86 | 0.99 |
|---------|------|------|------|------|------|------|------|
| Brand-B | 0.89 | 0.91 | 0.78 | 0.91 | 0.81 | 0.84 |      |

Do the data provide strong evidence that Brand-A measures high mercury concentration in the air?

*Solution.* Let $X$ and $Y$ be the mercury concentration of Brand-A and Brand-B, respectively. Assume that both the population follows a normal distribution. Also given $n_1 = 7$, $n_2 = 6$, $\bar{x} = 0.91$, $\bar{y} = 0.86$, $s_1 = 0.061$, $s_2 = 0.055$, and the pooled standard deviation according to the formula, $s = \sqrt{\frac{(n_1-1)s_1^2+(n_1-1)s_1^2}{(n_1+n_2^2-2)}} = 0.058$, and $\alpha = 0.10$. It is desired to test $H_0 : \mu_1 = \mu_2$ against $H_1 : \mu_1 > \mu_2$. This is a test of equality of two means. Since sample sizes are small and the variances are unknown, we use a $t$-test for equality of mean to test the hypothesis. The procedure is as follows:

1. $H_0 : \mu_1 = \mu_2$ against $H_1 : \mu_1 > \mu_2$
2. Significance level, $\alpha = 0.10$
3. Computation of test statistics

$$t_{cal} = \frac{\bar{x} - \bar{y}}{s\sqrt{\left(\frac{1}{n_1} + \frac{1}{n_2}\right)}} = 1.649.$$

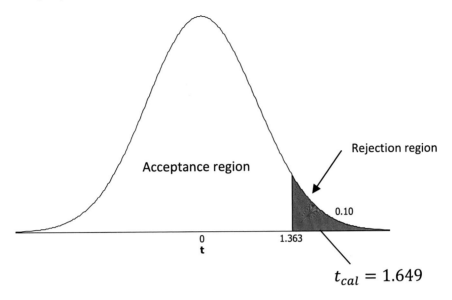

**Fig. 8.19** $t$ test statistic and critical region

4. For $\alpha = 0.10$, $t_{\alpha,(n_1+n_2-2)} = 1.363$. Since $t_{cal} > t_{\alpha,(n_1+n_2-2)}$, we reject the hypothesis.
5. Since the hypothesis is rejected, we conclude that there is no significant evidence to say that the effectiveness of the two drugs is the same.

The $p$-value of the test is computed as $p$-value $= P(t_{cal} > 1.649) = 0.064$. The graphical presentation of the critical region and the test value is shown in Fig. 8.19. It is evident that the $z_{cal}$ value falls under the acceptance region.

***Example 8.13*** Weights in kg of 10 students are as follows: 38, 40, 45, 53, 47, 43, 55, 48, 52, and 49. Can we say the variance of the distribution of weights of all students from which the sample of students was drawn equals 20 kg2? Use $\alpha = 0.10$.

*Solution.* Let $X$ be the weight of students in kg. Assume that the population follows a normal distribution. Also given $n = 10$, $s^2 = 31.1$, and $\alpha = 0.05$. It is desired to test $H_0 : \sigma^2 = \sigma_0^2 = 20$ against $H_1 : \sigma^2 \neq \sigma_0^2 \neq 20$. This is a test of variances, and hence, we use $x^2$-test. The table values of $x^2-$ distribution are given in Appendix Table 6. The procedure is as follows:

1. $H_0 : \sigma^2 = \sigma_0^2 = 20$ against $H_1 : \sigma^2 \neq \sigma_0^2 \neq 20$.
2. Significance level, $\alpha = 0.10$
3. Computation of test statistics

$$x_{cal}^2 = \frac{(n-1)s_1^2}{\sigma_0^2} \sim x_{(n-1)\alpha/2}^2$$

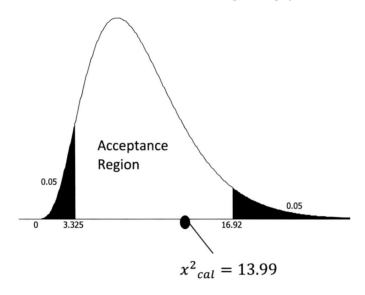

Acceptance
Region

0.05

0.05

0   3.325                    16.92

$$x^2{}_{cal} = 13.99$$

**Fig. 8.20** $x^2$ test statistic and critical region

$$= 13.99$$

4.   For $\alpha = 0.10$, $x^2_{(n-1)\alpha/2} = x^2_{(9)0.05} = 16.92$. Since $x^2_{cal} < x^2_{(9)0.05}$, we do not reject the hypothesis.
5.   Since the hypothesis is not rejected, we conclude that there is no significant evidence to say that the variance is not equal to 20.

The $p$-value of the test is computed as $p$-value $= 2\,P(x^2_{cal} > 13.99) = 0.24$. The graphical presentation of the critical region and the test value is shown in Fig. 8.20. It is obvious that the $F_{cal}$ value falls under the acceptance region.

## 8.3.7   Design of Experiments

Suggesting incremental changes and solutions, eliminating or reducing defects, costs, or cycle time without changing the basic structure of the process need statistically validated experiments and trials. Design of experiments or DOE, in short, is one such tool. The DOE is the process of planning experiments so that appropriate data will be collected, the minimum number of experiments will be performed to acquire the necessary technical information, and suitable statistical methods will be used to analyze the collected data. A DOE is carried out by researchers or engineers in all fields of an experimental study to compare the effects of several conditions or to discover something new.

The identification of the causal relationship between independent ($x$) and the response variable ($y$) is carried out through the DOE concept, where the experimenter uses three essential principles of experimentation. They are randomization, replication, and local control or blocking. As per the DOE terminology, *randomization* is used to assign treatments to experimental units so that each unit has an equal chance of being assigned a particular treatment, minimizing the effect of variation from uncontrolled noise factors. *Replication* means the repetition of the same treatment in many plots to estimate the error variance of the experiment. The purpose of *local control* or *blocking* is to increase the experiment's accuracy without a substantial increase in cost. To achieve local control, first, we detect some sources of variation in the experimental units and classify the units into homogeneous groups according to these sources of variations. The technical aspects of experimentation, models, and type of designs are beyond this book's scope. However, we put forth how DOE can be perceived as a quality tool for modeling and statistical reasoning of industrial experiments following the objective of this book. One may refer to Montgomery (2003) for comprehensive technical details and application on DOE. For the reader's interest, we give the following steps involved in the design of the experiment:

1. Identification of the problem
2. Choice of factors and levels
3. Selection of the response variable
4. Choice of experimental design
5. Performing the experiment
6. Data analysis
7. Conclusion and recommendations.

The relationship between input–output variables can be best established through the use of DOE techniques. There are many designs used for process optimization and design–redesign efforts of industrial processes. Table 8.20 compiles various such designs concerning their application contexts Montgomery (2003), Gryna et al. (2007), Muralidharan (2015). DOE also plays a significant role in engineering design activities, where new products are developed, and existing ones are improved. It is also interesting to note that DOE techniques remain a standalone tool because of its decision support capabilities. In short, DOE is also used for and not limited to:

1. Improving the performance of a complex manufacturing process. The application of DOE can economically determine the optimal values of process variables.
2. Development of new processes and designs. The application of DOE methods early in process development can result in reduced development time, reduced variability of target requirements, and enhanced process yields.
3. Screening important factors from among many competing input variables.
4. Engineering design activities include evaluation of material alternations, comparison of basic design configurations, and selection of design parameters so that the product is robust to a wide variety of field conditions.

**Table 8.20** Classification of designs

| Design | Appropriate when |
|---|---|
| Completely randomized | Only one experimental factor is being investigated |
| Randomized block | One factor is being investigated, and experimental material or environment can be divided into blocks or homogeneous groups |
| Balanced incomplete block | All the treatments cannot be accommodated in a block. |
| Partially balanced incomplete block | If a balanced incomplete block requires a large number of blocks, then it is practical |
| Latin square | One primary factor is under investigation, and results may be affected by two other experimental variables or by two sources of non-homogeneity. It is assumed that no interaction exists |
| Youden square | Same as Latin square, but the number of rows, columns, and treatments need not be the same |
| Factorial | Several factors are being investigated at two or more levels, and interaction of factors may be significant |
| Blocked factorial | The number of runs required for factorial is too large to be carried out under similar conditions |
| Fractional factorial | Many factors and levels exist, and funneling all combinations is impractical |
| Nested | To study relative variability instead of the mean effect of sources of variation |
| Response surface | To provide empirical maps (contour diagrams) illustrating how factors under the experimenters control to influence the response |
| Mixture designs | Same as factorial designs |
| Robust designs | To find a set of conditions for design variables, which are robust to noise, and to achieve the smallest variation in a product's function about the desired target value |

5.  Empirical model building is another process to determine the functional relationship between $x$ and $y$, leading to regression (prediction) and grouping of observations.

The experimental principle common to all the design is randomization. Since the process of randomization involves the allocation of treatments to experimental units in such a way that an experimental unit has equal chances of receiving any of the treatments, the personal (human) bias will be minimum. Also, randomization makes the experiment free from any systematic influences of environmental impacts. It helps to make the *experimental errors* (the error caused by the extraneous factors which are beyond the control of the individual approach) independent. This is very important for the analysis of designs. Since all designs are model-based, the mathematical (or statistical) theory underlying necessitates randomization for any tests and inferences to be valid.

**Table 8.21**  Models and their ranking

| Therapist | Models and their ranking | | |
|---|---|---|---|
| | A | B | C |
| 1 | 2 | 3 | 1 |
| 2 | 2 | 3 | 1 |
| 3 | 2 | 3 | 1 |
| 4 | 1 | 3 | 2 |
| 5 | 3 | 2 | 1 |
| 6 | 1 | 2 | 3 |
| 7 | 2 | 3 | 1 |
| 8 | 1 | 3 | 2 |
| 9 | 1 | 3 | 3 |

Let us consider a couple of examples to understand the working principles of experimental designs.

*Example 8.14*  A physical therapist conducted a study to compare three models (A, B, C) of low-volt electrical stimulators. Nine other physical therapists were asked to rank the stimulators in the order of preference. A rank of 1 indicates a first preference, 2 for second preference, and 3 for third preference. The stimulators were given to therapists randomly without actually revealing the manufactures identity. The rankings of three models of low volt electrical stimulators, according to the physical therapists, are shown in Table 8.21.

This is an example of a two-way classified design (randomized design) having two sources of variation to be analyzed. The most important question to answer is whether the three models are equally preferred or not. If not, which model of the electrical stimulator is preferred over the other two? Similarly, it is also interesting to see whether the therapists (provided they are also randomized) have consistently done their ranking objectively or not.

*Solution.* To see the overall significance of the three models ($H_0 : \mu_A = \mu_B = \mu_C$) irrespective of therapists ranking, we can carry out a one-way analysis of variance (ANOVA) and is shown in Table 8.22. In the table, the mean sum of squares (MSS) is obtained by taking the ratio of sum squares to its corresponding degrees of freedom

**Table 8.22**  ANOVA table for overall significance of the model

| Source of variation | SS | df | MSS | F | P-value | F crit |
|---|---|---|---|---|---|---|
| Between groups | 7.407407 | 2 | 3.703704 | 7.692308 | 0.002622 | 3.402826 |
| Within groups | 11.55556 | 24 | 0.481481 | | | |
| Total | 18.96296 | 26 | | | | |

(df). The $F$-value is calculated as the ratio of MSS to the ESS. The sums of squares are computed as follows:

We have the totals as $T_A = 15$, $T_B = 25$, $T_C = 15$, and grand total $= 55$.

$$\text{Between groups sum of squares(TrSS)} = \frac{1}{9}(T_A^2 + T_B^2 + T_C^2) - \frac{55^2}{27}$$
$$= 7.407407$$

$$\text{Total sum of squares(TSS)} = 2^2 + 2^2 + \ldots + 3^2 - \frac{55^2}{27}$$
$$= 18.96296$$

$$\text{Within(error)sum of squares(ESS)} = \text{TSS} - \text{TrSS}$$
$$= 11.55556$$

The analysis clearly says that the model's performance is not the same throughout as the *p-value* is very significant, corresponding to a significance level of $\alpha = 0.05$. On further investigation, it is concluded that the model B is preferred over the other two followed by A and C. Commonly, this can be arrived at by computing *coefficient of variation* (a relative measure of dispersion), which is given by the ratio of the standard deviation to the average of all the three models, and choose the one which offers the least. The other way is to calculate the critical difference between any two treatments and select those with minimum critical-difference.

*Example 8.15*  Here, we consider a situation where randomization is done both at treatment and block-wise. A field experiment on seeds of four different corn (here treatment) is planned in five blocks (here it is the soil type). Each block is divided into four plots, which are then randomly assigned to the four classes. The compiled data is presented in Table 8.23 shows the yield in bushels per acre. The experimenter is interested to see whether the yields vary significantly for the soil quality and the type of corn.

**Table 8.23**  Yield of corn

| Soil types | Type of corn | | | | Total |
|---|---|---|---|---|---|
| | I | II | III | IV | |
| A | 15 | 24 | 12 | 10 | 61 |
| B | 19 | 11 | 15 | 12 | 57 |
| C | 18 | 12 | 14 | 15 | 59 |
| D | 16 | 16 | 11 | 12 | 55 |
| E | 17 | 14 | 16 | 11 | 58 |
| **Total** | 85 | 77 | 68 | 60 | 290 |

*Solution.* We have grand total (G) = 290; therefore, the correction factor $(CF) = \frac{G^2}{20} = 4205$. The sums of squares are:

$$\text{Treatment(Soil)sum of squares} = \text{TrSS} = \frac{1}{4}(T_A^2 + T_B^2 + \ldots + T_E^2) - \text{CF}$$

$$= \frac{1}{4}(61^2 + 57^2 + \ldots + 61^2) - \text{CF}$$

$$= 5$$

$$\text{Block(Corn)sum of squares} = \text{BSS} = \frac{1}{5}(T_I^2 + T_{II}^2 + \ldots + T_{IV}^2) - \text{CF}$$

$$= \frac{1}{5}(85^2 + 77^2 + \ldots + 60^2) - \text{CF}$$

$$= 70.6$$

$$\text{Total sum of squares (TSS)} = 15^2 + 19^2 + \ldots + 11^2 - \text{CF}$$

$$= 219$$

$$\text{Within(error)sum of squares(ESS)} = \text{TSS} - (\text{TrSS} + \text{BSS})$$

$$= 143.4$$

The two-way ANOVA, as per MINITAB software, is shown in Table 8.24. Here, the yields do not vary significantly for the soil quality and corn type, as $F < F_{crit}$ in both soil and corn types. See also *p-value*.

**Example 8.16** This example is based on a Latin square design (LSD), where three sources of variation can be analyzed through a single experiment. An online taxi company wants to consider four types of cars (say: A, B, C, D) for its efficient management of realizing customer demand. This is tested using four-car drivers on four different routes. The efficiency of cars is measured in terms of time in minutes to complete a particular distance. The layout and time consumed by each vehicle for each driver on each route are given in Table 8.25.

*Solution.* First, we obtain the total corresponding to three sources of variation. They are:

**Table 8.24** ANOVA table for yields

| Source of variation | SS | df | MSS | F | P-value | F crit |
|---|---|---|---|---|---|---|
| Soil | 5.0 | 4 | 1.2500 | 0.10 | 0.979 | 3.26 |
| Corn | 70.6 | 3 | 23.5333 | 1.97 | 0.1732 | 3.39 |
| Within groups | 143.4 | 12 | 11.9500 | | | |
| Total | 219.0 | 19 | | | | |

**Table 8.25** Time is taken to cover a particular distance

| Routes | Drivers | | | |
|---|---|---|---|---|
| | 1 | 2 | 3 | 4 |
| 1 | 18(C) | 12(D) | 16(A) | 20(B) |
| 2 | 26(D) | 34(A) | 25(B) | 31(C) |
| 3 | 15(B) | 22(C) | 10(D) | 28(A) |
| 4 | 30(A) | 20(B) | 15(C) | 9(D) |

| Routes | | Drivers | | Car types | |
|---|---|---|---|---|---|
| 1 | 66 | 1 | 89 | A | 108 |
| 2 | 116 | 2 | 88 | B | 80 |
| 3 | 75 | 3 | 66 | C | 86 |
| 4 | 74 | 4 | 88 | D | 57 |

The grand total (G) $= 331$; therefore, the correction factor (CF) $= \frac{G^2}{16} = 6847.563$. The sums of squares are:

Treatment (Routes) sum of squares

$$= \text{TrSS} = \frac{1}{4}(66^2 + 116^2 + \ldots + 74^2) - \text{CF}$$

$$= 380.6875$$

Block (Drivers) sum of squares

$$= \text{BSS} = \frac{1}{4}(89^2 + 88^2 + \ldots + 88^2) - \text{CF}$$

$$= 93.6875$$

Column (Car types) sum of squares

$$= \text{CSS} = \frac{1}{4}(108^2 + 80^2 + \ldots + 57^2) - \text{CF}$$

$$= 329.6875$$

Total sum of squares (TSS)

$$= 18^2 + 26^2 + \ldots + 9^2 - \text{CF}$$

$$= 893.4375$$

Within (error) sum of squares (ESS)

$$= \text{TSS} - (\text{TrSS} + \text{BSS} + \text{CSS})$$

$$= 89.375$$

The analysis of variance of the table of efficiency of cars is shown in Table 8.26.

The ANOVA table shows that the routes and car types are significant, whereas the driver's effect is not very significant. This is evident from the main effect plot as well (see Fig. 8.21). Routes 3 and 4 look similar, and car types B and C have some commonalities. The plot also conveys the fact that driver 3 seems to be apart from the other three guys. A close look at the interaction plots given in Fig. 8.22 conveys the reality that Routes 3 and 4 are significant for the drivers, any type of car on route-1 will take marginally the same time, and on other routes, it can differ, and there is high interaction between driver and car types.

**Example 8.17** A $2^2$ (2-factors used at 2-levels each) factorial experiment was conducted to determine whether the type of glass and phosphor type affected the

**Table 8.26** ANOVA for the efficiency of cars

| Source of variation | SS | df | MSS | F | P-value | F crit |
|---|---|---|---|---|---|---|
| Routes | 380.69 | 3 | 126.90 | 8.52 | 0.014 | 4.76 |
| Drivers | 93.69 | 3 | 31.23 | 2.10 | 0.202 | 4.76 |
| Car types | 329.69 | 3 | 109.90 | 7.38 | 0.019 | 4.76 |
| Within groups | 89.38 | 6 | 14.90 | | | |
| Total | 893.44 | 15 | | | | |

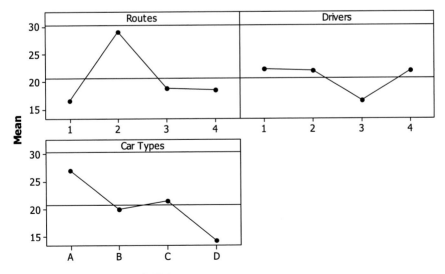

**Fig. 8.21** Main effect plots of efficiency parameters

## Interaction Plot for Efficiency
Data Means

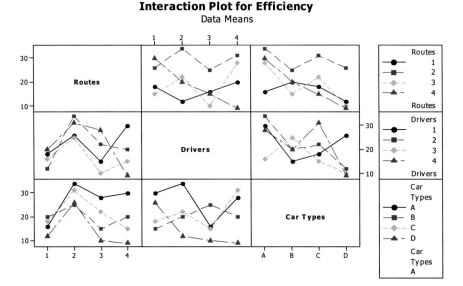

**Fig. 8.22** Interaction plot of efficiency parameters

brightness of a television tube. Table 8.27 presents three response measures each of the current necessary (in microamps) to obtain a specified brightness level.

*Solution.* The analysis of variance of the table of measurement of current is shown in Table 8.28.

The ANOVA table shows that both main effects A and B are significant, whereas there is no interaction effect between the two. This is evident from the main effect plot (Fig. 8.23) and the Pareto chart (see Fig. 8.24).

The quality of any design output is decided based on its performance features. For instance, the picture clarity of a camera, the alignment of wheels, the color density of a television set, the turning radius of an automobile, etc. are evaluated based on its performance as per customer's desires. To create such an output, engineers use engineering principles to combine inputs of materials, parts, components, assemblies, settings, etc. For each information, the engineer identifies parameters

**Table 8.27** Measurement of current (in microamps)

| Glass type | Phosphor type | |
|---|---|---|
| | A | B |
| 1 | 280 | 300 |
| | 290 | 310 |
| | 285 | 295 |
| 2 | 230 | 260 |
| | 235 | 240 |
| | 240 | 235 |

**Table 8.28** ANOVA table of measurement of current

| Source of variation | SS | df | MSS | F | P-value | F crit |
| --- | --- | --- | --- | --- | --- | --- |
| Main effects | 9066.67 | 2 | 453.33 | 64 | 0.000 | 3.55 |
| A | 8533.33 | 1 | 8533.33 | 120.47 | 0.000 | 4.41 |
| B | 533.33 | 1 | 533.33 | 7.53 | 0.025 | 4.41 |
| 2-way interactions | 33.33 | 1 | 33.33 | 0.47 | 0.512 | 4.41 |
| A*B | 33.33 | 1 | 33.33 | 0.47 | 0.512 | 4.41 |
| Residual error | 566.67 | 8 | 70.83 | | | |
| Total | 9666.67 | 11 | | | | |

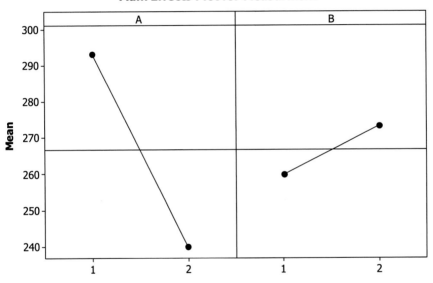

**Fig. 8.23** Main effect plots

and specifies numerical values to achieve the final product's required output. For each parameter, the specifications propose a target (or nominal) value and a tolerance range around the target. In selecting these target values, it is useful to set values so that the product's performance in the field is not affected by variability in manufacturing or field conditions. Such a design is called a *robust design*. Robust designs provide optimum performance simultaneously with variation in manufacturing and field conditions. Genichi Taguchi was the first person to develop a method for determining the optimum target values of product and process parameters that minimizes variation while keeping the mean on target (see Taguchi and Wu (1980), Taguchi (1987, 1991), Montgomery (2003), Muralidharan (2015)).

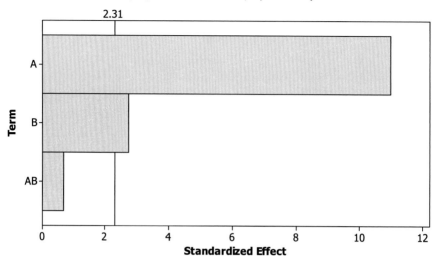

**Fig. 8.24**  Pareto chart of main effects

Although considered to be an expensive statistical tool, the technique of DOE is efficiently implemented in the identification of environmental factors associated with any industrial or organizational processes. To sustain business competitions, many organizations these days follow their customized experimentation method to reduce costs, resources, workforce, and time. It is also found that, for many organizations, the environmental performances are generally assessed based on the energy consumption and its impact on climate change issues, which may not require any kind of experimentation in real sense. But experimentation provides a basis for sustainable practices and user-friendly approaches. Some more performance-oriented quality monitoring techniques are discussed in the following sections.

## 8.4  Quality Function Deployment

At various places in this book, we have pointed out the importance of quality function deployment (QFD) as a useful lean management tool. QFD is a systematic way of documenting and breaking down customer needs into manageable and actionable detail. It is a planning methodology that organizes relevant information to facilitate better decision making and a way of reducing the uncertainty involved in product and process design. It promotes cross-functional teamwork and gets the right people together, early, to work efficiently and effectively to meet customers' needs. QFD is a structured methodology to identify and translate customer needs and wants into technical requirements and measurable features and characteristics (Rampersad and El-Homsi (2008), Muralidharan (2015)).

According to the Toyota Production System (TPS), for the development of new products or services, "built-in" tools like lean or Six Sigma must have experimented on a routine basis. This will ensure greater participation of the committed individuals on the cross-functional and multidisciplined development team works for the entire quality management of the organization. However, the DMAIC methodology of Six Sigma philosophy should be implemented cautiously with excellent expertise of quality professionals without compromising the quality essentials underlying the process. Otherwise, there is all chance for the derailment of the quality initiative. QFD is one such tool, where good subject knowledge and management expertise are jointly producing customer specifications into technical requirements. This is a powerful tool to capture *customer* voice (VOC) and translate them into critical product and process aspects at an early stage. In this approach, customer wishes are addressed with the help of matrices that use detailed technical parameters and project objectives (Rampersad and El-Homsi, 2008).

The basic building block of QFD is the house of quality (Sanchez et al. (1993), ReVelle et al. (1998), Pande et al. (2003), Muir (2006), Gryna et al. (2007), Kubiak and Benbow (2010), and Muralidharan (2015)). It is represented in a multidimensional matrix called *L*-matrix and shows the correlation between "What's" and "How's" of the process stages. Figure 8.28 shows the structure of the house of quality. A full QFD product design project will involve a series of these matrices, translating from customer and competitive needs down to detailed process specifications. It helps to identify critical quality characteristics. The characteristics critical to quality (CTQ) are nothing but a product feature or process step that must be controlled to guarantee that you deliver what the customer wants.

As seen in Fig. 8.25, the QFD matrix consists of several parts. The matrix is formed by first filling in the customer demands (zone 1), developed from the analysis of VOC. The technical requirements are established in response to the customer requirements and placed in zone 4. The relationship (interaction) zone 5 displays the connection between the technical specifications and customer requirements. The comparison between the competitors for the customer requirement is shown in zone 2, and zone 3 provides an index to documentation concerning improvement activities, also called actionable items. Zone 7, like zone 2, plots the comparison with the competition for the technical requirements. Zone 8 lists the target values for the technical specifications. Zone 6 shows the correlations (degree of relationship) between the technical specifications. A positive correlation indicates that both technical specifications can be improved at the same time. A negative correlation suggests that improving one of the requirements will worsen the other.

The step used for constructing the first quality house, which is aimed at product development, according to Rampersad and El-Homsi (2008) is:

1. Define who the customer is, make an inventory of customer wishes, and measure the importance as per the priority of these wishes with the help of weighing scores.

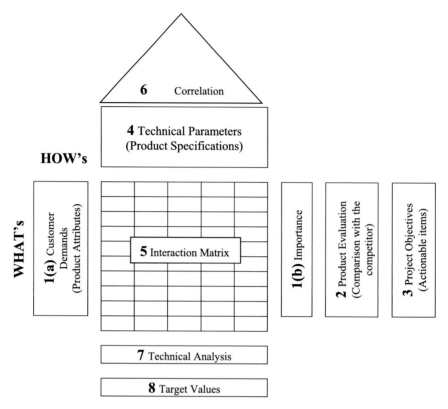

**Fig. 8.25** Structure of house of quality

2.  Compare your product performance to that of competitive products. Evaluate your product and identify the strength and weaknesses according to the customer.
3.  Identify and quantify the improvement objectives. Determine which customer wishes need to be improved about the competitive product and indicate this in a score.
4.  Translate customer wishes into quantifiable technical parameters that are product specifications. State how client wishes can be used to your advantage. Examples of technical parameters are dimensions, weight, number of parts, energy use, and capacity, and so on.
5.  Investigate the relationship between customer wishes and technical parameters. Indicate the same in the matrix to what extent technical parameters influence customer wishes, and indicate this relationship in a score.
6.  Identify the interactions between the individual technical parameters. Make the relationship between these parameters explicit in the roof of the quality house.
7.  Record the unit of measurement for all technical parameters. Express these parameters in measurable data.

8.  Determine the target values of the new product design or indicate the proposed improvements of the technical parameters.

As discussed, the objectives of QFD are to allow the VOC to be heard more frequently and clearly in the development process of new products and related processes and to comply with the *"do it right the first time"* principle (Rampersad and El-Homsi, 2008). All the customer wishes are established employing brainstorming sessions, where the team members should include a team leader, Six Sigma Black Belts, product/process engineers, market engineer, customer facilitator, few customers (preferable), and other executives as per the hierarchy of the project. Since most of the technical specifications and customer choices are expressed in qualitative characteristics (or attributes), it is better to use suitable scaling techniques to rank the specifications. Preferably a five-point scale (say: 5 = Very important, 4 = important, 3 = Less important, 2 = not so important, 1 = not important) may be used. One can also use a ten-point scale to describe the qualitative variable. In this case, the customer is free to express his wishes in some numbers explicitly.

As per the terminology used earlier, the *y*-variable or output variable is realized through the "WHATs" or by asking what the customer wants. So the use of *Voice of Customer* (VOC) is crucial here. This is further realized by comparing competitors' products and the advantage it is going to give. The rankings of these wishes are then done by identifying the functions or processes that impact customer wants. The strength of the interrelation between the WHATs and the HOWs is usually found using the correlation technique. Here also some suitable ranking procedures can be used. For instance: 5 = very strong, 4 = moderately strong, 3 = strong, 2 = weak, 1 = very weak, etc.

A partially filled house of quality is shown in Fig. 8.26. Note that the house of quality matrix constitutes an assemblage of the results of the benchmarking and cause and effect matrices methods together with additional information. Therefore, information is included on what makes customers happy and on measurable quantities relevant to engineering and profit maximization (Allen, 2008).

This way, a series of interlocking "house of quality" matrices that translates customer needs into product and process characteristics will be constructed (see Fig. 8.27). The completed matrix can provide a database for product development, serve as a basis for planning product or process improvements, and suggest opportunities for new or revised products or process introductions (ReVelle et al., 1998). If a matrix has more than 25 customer voice lines, it tends to become unmanageable. In such a situation, a convergent tool such as the *affinity diagram* and *Kano model* may be used to condense the list.

An affinity diagram is a tool to group a large amount of similar or related ideas generated through brainstorming. This is considered to be one of the seven Japanese management and planning tools. The Kano model introduced by Noriaki Kano (1996) is a model of customer satisfaction, which has been used in manufacturing, as well as service industries to identify the basic (or must be), expected, and attractive or unexpected need of customer or client (Rampersad and El-Homsi (2008). Whether

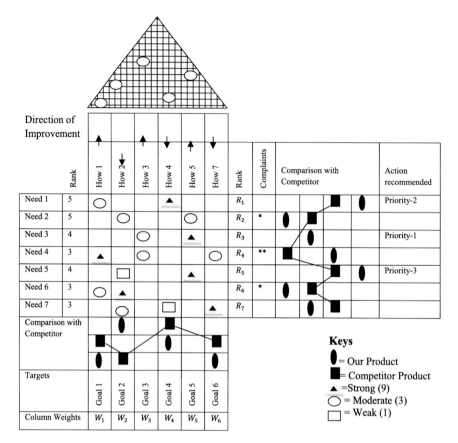

**Fig. 8.26**  A QFD matrix

these needs are fulfilled or not will determine whether the customers are satisfied, dissatisfied, neutral, or delighted (Kano and Gitlow, 1986).

Thus, QFD is a valuable decision support method to understand customer needs and competitors' strengths in one stretch. It helps identify the most correlated link between customer demands and technical requirements in coordination with the voice of the customer's process and voice. Hence, QFD facilitates the design–redesign effort to a new level through each stage of product development.

## 8.4.1   Dashboards

The monitoring of organizational functions can be effectively done using a dashboard. It works as a control panel for all activities in a nutshell. A *dashboard* is an

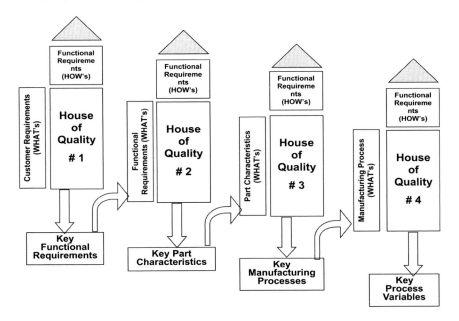

**Fig. 8.27** House of quality: relationship building

easy to read, often single page, real-time user interface, showing a graphical presentation of the current status (snapshot) and historical trends of an organization's key performance indicators to enable instantaneous and informed decisions to be made at a glance.

Management is about efficiency. Doing the right things at the right times. For organizational excellence, we need a supportive corporate environment, result-oriented programs, and ethical leadership. Successful design and implementation would lead to the creation of a fully functioning and active organization capable of a dynamic balancing among conflicting concerns, challenges, and paradoxical demands. A dashboard facilitates state-of-the-art performance measures related to process variables and their relevance. Dashboards typically are limited to show summaries, key trends, comparisons, and exceptions (Muralidharan, 2015). The summaries could be anything like the status of employees, types of machinery available, SOPs, ongoing projects, and training, etc. The vision and mission statements could also be a part of dashboards.

There are many types of dashboards. Among all, digital dashboards dominate the business environment. Digital dashboards may be laid out to track the flows inherent in the business processes that they monitor. Digital dashboard projects involve business units as the driver and the information technology department as the enabler. The success of digital dashboard projects often depends on the metrics that were chosen for monitoring. Key performance indicators, balanced scorecards, and sales performance figures are some of the content appropriate on business dashboards.

Benefits of using digital dashboards include

- Visual presentation of performance measures
- Ability to identify and correct negative trends
- Measure efficiencies/inefficiencies of process/product/services/transactions, etc.
- Ability to generate detailed reports showing new trends
- Ability to make more informed decisions based on collected business intelligence
- Align strategies and organizational goals
- Saves time compared to running multiple reports
- Gain total visibility of all systems instantly
- Quick identification of data outliers and their associations.

The outliers mentioned in the last bullet point can be unique to the organization sometimes. So highlighting the same will add value to the organization and can increase brand value and image. A customer who visits the premises will immediately develop an emotional bonding with the organization, an add-on for future performances and reputation.

## 8.4.2   Prioritization Method

An organization works on many competence areas, where quality criteria may often be weighted and ranked. The systematic approach of identifying, weighing, and applying rules to the options is called *prioritization*. Usually, this job is entrusted to the quality department, who can objectively prioritize the core areas for improvement. A measurement-based study fully supported by prioritization can, through light on the significance of each variable, will be the part of the analysis. The prioritization matrix is used when; there are (i) too many variables that might have an impact on the output of the process, (ii) collecting data about all possible variables would cost too much time and money, and (iii) team members have different theories and perceptions about what happens in the process.

Consider an example to illustrate the prioritization method on the academic performance of students of a university. The first step toward preparing a prioritization matrix is to list all the input variables (say, $x$) and output variables (say, $y$) associated with the study. For instance, the standard inputs will be the amount of time invested for education ($x_1$), the number of lectures attended ($x_2$) attended, hours of sleep ($x_3$), the ambiance at home ($x_4$), the ambiance at the campus ($x_5$), quality of food intake ($x_6$), and so on. The standard outputs will be anything like grade obtained in the examination ($y_1$), distinction got ($y_2$), promotion to higher studies ($y_3$), and leadership traits acquired ($y_4$), value-added services received ($y_5$), and so on. The identification of $x$'s and $y$'s can be made through some brainstorming sessions or some opinion-sharing process.

The next step is to rank the importance of all these output variables. We know that all output variables may not be equally important, and hence, they may have varying significance. Accordingly, the output variables will be weighted. This is what is shown in Table 8.29. To understand the strength of the relationship between output

**Table 8.29** A prioritization matrix

| Output variables | | $y_1$ | $y_2$ | $y_3$ | $y_4$ | Total |
|---|---|---|---|---|---|---|
| Weight | | 8 | 8 | 7 | 5 | |
| Process input variables | $x_1$ | 2 | 4 | 5 | 2 | 93 |
| | $x_2$ | 9 | 8 | 2 | 5 | 175 |
| | $x_3$ | 5 | 5 | 2 | 1 | 99 |
| | $x_4$ | 5 | 5 | 4 | 2 | 128 |
| | $x_5$ | 8 | 7 | 7 | 2 | 179 |
| | $x_6$ | 5 | 5 | 5 | 3 | 130 |
| | ... | ... | ... | ... | ... | ... |
| | ... | ... | ... | ... | ... | ... |
| | ... | ... | ... | ... | ... | ... |

and input/process variables, we cross-multiply the weight and factor scores and place them in the total column. The factor scores of each input variable are decided in association with the output variables, and the weightage of each output variable is agreed through a brainstorming section. The weightage can also be calculated in terms of its probability of occurrence and its significance in comparison with the input variables. In this process, we can identify those critical quality characteristics affecting the performance in the examination and evaluation procedure. As per the table, and based on the proposed evaluation method, the critical-to-quality variables for poor performance may be the ambiance at the campus ($x_5$) and the number of lectures (or sessions) attended ($x_2$) in that order.

There are plenty of ways to attempt the exercise of prioritizing the variables. We know that the quality of a process can be improved if its input variables are improved. Therefore, the idea here is to prioritize the significance of input variables connected with the output requirement. At the same time, one can also prioritize the importance of output variables that need maximum attention. Preparing a prioritization matrix is very common during the DOE, root cause analysis, and failure mode effect analysis. Interestingly, all these techniques are part of lean manufacturing and Six Sigma project management. Figure 8.28 shows the supplier scorecard prepared through a prioritization exercise for selecting vendors of particular business activity. The vendor has a high score will be given preference for order and sales.

## 8.4.3 Root Cause Analysis

According to Ishikawa, "Conclusions based on facts and data are necessary for any improvement." So developing performance metrics and track them every day to monitor for quality are the best thing one should do. To identify those metrics, one should get into the root cause of the problem. The highest-level cause of a problem is

| LGC Criteria | Weight | Supplier/Vendor | | | | | |
|---|---|---|---|---|---|---|---|
| | | A | B | C | D | E | E |
| Brand | 2 | 5 | 8 | 5 | 7 | 7 | 6 |
| Years in Service | 1 | 7 | 7 | 6 | 6 | 5 | 5 |
| Quality history | 4 | 5 | 9 | 10 | 9 | 9 | 6 |
| Material cost | 3 | 8 | 8 | 8 | 7 | 8 | 6 |
| Timeliness of delivery | 5 | 8 | 8 | 8 | 9 | 10 | 7 |
| Technical specifications | 3 | 7 | 7 | 7 | 7 | 7 | 7 |
| Design and Features | 3 | 9 | 10 | 10 | 8 | 8 | 9 |
| Energy specifications | 2 | 8 | 4 | 6 | 7 | 8 | 6 |
| Capacity & load features | 2 | 8 | 9 | 8 | 6 | 6 | 5 |
| Commitment for maintenance | 4 | 6 | 6 | 8 | 8 | 9 | 5 |
| Warranty & Specifications | 4 | 8 | 6 | 9 | 10 | 6 | 7 |
| **Total Score** | | **237** | **248** | **273** | **265** | **262** | **212** |

**Fig. 8.28**   Vendor selection method

called the root cause. *Root cause analysis* (RCA) is the process of identifying causal factors using a structured approach with techniques designed to provide a focus for identifying and resolving problems (risks).

Problem (or failure or variation) is a part of any process. A problem can have one or more causes, and all the causes may not be root causes. A problem can be due to mechanical, technical, managerial, material, service-related, product-related, customer-related, and personal-related, failure-related, and so on. Hence, causes and its method of identification are context-based and cannot be due to the same nature and identical behavior. So root causes explain the way we need to look at the problem and its solution.

Interestingly, many quality tools discussed in this chapter become a part of the root cause analysis. Apart from those tools, process flow diagram, 5-Why's technique, fault tree analysis (FTA), FMEA, reality charting, etc. are also used as supporting tools for RCA. Five whys (*5-Why*) techniques are a technique designed to examine the root cause of a problem at different granularity levels. It involves asking the simple question "*why*" five times in succession. This is done to drill through the layers of cause and effect to arrive at the root cause. It consists of looking at an undesirable result and asking, why did this occur? When the question is answered, the next question asked is how, when, what, where, and so on in an arbitrary way. This process will finally lead to the actual cause of the problem. The following quote explains this well: "*If you don't ask the right questions, you don't get the right answers. A question asked in the right way often points to its answer. Asking questions is the ABC of diagnosis. Only the inquiring mind solves problems.*"—Edward Hodnett.

Let us find out the root cause of the problem of an abrupt halt of air conditioner used in the power generators assembly:

- Why did the air conditioner stop working?

- – Because there was a high-voltage problem.
- • Why did this happen?
  - – Because there was a sudden stoppage of the transformer.
- • Why did this happen?
  - – Because the fuse wire used was broken.
- • Why did this happen?
  - – Because poorly welded coils used for fuse connection.
- • Why did this happen?
  - – Because the service provider delivered poorly welded coils for the fuse.

There doesn't need to be only a single answer for each "why." There can be multiple answers which cause a particular quality characteristic. For instance, suppose a marketing manager is interested in knowing the poor marketing effort of a product. As per a serious brainstorming discussion with the executives, it was found that the problem is due to three dimensions of quality, viz. price, promotion, and product quality itself. Now each aspect is causing some problem with the acceptability of the product, and they, in turn, associate with each other. This situation can be represented in the following 5-Why diagram (see Fig. 8.29).

The first thing toward finding the root cause is the identification of a clearly stated problem. If the problem statement itself is not clear and straightforward, then finding the root causes will be difficult and can lead to misleading results. The problem statement should be a concise and focused description of "what's wrong"—either the pain arising from the problem or the opportunity that needs to be addressed. A precise problem statement should include characteristics like what is the problem? When did it happen? Where did it happen? How were the overall goals impacted? The RCA begins with an impact on the overall goals, and it answers the "WHY" by addressing each following question explicitly. A problem statement generally serves the purposes like:

- • Validating the project rationale
- • Consensus formation of project teams
- • Assessing the clarity of the supporting data
- • Establish a baseline measure against which progress and results can be tracked.

As a course of corrective or preventive action, the quality professionals involved with the RCA should anticipate probable causes in advance for enabling the deployment of quality. A re-exercise of quality tools and reconfirmation of performance index in tandem with available resources at disposal should be practiced frequently for any changes or monitoring. For this the following checklists may also be used:

- • Have you drawn the process flow, FMEA, stratification groups, C/E diagrams and identified all sources of variation?
- • Have all sources of information been used to define the cause of the problem?

| Why? | Why? | Why? | Why? | Why? |
|---|---|---|---|---|
| Poor customer acceptance | Price | The high cost of materials | Less material availability | Materials outdated |
| | | | | … |
| | | | Fewer vendors | Local vendors not interested |
| | | | | … |
| | | | High GST and taxes | New Tax system |
| | | High labor cost | Labors imported | Local labors not interested |
| | | | Labors amenities | No local hospitalities |
| | | | Skilled labors | Experts not available |
| | | | | … |
| | | Competitor price equal | Established competitor | Established brand |
| | | | New technology used | Automated machines in use |
| | | | | Technocrats available |
| | | | … | … |
| | Promotion | Brand image | New player | … |
| | | | Poor advertisements | Poor  logistics |
| | | | … | |
| | | Low awareness | Poor coordination | Unexpected changes |
| | | | Customer base unknown | VOC not performed |
| | | | … | … |
| | | Time management poor | Ground reality unknown | VOP not performed |
| | | | Wrong time of launching | Unexpected shutdown |
| | Product | Poor quality | Poor materials | Poor selection of vendors |
| | | | Poor office keeping | 5S not practiced |
| | | | No TQM practices | Kaizen not practiced |
| | | Poor packaging | Untrained and unskilled labors | No random checking, No Acceptance Sampling |
| | | Poor design | No GAGE R&R practiced | No measurement plan followed. |
| | | … | … | … |

**Fig. 8.29**  5-Why drilled down

- Do you have the physical evidence of the problem, if possible?
- Can you establish a relationship between the problem and the process?
- Is this a unique situation, or is the likely problem similar to experience?
- Has a comparative analysis been completed to determine if the same or similar problem existed in related products?
- What are the experiences of recent actions that may be related to this problem?
- Why might this have occurred?
- Why haven't we experienced this before?
- What changed finally?

## 8.5  Exercises

8.1.  What are the eight dimensions of quality?
8.2.  Distinguish between quality assurance and quality council.
8.3.  What is statistical process control? How does it differ from statistical quality control?
8.4.  What are the significant sources of variation encountered in the manufacturing process?
8.5.  Distinguish between specification limits, tolerance limits, and control limits
8.6.  Describe the procedure of constructing a control chart.
8.7.  The data below are $\overline{X}$ and R values for 24 samples of size n = 5 taken from a process producing bearings. The measurements are made on the inside diameter (in mm) of the bearing, with only deviations from 0.50 in multiples of 10,000 are recorded.

| Sample no. | $\overline{X}$ | R | Sample no. | $\overline{X}$ | R |
|---|---|---|---|---|---|
| 1 | 34.5 | 3 | 13 | 35.4 | 8 |
| 2 | 34.2 | 4 | 14 | 34.0 | 6 |
| 3 | 31.6 | 4 | 15 | 37.1 | 5 |
| 4 | 31.5 | 4 | 16 | 34.9 | 7 |
| 5 | 35.0 | 5 | 17 | 33.5 | 4 |
| 6 | 34.1 | 6 | 18 | 31.7 | 3 |
| 7 | 32.6 | 4 | 19 | 34.0 | 8 |
| 8 | 33.8 | 3 | 20 | 35.1 | 4 |
| 9 | 34.8 | 7 | 21 | 33.7 | 2 |
| 10 | 33.6 | 8 | 22 | 32.8 | 1 |
| 11 | 31.9 | 3 | 23 | 33.5 | 3 |
| 12 | 38.6 | 9 | 24 | 34.2 | 2 |

(1)  Set up $\overline{X}$ and R-charts on this process. Does the process seem to be in statistical control? If necessary, revise the trail control limits. Estimate process means and process standard deviation.

(2)  If specifications on this diameter are $0.5030 \pm 0.0010$, find the percentage of non-conforming bearings produced by this process. Assume that the diameter is normally distributed.

8.8.  Discuss in brief the control chart for attributes. Propose the control limits for p, np, u, and c-charts.

8.9.  A personal computer manufacturer wishes to establish a control chart for non-conformities per unit on the final assembly line. The sample size is selected as five computers. Data on the number of non-conformities in 20 samples of 5 computers each are shown in the table.

| Sample No. | Total number of non-conformities | Sample No. | Total number of non-conformities |
|---|---|---|---|
| 1 | 10 | 11 | 9 |
| 2 | 12 | 12 | 5 |
| 3 | 8 | 13 | 7 |
| 4 | 14 | 14 | 11 |
| 5 | 10 | 15 | 12 |
| 6 | 16 | 16 | 6 |
| 7 | 11 | 17 | 8 |
| 8 | 7 | 18 | 10 |
| 9 | 10 | 19 | 7 |
| 10 | 15 | 20 | 5 |

Set up a control chart for non-conformities per unit. Does the process in statistical control?

8.10.   A manufacturer of paper used in copy machines and laser printers monitors various aspects of its production using control charts. Paper is produced in large rolls 10 ft long and 5 ft. Inch in diameter. Each shift, a sample is taken from each completed roll and is checked in the testing laboratory for non-conformities such as particles of dirt and discoloration. All these non-conformities are given the same weight in forms of importance. Below are data for 24 consecutive production shifts.

| Shift | Total no. of non-conformities | No. of rolls produced | Shift | Total no. of non-conformities | No. of rolls produced |
|---|---|---|---|---|---|
| 1 | 52 | 10 | 13 | 36 | 6 |
| 2 | 59 | 10 | 14 | 39 | 8 |
| 3 | 44 | 7 | 15 | 63 | 9 |
| 4 | 50 | 7 | 16 | 41 | 7 |
| 5 | 35 | 8 | 17 | 50 | 10 |
| 6 | 49 | 10 | 18 | 43 | 7 |
| 7 | 60 | 9 | 19 | 42 | 9 |
| 8 | 55 | 9 | 20 | 32 | 7 |
| 9 | 48 | 9 | 21 | 41 | 7 |
| 10 | 56 | 10 | 22 | 46 | 10 |
| 11 | 57 | 9 | 23 | 48 | 20 |
| 12 | 46 | 8 | 24 | 34 | 6 |

(1)   Define an inspection unit as one roll of paper. Construct u-chart with exact limits for the data. Does the process appear to be in control?

(2) Based on the results of part(a), give the specific details of the design of control limits for monitoring future data on total no. of non-conformities.
(3) Construct u-chart based on average inspection unit size.
(4) Construct a standard u-chart for the data. Applying all the supplementary runs rules to see if there is any evidence of unusual runs of the data.

8.11. The following table shows the total annual sales of scooters at a local store. Draw a time series plot. Estimate the linear trend for the sales data, find R-square, and state the goodness of the fit. Forecast sales for the years 2014 and 2015.

| Year | 2004 | 2005 | 2006 | 2007 | 2008 | 2009 | 2010 | 2011 | 2012 | 2013 |
|------|------|------|------|------|------|------|------|------|------|------|
| Sale | 814 | 818 | 869 | 895 | 908 | 956 | 974 | 996 | 1025 | 1100 |

8.12. What are the uses of the Pareto chart?
8.13. The following defects are noted for respirator masks inspected during a given period:

| Dent | Pinhole | Strap | Dent | Pinhole |
|------|---------|-------|------|---------|
| Pinhole | Discoloration | Dent | Discoloration | Strap |
| Strap | Dent | Strap | Strap | Pinhole |
| Strap | Pinhole | Discoloration | Dent | Discoloration |
| Discoloration | Strap | Discoloration | Dent | Pinhole |
| Dent | Pinhole | Discoloration | Pinhole | Discoloration |
| Dent | Discoloration | Strap | Pinhole | Dent |
| Discoloration | Strap | Dent | Discoloration | Strap |

Prepare a Pareto chart and conclude.
8.14. Distinguish between correlation and regression. State the purpose of each.
8.15. The data for the consumption (per month) of petrol of a medium-class family and the price of petrol (Rupees/liter) were found as per the following.

| Consumption (in liter) | 65 | 62 | 56 | 52 | 47 | 42 | 40 | 37 | 33 | 30 |
|------------------------|-----|-----|-----|-----|-----|-----|-----|-----|-----|-----|
| Price of petrol ('/liter) | 40 | 45 | 50 | 54 | 57 | 60 | 62 | 65 | 68 | 70 |

(1) Identify the independent and dependant variable.
(2) Draw a scatter diagram and suggest appropriate conclusion
(3) Fit a suitable regression line
(4) Test the significance of the regression coefficient
(5) Compute correlation coefficient using the Pearson's formula.
(6) Compute correlation coefficient using the Spearman's formula

8.16.   The following table shows the percentage marks of 25 students of BCA class secured in Semester 1 through Semester 4 and final examinations.

| Student no. | Sem. 1 ($X_1$) | Sem.2 ($X_2$) | Sem. 3 ($X_3$) | Sem. 4 ($Y$) |
|---|---|---|---|---|
| 1 | 72 | 78 | 76 | 76 |
| 2 | 91 | 84 | 90 | 93 |
| 3 | 88 | 91 | 90 | 90 |
| 4 | 94 | 90 | 90 | 93 |
| 5 | 70 | 62 | 70 | 72 |
| 6 | 55 | 49 | 56 | 52 |
| 7 | 68 | 73 | 76 | 74 |
| 8 | 48 | 57 | 62 | 58 |
| 9 | 85 | 78 | 88 | 83 |
| 10 | 74 | 70 | 85 | 82 |
| 11 | 67 | 72 | 77 | 70 |
| 12 | 71 | 61 | 72 | 71 |
| 13 | 92 | 92 | 92 | 92 |
| 14 | 76 | 82 | 77 | 76 |
| 15 | 72 | 76 | 78 | 79 |
| 16 | 91 | 89 | 90 | 91 |
| 17 | 75 | 74 | 70 | 74 |
| 18 | 80 | 88 | 86 | 87 |
| 19 | 85 | 93 | 88 | 88 |
| 20 | 78 | 83 | 77 | 59 |
| 21 | 82 | 86 | 90 | 77 |
| 22 | 86 | 82 | 89 | 75 |
| 23 | 78 | 83 | 85 | 75 |
| 24 | 76 | 83 | 71 | 49 |
| 25 | 96 | 93 | 95 | 92 |

(1)   Estimate the impact of all semester scores on end-semester exam scores.
(2)   Compute correlation matrix
(3)   Check whether there is any impact of multicollinearity.

8.17.   Discuss the significance of the cause and effect diagram in identifying the root causes.

8.18.   What is the testing of the hypothesis (TOH)? Write down the steps involved in TOH.

8.19.   Define the following terms:

(1)   Significance level

      (2)    Power of a test
      (3)    Distribution-free tests
      (4)    Composite hypothesis

8.20.    A fast-food product producer claims that the percentage of fat in its cheese-burger is not more than 15%. The food inspector took 40 specimens of the cheeseburger and found the average fat percentage of 15.5% with a standard deviation of 5%. Assuming the percentage fat as a normal random variable test the validity of the producer's claim at a 5% level of significance.

8.21.    A maternity nursing home observes that the average newborn baby will be about 20 in. long. On a survey, it is believed that the newborn babies are longer than normal for a particular community. The heights of 20 newborn babies born to this particular community woman are recorded as

| 23 | 15 | 18 | 21 | 25 | 21 | 25 | 19 | 24 | 23 |
|----|----|----|----|----|----|----|----|----|----|
| 23 | 19 | 20 | 18 | 22 | 21 | 20 | 15 | 22 | 25 |

Do the data agree with the belief about the community at a significance level 0.025?

8.22.    In a city, there are two sources of water through which the residents get water at home. The alkaline level in water in both sources is normally distributed with the same mean pH but with standard deviations of 0.5 pH in source one and of 1 pH in source 2. Municipal engineer periodically checks the pH level of samples from both the sources. The engineer recently took ten random samples from sources 1 and 15 random samples from sources 2. The readings of alkaline pH are shown in the following:

| Source 1: | 9.0 | 9.7 | 9.1 | 9.4 | 8.5 | 9.3 | 9.6 | 10.0 | 8.7 | 8.8 | | | | | |
|-----------|-----|-----|-----|-----|------|-----|------|------|-----|------|-----|------|------|-----|-----|
| Source 2: | 8.1 | 9.1 | 9.3 | 8.4 | 10.2 | 9.4 | 10.0 | 7.8 | 8.9 | 10.7 | 10.8 | 10.9 | 9.8 | 8.7 | 8.0 |

      (1)    Does the data indicate that the alkaline level in water in both sources have no same mean pH? Test using $\alpha = 0.01$.
      (2)    Suppose the population standard deviations 0.5 and 1 are not known, how will you carry out the test assuming unequal population variances? What would be the result of the test?

8.23.    Suppose each of the three different medicines was given to randomly selected eight patients with identical conditions. Total cholesterol was recorded after three months, and the results were as follows.

| Medicine 1 | 221 | 209 | 182 | 230 | 188 | 232 | 203 | 206 |
| Medicine 2 | 202 | 209 | 180 | 196 | 217 | 185 | 220 | 192 |
| Medicine 3 | 228 | 208 | 202 | 181 | 194 | 214 | 198 | 217 |

Is the sample data evidence that the average effect of all three medicines is the same?

Write the necessary assumptions required for your test. Assume a 5% level of significance.

8.24.   The following data shows the number of calls received at the control room in pick hours 10 am to 2 pm. Test whether the sample data have come from the exponential population with the mean number of calls 80. Use a 5% level of significance for the test. Calculate p-value.

| No. of calls | 0–100 | 100–200 | 200–300 | 300–400 | 400–500 |
| Frequency | 360 | 100 | 27 | 10 | 3 |

8.25.   What are the principles of experimentation? State their importance in designing an experiment.

8.26.   Describe in brief the uses of various designs used for experimentation.

8.27.   Write short notes on quality function deployment.

8.28.   Write down the practical significance of root cause analysis.

8.29.   Ten subjects with exercise-induced asthma participated in an experiment to compare the protective effect of a drug administered in four dose levels. Saline was used as a control. The variable of interest changed in FEV1 after the administration of the drug or saline. The results were as follows.

| Subject | Saline | The dose level of the drug (mg/ml) | | | |
| | | 2 | 10 | 20 | 40 |
| 1 | −0.68 | −0.32 | −0.14 | −0.21 | −0.32 |
| 2 | −1.55 | −0.56 | −0.31 | −0.21 | −0.16 |
| 3 | −1.41 | −0.28 | −0.11 | −0.08 | −0.83 |
| 4 | −0.76 | −0.56 | −0.24 | −0.41 | −0.08 |
| 5 | −0.48 | −0.25 | −0.17 | −0.04 | −0.18 |
| 6 | −3.12 | −1.99 | −1.22 | −0.55 | −0.75 |
| 7 | −1.166 | −0.88 | −0.87 | −0.54 | −0.84 |
| 8 | −1.15 | −0.31 | −0.18 | −0.07 | −0.09 |
| 9 | −0.78 | −0.24 | −0.39 | −0.11 | −0.51 |
| 10 | −2.12 | −0.35 | −0.28 | +0.11 | −0.41 |

Can one conclude based on these data that different dose levels have different effects?

8.30. The sample data in the following Latin square are the scores obtained by nine college students of various ethnic backgrounds and different professional interests in an American history test.

| Professional interest | Ethnic background | | |
|---|---|---|---|
| | Mexican | German | Polish |
| Law | A<br>75 | B<br>86 | C<br>86 |
| Medicine | B<br>95 | A<br>79 | B<br>86 |
| Engineering | C<br>70 | A<br>83 | B<br>93 |

In this table, A, B, and C are the three instructors by whom the college students were taught. Identify the design and analyze it at a 5% level of significance.

# References

Allen, J. P. (2008). The attachment system in adolescence. In J. Cassidy & P. R. Shaver (Eds.), *Handbook of attachment: Theory, research, and clinical applications* (pp. 419–435). The Guilford Press.

Bower, K. M. (2000). *Using Exponentially Weighted Moving Average.* Asia Pacific Process Engineer: EWMA) Charts.

Champ, C. W., & Woodall, W. H. (1987). Exact results for shewhart control charts with supplementary runs rules, *Technometrics 29.*

Chakraborti, S., Van der Laan, P., & Bakir, S. T. (2001). Nonparametric Control Charts: An Overview and Some Results. *Journal of Quality Technology, 33,* 304–315.

Chakraborti, S., & Graham, M. A. (2007). Nonparametric control charts. *Encyclopedia of Statistics in Quality and Reliability, 1,* 415–429. Wiley.

Chakraborti, S, Human, S. W., & Graham, M. A. (2011). *Nonparametric (Distribution-free) quality control charts.* In: N. Balakrishnan (ed.) Handbook of methods and applications of statistics: Engineering, quality control, and physical sciences (pp. 298–329). Wiley.

Chandra, M. J. (2016). *Statistical quality control,* CRC Press.

Garvin, D. A. (1987). Competing on the eight dimensions of quality. *Harvard Business Review, 65,* 101–109.

Gryna, F. M., Chua, R. C. H., & Defeo, J. (2007). *Juran's quality planning and analysis for enterprise quality.*

Kano, N. (1996). *Business strategies for the 21st century and attractive quality creation* (p. 105). ICQ.

Kano, N., & Gitlow, H. (1986). *The kano program,* Notes from seminar on May 4–5, University of Miami.

Kubiak, T. M., & Benbow, D. W. (2010). *The certified six sigma black belt handbook* (2nd ed.), Dorling Kindersley (India) Pvt. Ltd.

Lehman, E. L. (1998). *Testing of statistical hypothesis,* Springer

Montgomery, D. C. (2003). *Introduction to statistical quality control* (4th ed.). Wiley.

Muralidharan, K. (2015). *Six sigma for organizational excellence: A statistical approach*, Springer Nature.

Muralidharan, K., & Syamsunder, A. (2012). *Statistical methods for quality, reliability and maintainability*. PHI India Ltd.

Muir, A. (2006). *Lean six sigma way*. McGraw Hill.

Oakland, J. S. (2005). *Statistical process control*, Elsevier.

Page, E.S. (1954). Continuous Inspection Schemes, *Biometrika*, 41.

Pande, P. S. Newuman, R. P., & Cavanagh, R. R. (2003). *The six sigma way*. Tata McGraw Hill

Rampersad, H. K., & El-Homsi, A. (2008). *TPS-lean six sigma: Linking human capital to lean six sigma*, SARA Book Pvt. Ltd.

ReVelle, J. B., Moran, J. W., & Cox, C. A. (1998). The QFD Handbook, John Wiley & Sons.

Sanchez, S. M., Ramberg, J. S., Fiero, J., & Pignatiello, Jr., J. J. (1993). Quality by Design, Chapter 10, in: A. Kusiak (ed.) *Concurrent engineering: Automation, tools, and techniques*. Wiley.

Taguchi, G., & Wu, Y. (1980). *Introduction to off-line quality control*, Japan Quality Control The organization.

Taguchi, G. (1987). A system for experimental design, UNIPUB.

Taguchi, G. (1991). *Taguchi methods: Research and development*. In: S. Konishi (ed.) Quality engineering series, Vol. 1. The American Supplier Institute.

Wardsworth, H. M., Stephens, K. S., & Godfrey, A. B. (2001). *Modern methods for quality control and improvement*. Wiley.

Wardsworth, H. M., Stephens, K. S., & Godfrey, A. B. (2004). *Modern methods for quality control and improvement*. New York: Wiley & Sons.

Womack, J. P., & Jones, D. T. (1996). *Lean thinking: Banish waste and create wealth in your organization*. New York: Simon and Schuster.

Woodall. (1997). Control charts based on attribute data: Bibliography and review (pp. 172–183), https://doi.org/10.1080/00224065.1997.11979748.

https://www.inmo.ie/MagazineArticle/PrintArticle/11397

http://www.hse.ie/eng/about/Who/qualityandpatientsafety/

https://www.hse.ie/eng/about/who/qid/governancequality/qswalkrounds/quality-and-safety-walk-rounds-a-co-designed-approach-toolkit-and-case-study-report.pdf

https://thinkers50.com/blog/walk-the-talk-of-quality/

# Chapter 9
# Lean, Green, and Clean: Some Case Studies

## 9.1 Introduction

This chapter presents three practically relevant case studies to make the readers comfortable working with LGC projects for quality improvement. Case studies are valuable for several reasons: (a) It provides hands-on experience with organizational problems, (b) it facilitates confidence in working with projects, and (iii) the concept learned in the chapters can be implemented appropriately. All the case studies are presented from an academic perspective and cannot be used directly for implementation. Information for the case studies was obtained from company consultations and personal contacts. The confidentiality of the company's name and the people associated with the projects are kept confidential on request.

All three case studies are different from each other. The objective here is to explain the concepts discussed in various chapters of this book through live projects. There will be many changes in its conceptual development and presentations while applying to a case study. So, readers are advised to consult standard books and references for clarifying the concepts. The details of facts and figures are not presented in every aspect but tried to give in a precise form to increase the readability.

## 9.2 Case Study Based on Lean Six Sigma Project

Here, we discuss a LSS Black Belt project to improve the product's quality by reducing the number of defects in a manufacturing product. This case study has strictly followed the DMAIC procedure in its entirety. The project objective is to "reduce quality diversion problem due to edge cracks in steel coil."

The company is based in India, located at Kolhapur, and is one of the largest single-location flat steel producers globally. The company has a strategically located site with dedicated infrastructure and installation of state-of-the-art steelmaking and

© The Author(s), under exclusive license to Springer Nature Singapore Pte Ltd. 2021    291
K. Muralidharan, *Sustainable Development and Quality of Life*,
https://doi.org/10.1007/978-981-16-1835-2_9

**Table 9.1** Prioritization of causes

| Causes | Sub-causes ($X$'s) | Prioritization score | Remarks |
|---|---|---|---|
| Man | SOP violation | 330 | 2nd hit list |
| | Lack of expertise | 220 | Discarded |
| Machine | Physical damage | 320 | 3rd hit list |
| | Segment life | 290 | 2nd hit list |
| | Mold life | 410 | 2nd hit list |
| | Roll alignment | 545 | 1st hit list |
| | Centering of bar | 465 | 2nd hit list |
| Method | Taper | 320 | 3rd hit list |
| | Speed curve | 520 | 1st hit list |
| | Spray curve | 480 | 1st hit list |
| | Thermal taper | 495 | 1st hit list |
| | Cross-spray water | 250 | Discarded |
| Materials | Superheat | 510 | 1st hit list |
| | Mold powder | 330 | 2nd hit list |
| | Total AI | 470 | 1st hit list |
| | Ladle-free opening | 230 | Discarded |

rolling technology. It is considered to be the best in the country in terms of sustainability indicators. The company is accredited for both processes and products and is seriously contributing to society through CSR activities. This project is addressed to one of the flat steel products named QRT-7 DSP Caster, where the cracks are generally found. The problem statement is: (i) 760 coils had defects due to edge crack in DSP Caster, and (ii) 9800 tons of total slab production was found to have imperfections due to edge crack 3.56% of the total output. This causes a number of customer diversion and dissatisfaction problems for the company. So, the immediate concern of the company is to reduce diversion of the edge crack slabs in QRT-7 DSP Caster to a zero level.

Since there are two trained Black Belts and nine other executives available with the project, a brainstorming session was conducted to identify the initial bottlenecks and a possible solution to get rid of customer diversion. A close examination of a few samples of the coil has resulted in two critical reasons for the crack. They are: (i) The pickled narrow face edge revealed that there were cracks in the slab corners or oscillation marks and (ii) edge crack in the coil due to preexisted cracks at the slab stage. These two causes lead to a majority of variations and are the primary concern for the process owners. Through a Pareto chart, it was also ascertained that the primary sources of variation are coming from man (5%), machine (15%), method (60%), and materials (20%). Through the C&E diagram, the sub-causes were identified and ranked according to its impact on the edge crack. See Table 9.1 for details.

Since many variables impact others concerning the changes in dimensions and design, the degree of association was found using the correlation method. The highly correlated variables were further tested for its statistical significance. Simultaneously, a failure mode effect analysis (FMEA) was also carried out to understand the effects of failure causes. They are summarized in Table 9.2.

Some improvements and controls suggested are:

- Free ladle open to avoid open casting
- Controlling TlN to prevent the loss of ductility trough
- Controlling speed and spray water to reduce stress nodes
- Controlling of superheat to reduce overcooling of slabs
- Mold geometry control to avoid erosion of copper plates

**Table 9.2** Variable association and failure effects

| Variables | Reasons for variations | Failure effects |
|---|---|---|
| Speed curve | This changes as per the casting speed. Mainly due to slab bulging, high presence of silicon in steel grade | Low unbending temperature lies in ductility trough causing TlN precipitation along austenite grain boundaries and leads to edge crack defect |
| Total PI | Higher addition of Pl, higher tapping oxygen | Edge crack defect in slabs and heat flux deviation, SEN clogging, bias flow in mold and defects in plates |
| Casting speed | Casting speed delays due to upstream, downstream, and internal reasons | To set speed such that avoid ductility trough zone |
| Superheat | Ladle processing at LRF | Low superheat leads to insufficient mold powder lubrication at the edges resulting in hooking, high superheat points to slab bulging, surface cracks and casting speed reduction |
| Nitrogen | High nitrogen in steel from LRF, high nitrogen pickup in the edge | Brittle cracks, the low unbending temperature lies in ductility trough causing AlN precipitation along austenite grain boundaries and leads to edge crack defect |
| Mold condition | Poor machining, and erosion or damage of Cu plate | Entrapment of copper leads to hot shortness resulting in edge cracks, and geometry of Cu plate causes hooking |
| Cross-spray | Cross-spray water accumulated within Bunkers Bottom and then goes into the flume. In case choking of Bunkers Bottom, the overflow of water occurs on one side/both sides of slab | In case choking of Bunkers Bottom, overflow of water takes place on one side/both sides of slab causing edge crack |
| Grid alignment | Wear-out rolls | Development of misalignment strains in slab and leads to surface cracks |

**Table 9.3** Control plans

| Action deliverables | Plans |
|---|---|
| SOP established for cast speed versus TIN factor | Need strict adherence |
| SOP revised spray table for cast speed | Highly correlated, hence need close monitoring |
| Cross-spray put off in all grades | Continue to monitor |
| Segment life monitoring and changing | Responsibility assigned |
| Heat flux ratio control by SOP | Necessary changes have done |
| Thermal taper changed as per specifications | Specifications changed |
| Daily auditing process parameters with quality control | Compulsorily followed |

- Segment alignment control to prevent microsegregation in slabs
- Cross-spray water control to avoid overcooling and overstress
- Heat flux control to prevent chilling of the shell.

The sustainable control plan is summarized in Table 9.3. A daily auditing process parameters with quality control and strict adherence of standard operating procedures are made compulsory, to monitor the deviations of the other quality parameters.

Finally, a performance comparison for the four main process activities is made based on the sigma-level calculation. They are presented in Table 9.4. It was very clear from the table that the process was not running according to the normal operation before the start of the project, but performance during and after the project was tremendous. The company can now assess the profit in financial terms (cost of quality) and the loss in terms of cost of poor quality by translating the sigma level in defects per million opportunities.

**Table 9.4** Sigma level of the processes

| Process/activity | Sigma level | |
|---|---|---|
| | Before | After |
| Solidification | 1.83 | 3.03 |
| Liquid steel | 1.76 | 3.21 |
| Slab formation | 2.01 | 4.15 |
| Segment alignment | 2.43 | 3.98 |

## 9.3   Case Study Based on 5S and Sustainable Practices

This project focuses on active workplace organization and implementation of standard work procedures through 5S (*sort–set in order–shine–standardize–sustain*) and sustainable business practices. The proposed project was shared for discussion by the mentor of the project, Mr. Abhinav (name changed) of TXT Cables LTD, Hassan. The company specializes in producing HT cables and is a large supplier across India. They have a large establishment at Hassan, a southern city in Karnataka, and house sub-units in many major cities in India. The company feels that its infrastructure management needs a complete makeover, as it is scattered over a large work area and production area. Since there is no competition for the products and services, their focus is now on improving these two areas to have new production lines added for sustaining the demand for the cables.

Upon an informal brainstorming session, the company has identified some gray areas for sustainable improvement. They include materials and tools disorganization, poor space management within the work area and production area, safety lapses in handling the product, and disarrayed waste (scrape) material disposal. The company feels that a slight improvement in these areas will have a tremendous impact on work efficiency and output generation. Hence, finally, it was decided to go for a 5S audit and, after that, a continuous improvement through Kaizen approach. A timeline chart (Gantt chart) for six months was prepared to assign each activity's responsibility systematically. It is presented in Fig. 9.1.

The processes identified to address the improvement in both the areas are wire drawing, stranding, vulcanization, copper tapping, inner sheathing, armoring, and testing. Improvement in these areas will directly optimize the management of warehouse, engineering, utility, and scrap yard area effectively. Further, for each identified process, 5S implementation rules were made. Correctly, the reasons to determine the excess utilization of wire drawing machine area are identified through a C&E diagram (fishbone diagram). This is shown in Fig. 9.2. Similar C&E diagrams are drawn for other core areas as well.

| Process Steps | Status | May | June | July | August | September | October |
|---|---|:---:|:---:|:---:|:---:|:---:|:---:|
| Problem identification | Completed | √ | | | | | |
| Data Collection for Quantifying Problem | Partially completed | √ | √ | | | | |
| Process Mapping | Initiated | | √ | √ | | | |
| Identification of VA/NVA Activities | Initiated | | √ | √ | | | |
| Root Cause Analysis | Initiated | | | √ | | | |
| Potential Solution identification | Initiated | | | √ | √ | √ | |
| Actions to be taken to eliminate NVA | Initiated | | | | | √ | |
| Control Plan | Initiated | | | | | | √ |

**Fig. 9.1**  Timeline chart

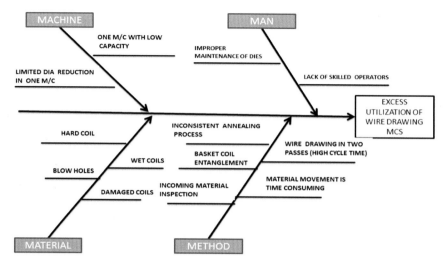

**Fig. 9.2**  C&E diagram for excess utilization of wire drawing area

As usual, the C&E diagram was further examined through some analytical study for earmarking the causes of the excess utilization of the wire drawing machine area. Upon some prioritization and statistical investigation, it was found that

- Documentation of detailed SOPs for all new wire drawing process is to be initiated with immediate effect.
- Training to shop floor associates is a must and may need frequent updations.
- Real-time monitoring and archiving of process record charts should be assigned to the QC department.
- Use of mild steel strips to tie heavy coils before annealing can avoid the use of excess materials and spaces.

Real-time monitoring of the above was made mandatory for quality improvement, and the responsibility was entrusted to the project leader himself. A dedicated team of employees for implementing Kaizen activities was selected for every process, and the process of sensitization of 5S and Kaizen quality improvement were also started immediately. The team member responsible for each operation was entrusted with three primary responsibilities for thirty days. They include.

- Creating a highly productive and well-organized workplace
- To experience the direct benefits of improvements through 5S techniques practically, and
- To identify potential wastes on the line and to make rapid improvements.

The focuses of 5S activities for each area were centered on:

- *Sort*: Sorting of items, labeling of things, and tagging of items.
- *Set in order*: Items to reorganize as per the size, dimension, and location.

**Table 9.5**  Processes and 5S scores

| Process/area | Scores | | | | | Average scores (50) | Comments |
|---|---|---|---|---|---|---|---|
| | Sort | Set in order | Shine | Standardize | Sustain | | |
| Wire drawing machine | 7 | 6 | 7 | 7 | 9 | 36 | > 35 |
| Stranding | 9 | 9 | 7 | 8 | 7 | 40 | > 35 |
| Testing | 8 | 8 | 8 | 8 | 8 | 40 | > 35 |
| Vulcanization | 7 | 7 | 8 | 9 | 9 | 40 | > 35 |
| Copper tapping | 7 | 7 | 6 | 8 | 8 | 36 | > 35 |
| Inner sheathing | 8 | 9 | 8 | 9 | 8 | 42 | > 35 |
| Armoring | 8 | 9 | 9 | 9 | 8 | 43 | > 35 |

- *Shine*: Items of large sizes were cleaned; small items were made mobile, and removal of toiled units as per reuse potential segregated.
- *Standardize*: Items were segregated and coded with necessary colors and dashboards prepared.
- *Sustain*: Reduce, reuse, and recycle policy adopted as a result of the previous 4S with a complete audit of each activity.

A scorecard based on 5S was prepared simultaneously, involving all those involved with the wire drawing machine areas. As a result, a suitable plan for the wire drawing area was realized. Simultaneously, other process teams also compiled their improvement plan based on each 5S and Kaizen activity. Also, members common to other processes were made available for evaluation and comments. To understand the real improvement, the team members and other officials in the company were asked to score (larger the better) each 5S activity according to a scale of 1–10 (10 being the highest) for each process. An average score of more than 35 was decided as a measure of improvement as per the pre-approved 5S audit's recommendation. The compiled summary is shown in Table 9.5.

There is no doubt that the 5S audit has paid a high dividend for the teams as the average score for each category is above 35. The suggested control plan according to 5S checklist for wire drawing area was prepared according to the following action:

- 5S list (visual plan) for all regions was made mandatory for day-to-day assessments and review.
- Team leaders of responsible processes to do the daily 5S audit in their assigned area.
- Once in a day, the area needs to be checked as per the checklist by a centralized team.
- Any abnormality found to be corrected immediately, otherwise recommended for intervention from the respective process leaders.

- Actions for which support is required to be placed on team board as concerns and discussed in daily empower committee meeting.
- It was decided to promote KANBAN (workplace management) strictly and audit.

Every improvement project is an experience. There are many things to learn from each project, even if it is a success or failure. The learning outcome of this project according to the improvement team is:

- Kaizen was found to be a robust team-building activity and brought colleagues together to get even better.
- Transparent communication was available from the top management with sufficient resources.
- Participation by managers in 5S and Kaizen events was thrilling.
- The path toward improvement is a natural process and possible.
- The change from an "emotional world" into a "facts and figures world" is difficult, but will bear the fruits.
- Team spirit increases confidence and removes bias in decisions.

## 9.4  Case Study Based on the Quality Walk

In Chap. 2, we had introduced the concept of the Quality Walk as a method of finding instant solutions to problems by communicating directly with the people and process with a purpose to detect, prevent, and mitigate the quality issues. Based on this, a quality improvement project was carried out in an educational institute. Specifically, the proposed project is on "Constructing a student facilitation center for examination-related activities." Currently, the center is working at a makeshift place aligned with a temporary establishment. According to the hierarchy of officials and their job function, the areas of quality improvements are identified and are shown in Table 9.6. The quality improvement components are not wholly mutually exclusive and, hence, not restricted to a single job function.

A concept note was prepared and circulated in advance to all the people involved with the exercise. A briefing was also carried out to make them comfortable filling the checklist. The sixteen-member Quality Walk includes four students, two each of teachers, non-teachers, and technicians. The rest of them are higher officials from the same organization of varying designations. A specimen of the quality checklist is shown in Table 9.7.

The Quality Walk resulted in many surprising conclusions. One of the suggestions unanimously supported by the members was to have a permanent student facilitation center. Members were the opinion that the enrollment of students is increasing every year, and catering to their demand, such a suggestion is necessary. A majority of the members suggested having increasing green and clean components included in the center. The other vital ideas are.

- To develop a new-age technology-enabled support system

**Table 9.6** Job profile versus improvement components

| Job function | Quality improvement |
| --- | --- |
| Teaching staff | - Improve self-discipline<br>- Improve commitment<br>- Manage time and lectures<br>- Apply self-control<br>- Establish workflow<br>- Continuous updating of knowledge<br>- Promote Kaizen<br>- Maintain quality standards in education |
| Non-teaching staff | - Improve self-discipline<br>- Improve commitment<br>- Adhere to rules and laws<br>- Respect the work processes<br>- Promote 5S and Kaizen |
| Faculty administrators | - Manage the workforce efficiently<br>- Apply self-control<br>- Be transparent on information dissemination<br>- Communicate with stakeholders<br>- Make use of mistake proofing<br>- Promote zero-defect concept<br>- Promote 5S and Kaizen<br>- Maintain process standards |
| Technical and support staff | - Respect the system<br>- Respect the people<br>- Improve knowledge in IT<br>- Continuous use of Poka-Yoke<br>- Promote zero-defect concept<br>- Promote 5S and Kaizen |
| Examination and evaluation | - Improve self-discipline<br>- Improve commitment<br>- Adhere to rules and laws<br>- Improve efficiency in work<br>- Promote zero-defect concept<br>- Maintain confidentiality<br>- Coordinate with teaching staff |
| Accounts and audit | - Have good knowledge of accounting and audit<br>- Improve commitment<br>- Improve efficiency in transactions |

(continued)

**Table 9.6**  (continued)

| Job function | Quality improvement |
|---|---|
| Upper management | - Provide adequate facilities<br>- Provide sufficient training and expertise<br>- Promote zero-defect concept<br>- Promote 5S and Kaizen<br>- Instill SOPs as common practice<br>- Update curriculum and general practice<br>- Promote organizational excellence |

**Table 9.7**  Quality checklist

| Waste types | Details and observations specific to | | | Ideas to eliminate | Ideas to improve | Actions and recommendations |
|---|---|---|---|---|---|---|
| | **Lean** | **Green** | **Clean** | | | |
| Overproduction | | | ✓ | Space minimization | Construction under the vicinity | Close to administrative office |
| Inventory/materials | ✓ | ✓ | ✓ | Real-time use of materials | Heavy load transportation | Construct a transit approach road |
| Defects | | ✓ | | – - | – - | Use of standard items |
| Processing | | | ✓ | Use nearby ongoing site | Use of manpower | Use of nearby ongoing site |
| Waiting | ✓ | | | – - | Prepare a timeline chart | Strict adherence to timeline |
| Motion | ✓ | | | – - | – - | Maximum use of existing inventory |
| Transportation | | ✓ | ✓ | – - | Use of existing vehicles | Vehicle vendor selection advised |
| People | ✓ | ✓ | ✓ | Use of skilled employers' utilization | Constraint evaluation suggested | Trained consultant appointment suggested |
| Energy | | ✓ | | – - | Installation of monitoring equipments | Procurement of energy-efficient tools and equipments |

- To turn all the examinations online
- Multilayer floor components to facilitate the movement of examination answer books and materials
- Centralized evaluation of all courses at one place
- Student amenities of all sorts including a food court
- Surveillance cameras in all floors and examination rooms
- Dashboards on examination schedules to install at all entry points.

Before compiling the checklist and the recommendations, student feedback was also carried out to know their perception about the facilitation center. Finally, the suggestions from all participants were compiled to understand various aspects of LGC to include in the formation of the facilitation center. The exercise was followed by a brainstorming session to initiate the functioning of tendering and procurement of inventories. For this, a high-power facilitation committee was formed, and a *Responsible–Accountable–Consulted–Informed* (RACI) matrix to understand the areas of concern and responsibility was prepared. See Table 9.8 for details. Table 9.9 presents the LGC tools according to some select processes (functions).

The process of constructing the facilitation center is now on. The use of LGC components and its acceptance is well recognized by the stakeholders. The involvement of futuristic technology and sustainable items is highly recommended to include in every process of construction. It was also suggested to repeat the Quality Walk once in every month to assess the improvement and progress.

This author feels that the kind of exercise discussed above should be tried at every organizational level if the quality and sustainability are the major concerns. However, for reaping the benefits of LGC, one should follow a holistic approach with the right attitude and patience. The importance of leading indicators over trailing indicators is a crucial concept of balanced score carding relevant to LSS deployment (Butler et al., 2009). The Quality Walk exercise facilitates the identification of those indicators affecting quality very efficiently. Although discussed at many places in this book, Quality Walk as a part of the LSS initiative ensures a solution set for enhanced quality improvement with long-term sustainability. They include.

- On the table agreement on problems and its solutions
- Eliminates redundancy in the system by eliminating things that are not required
- Facilitates consolidation of job functions and responsibilities
- Promotes the best use of infrastructure and logistics
- Identifies the reduce, reuse, and recycle potential items and activities from the clutter
- Improves clarity and vicinity of problems and issues
- Minimizes handoffs and influences
- Removes bottlenecks and constraints
- Improves transparency in the system
- Nullifies the effect of bossism and authority.

**Table 9.8** RACI matrix

| Functions | Departments | | | | | | | | | |
|---|---|---|---|---|---|---|---|---|---|---|
| | Legal | Finance | Sales | R&D | Accounts | Purchase | Consultant | Construction | Design engineer | Structural engineer |
| Advertisement and campaign | I | R | I | A | I | I | I | C | C | I |
| Preparation of tender | R | C | C | I | C | I | C | I | I | I |
| Preparation of site plan | C | C | I | I | C | I | I | I | A | R |
| Appointment of site officer | C | C | I | I | C | I | I | R | I | I |
| Preparation of budget | I | C | C | C | R | C | I | I | I | I |
| Procurement of inventory | C | C | R | C | C | A | I | I | C | C |
| Procurement of logistics | I | I | C | I | C | A | I | R | C | C |
| Inventory control | I | C | A | I | C | R | I | C | I | I |
| Security and surveillance | C | C | I | R | C | I | I | I | C | C |

**Table 9.9**  Process versus LGC components

| LGC components | | | |
|---|---|---|---|
| Process | Lean | Green | Clean |
| Site establishments | - Mobility<br>- Accessibility<br>- Features<br>- Size<br>- Design | - Sustainable<br>- Eco-friendly<br>- Aesthetics<br>- Relaxing | - Ambience<br>- Technology<br>- Coverage<br>- Capability<br>- Safety |
| Construction | - Reliability<br>- Acceptability<br>- Adaptability<br>- Perceived quality<br>- Size<br>- Design | - Sustainable<br>- Eco-friendly<br>- Serviceability<br>- Coverage<br>- Energy efficient<br>- Structure | - Technology<br>- Capability<br>- Safety<br>- Relaxing<br>- Freshness<br>- Modern |
| Purchase | - Conformance<br>- Satisfaction<br>- Durable | Utility<br>Sustainable<br>Variety | - Large user size<br>- Instant<br>- Aesthetics |
| Administration | - Accessibility<br>- Size<br>- Modern<br>- Technology<br>- Capability<br>- Perceived quality | - Communication<br>- Features<br>- Eco-friendly<br>- Serviceability<br>- Aesthetics<br>- Variety | - Simple<br>- Flexible<br>- Ambience<br>- Relaxing<br>- Energy-efficient |
| Inventory and logistics | - Security<br>- Reliability<br>- Durability<br>- Acceptability<br>- Conformance<br>- Capability | - Minimum space<br>- Sustainable<br>- Eco-friendly<br>- Serviceability<br>- Energy efficient<br>- Variety | - Textures<br>- Space<br>- Trust<br>- Color<br>- Safety |

# Reference

Butler, G., Caldwell, C., & Poston, N. (2009). *Lean Six Sigma for Healthcare: A senior leader guide to improving cost and throughput*. ASQ Quality Press.

# Chapter 10
# Moving Toward Sustainable Quality Life

"The reality today is that we are all interdependent and have to
coexist on this small planet. Therefore, the only sensible and
intelligent way of resolving differences and clashes of interests,
whether between individuals or nations, is through dialogue",
The XIV Dalai Lama.

## 10.1 Introduction

Lean is about adopting a new paradigm—it is a new way of thinking about quality, productivity, efficiency, and value creation. Similarly, lean manufacturing is about questioning, challenging, and changing the traditional methods of working. Those lean practitioners who have achieved more than fleeting success with lean will know that to achieve economic success, you require a significant focus on your people and planet. A business that is carbon-positive and wastes nothing is regenerative to environments and has a positive effect on people's lives. Clean, breathable air is not a luxury; it is a fundamental human right (Swami Chidanand Saraswati, Speaking Tree, TOI dated November 20, 2018). According to the author, some solutions are possible, available, and doable. It merely requires that we wake up, pay attention, and make finding a solution to pollution as our priority. One way to counter air pollution is to plant more trees. Scientific studies support that the amount of oxygen produced by an acre of trees per year equals the amount consumed by 18 people annually.

According to the daily newspaper, The *Times of India* report, one tree produces nearly 260 lb of oxygen each year, and one acre of trees removes up to 2.6 tons of carbon dioxide each year. Not only do trees provide us with oxygen, but they also give us shade that makes the planet cooler by several degrees in summer. Toward this, Hero-MotoCorp-Times of India's Green drive is commendable, as they have been facilitating the planting of saplings since 2015 (The Speaking tree, TOI, October 20, 2018). The "Swatch Bharat Abhiyan" or "clean India mission" initiated by the current government at the center is another landmark program for sustainable practices. The program aims to clean up the streets, roads, and the infrastructure of India's cities, towns, and rural areas besides improving solid waste management and eliminate open defecation.

Notably, in our economic system, the monetary value of goods and services is set by the balance of supply and demand in markets. Yet prices set in markets traditionally omit individual costs and benefits that do not have a market price—the so-called externalities. Externalities often come from stakeholders lacking the power to make their value or cost recognized, such as the detrimental impacts of a large corporation on its employees' health or the natural resources used and the pollutants produced. As a consequence of externalizing the environmental costs, our industrial practices are profoundly disturbing the ecological balance of the planet in many different ways. At the same time, disparities are widening in human society. Some of these challenges are climate change, eutrophication (enrichment of water by nutrient salts that causes structural changes to the ecosystem), ozone layer depletion, loss of biodiversity and deforestation, extreme poverty, malnutrition and undernourishment, and acute social inequality, to name but a few. It is recommended that the human economic system needs to internalize such externalities (Zokaei et al., 2013).

A quality life addresses all concerns of society and environment suitably incorporating all types of externalities discussed above. Establishing a close watch on the externalities should be the primary concern of all human beings. In this process, human beings should forgo their egos and priorities for the betterment of other living people, if it matters the livelihood of others. By changing the preferences, one becomes closer to the ecosystem they belong to. At least, if we can improve our thinking process to create a better world, then that itself becomes a unique contribution of each individual. A positive thought process can eliminate all kinds of negative thoughts (wastes) from the human mind. Most of the time, negative thoughts dominate positive impressions and lead to chaos, tensions, and crisis. Note that, a layman typically relates quality to feature-based measures as determined by design and performance-based measures as determined by conformance to design or ideal value. All goods and services are produced in various grades or levels of quality. These variations in grades or levels of quality are intentional and are due to the design of excellence. Therefore, adding value to any process should be the priority of every management. Let us now articulate how to add value to the system (process) and improve the process (method) for a sustainable quality life.

## 10.2  Identifying Value-Added Activities

Quality improvement comes from reducing variation by eliminating special causes, improving the consistency of measurement systems, controlling the product by controlling the processes, and reducing losses and eliminating waste. The critical place to begin any quality improvement effort is with a precise specification of the value of a product or service, as perceived by the customer. The demand for a product with the right capability, at the right price, and the right time determines value. Identification of value-added (VA) and non-value-added (NVA) component is the backbone of any improvement activity. *Value-added activities* transform the product, service, or transaction in a way that is meaningful to the customer (layman). In other words, value-added activities are those activities that are valuable from the external point of view of the customer. Activities that add no value to the process are called NVAs. *Non-value-added activities* are those activities that are required by a business to execute work that adds no value to the customer, and that customer is not willing to pay for. Note that waste points us to problems within the system. The more the value-added less the waste it contributes. This is what exactly, the lean, green, and clean principles promote. The involvement of more value-added activities in any process ultimately leads to sustainable products, services, and information. Some other perceptions of VA/NVA activities from various authors are as follows:

> VA activity is an activity that is being performed as efficiently as possible, or that supports the primary objectives of producing output—Kaplan and Cooper (1998).
>
> VA in economics defined as the difference between the costs of purchases and the revenue from the sales of goods and services—Koskela (1992).
>
> NVA activity is an activity which absorbs resources but creates no value—Womack and Jones (1996).
>
> A VA activity contributes to the customer's perceived value of the product or service. An NVA activity is one which, if eliminated, would not detract from the customer's perceived value of the product or service—Convey (1994).
>
> VA step directly impacts the customer's perception of the product's value—George (2003).

The mapping of all process activities is a useful communication tool and identifies vital input variables and the resulting output. It tends to show the complexity of the process by systematically displaying the non-value-added activities. Some examples of inputs are raw materials, equipment, process, people, methods, and components. All inputs can be subject to errors and variations, and they ultimately decide the quality of the output. Inputs that are green and clean contribute to the lean value streams to establish the flow in the production. It guarantees to provide sustainable outputs which have a high survival rate and endurance potential. Sustainable outputs generally contribute to less waste and pollution. It is believed that the ratio of VA to NVA is 5–95% most of the time. For example, ordering raw materials, shipping to customers, preparing engineering drawings, processing customer deposits, assembling products, filing claims, etc., are value-added activities, and inspection, recording, testing, revising, counting, approvals, checking, etc., are non-value-added activities, as they do not contribute anything directly to the customer. Hence, they are all wastes. That is, all NVAs contribute wastes. Figure 10.1 makes a comparison of VA/NVA concerning eight deadly wastes.

Womack and Jones (2003) describe the value of creating tasks as the immediate need for a modern business environment to sustain in a competitive environment. Taiichi Ohno, who introduced the Toyota Production System (TPS), suggested that the NVA activities accounted for approximately 95% of all costs in a non-lean manufacturing environment. When applied to a company or an aggregate of companies, the concept of VA is clear and quantitative: It is the difference between the price at which goods are sold and the cost of the materials used to make them. It can be

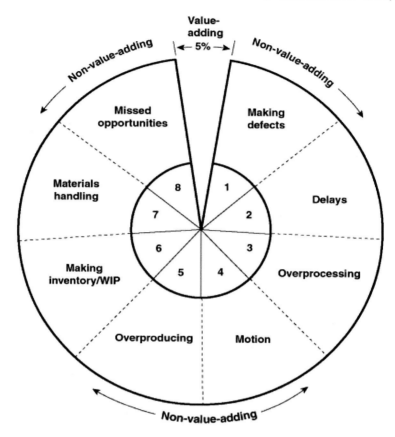

**Fig. 10.1** Comparison of VA/NVA activities and eight deadly wastes. ( *Source* Harry & Mann, 2010)

calculated easily from accounting data and is used as a basis for the value-added taxes paid in many countries. Comparisons of value-added activities among manufacturing industries provide insights into their structures. For example, in metal smelting, the value-added maybe 20% of sales; in prestige cosmetics, 80%; in software, 99%, etc. Such differences impact wages, the level of automation, and competitive strategies, and hence, the assessment of VA/NVA activities is quite natural for performance check. Value-added can also be calculated for individual products and then enhanced by finding cheaper materials. In this process, one should be careful not to diminish the value of the product to customers by compromising the quality of the product.

The lean methodology first pays attention to NVA activities in the form of *waste* in all functional movements. No customers, whether they are internal or external, want to pay for any kind of wastes. Companies and organizations that fail to acknowledge this fact are not lean. Lean thinking is driven by the precise definition of VA and NVA activities. The concept of waste may be less elegant, but it is more operational. It is possible to identify what is waste and what is not, and, by definition, waste needs to be eliminated. By saying so, we cannot avoid some NVA activities like training and appointment of personal, accounting and tax filing, sales and marketing, repair and maintenance, and so on, which are adding value to the organization. May be by reducing the frequency and by the use of the right expertise, the impact of NVA activities can be minimized. Therefore, for any type of quality improvement, one need to identify the source of variation first, its contribution to waste generation secondly and then look for all methods of its elimination. In various chapters, we have already articulated the type of waste, their origin, and the importance of eliminating them. Some sustainable inputs to overcome the waste menace will be discussed further here.

There are many limitations of classifying VA and NVA activities, as discussed by the authors Kaplan and Cooper (1998) and Koskela (1992). Koskela (2000) points out that classifying activities into VA and NVA will sometimes be problematic as we cannot consistently define what constitutes a VA or NVA activity. Few authors provide an alternative classificatory approach to deal with such a problem. Womack and Jones (2003) proposed three categories as described below to identify the proper scope of improvement in process line:

- Those who create value as perceived by the customer
- Those who create no value but are currently required by the process (processing, training, inventory, etc.)
- Those who do not create value as perceived by the customer (meetings, appointments, transportation, etc.).

George (2002) proposed the following set of questions to identify whether the activity is VA or NVA activity:

- Does the task add a form or feature to the product or service?
- Does the task enable any competitive advantage?
- Would the customer be willing to pay extra or prefer us over the competition if he or she knew we were doing this task?

Note that the customer who wants to buy the right product with the right capabilities at the right price determines the amount of VA in the product. According to Womack and Jones (2003) and Muralidharan (2015), the value can only be defined by the ultimate customer. It is meaningful only when expressed in terms of a specific product that meets the customer's needs at a particular price at a particular time. Hence, the classification of VA/NVA activities from the eye of the customer plays an important role.

## 10.2.1 Value Stream Mapping

The best way to identify the non-value-added activities in a process is by drawing a value stream mapping (VSM). VSM is a hands-on process to create a graphical representation of the process, material, and information flow within a value stream. It facilitates.

- The means to see the material, process, and information flows as per requirement
- The prioritization of continuous improvement activities at the value stream
- The basis for facility layout and
- The elimination of waste drastically.

VSM also helps the customer's significant process variation and protects the considerable demand variation as perceived by the manufacturing and operation entities. Every effort should be made to minimize these variations. These variations are driven by one of the eight forms of waste (already discussed at many places in this book) and should be eliminated. A typical value stream map (VSP) is shown in Fig. 10.2.

The non-value-added activities in a process can be eliminated in the following ways:

- identify and eliminate bottlenecks,
- decrease wait time or queue time,
- increase process frequency,
- simplify the process,
- allow more time for conducting value-added work.

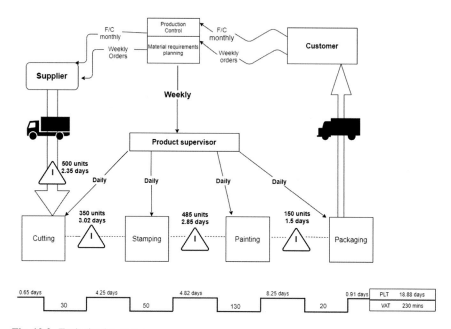

**Fig. 10.2** Typical value stream map

Non-value-added activities are roadblocks for improvement, whether manufacturing, nonmanufacturing, transactional, service, and administrative. Everybody experiences the burden of non-contributing events in the process. Often the cost of investment is spent more on NVA activities. These costs include raw material and disposal costs, as well as costs for compliance management activities and pollution control processes and equipment. The costs associated with the improvement process are plans, goals, budgets, and review (Muralidharan and Raval, 2015). The NVA activities can be perceptional, qualitative, quantitative, and purely managerial.

The perceptional-type NVA activities are.

- Not doing things right the first time
- Bottlenecks on production lines
- Cannot find things or "lost" or misplaced material or equipment
- Human NVA activities like mistakes, forgetfulness, poor communication, indecisiveness, no authority to take action, etc.

The qualitative-type NVA activities are.

- Unavailability of information from the management
- Using ineffective equipment, e.g., equipment not intended for use
- Unskilled and untrained personnel
- Mistakes like using expired material
- Excess or unnecessary use of chemicals
- Use of hazardous chemicals that could harm human health, worker safety, and the environment
- Hazardous wastes generated from production and the disposal of products

The managerial-type NVA activities are.

- Unorganized work areas and lack of housekeeping
- Reworks and sorting defect material
- Unnecessary equipment stoppages or manufacturing downtime
- Lack of materials or human resources hindering operation flow
- Unlabeled or unidentifiable material
- Running out of consumables (order in advance)
- Too long or insufficient setup times
- The continuous state of "fire fighting" or always performing emergency fixes to problems
- Ineffective or lack of planning
- Generating too much waste and scrap material
- Untapped personnel or not using people's expertise
- Overprocessing of materials, methods, and machinery
- Overproduction and holding too much inventory especially where shelf life is important
- Unscheduled machine maintenance or repairs

- Transportation problems, etc.

The quantitative-type NVA activities are.

- Too much sampling and testing
- Too much "travel distance/time" due to poor layout on the shop floor
- Incorrect storage conditions where first in first out (FIFO). Point of use storage might eliminate warehousing and non-value-added handling or waiting time
- Queue or waiting: that is waiting to be processed, waiting to move, waiting for parts or material
- Unreliable suppliers, who do not deliver on time or deliver poor quality or incorrect material
- Unnecessary machine changeover—aims to do "same" size or product when equipment has been set up, then change to another process
- No safety stock
- Producing incorrect yields
- All kinds of transportation, inventory, movement, waiting, over-processing, overproduction, and defects.

Some lean tools specific to achieve VA goals are as follows:

- For stability (stable and capable workshop producing high quality, consistent production outputs), the tools can be 5S + Safety, standardized work procedures or SOPs, error proofing, Six Sigma.
- For continuous flow (products flow smoothly through all operations without stopping), one can use set up reduction or SMED, TPM, and continuous flow.
- For standardized work (synchronize the products' rate of flow with the demand of the customer—TAKT), it is better to follow standardized work procedures or SOPs.
- Pull material (where continuous flow is not possible, *pull* is used to control production between streams) is better to consider pull (Kanban) system.

Most of the methods mentioned above are very powerful in yielding excellent process flow and short- and long-term gains in process efficiency and speed. Keeping the lean characteristics in place, these methods promote best practices for sustaining business growth and reputation. For new generation business ventures and start-up initiatives, the development of lean culture is a must to get speedy results quickly.

One can also use various statistical principles to identify, classify, and control value-added activities in a process. DOE is a powerful method used to identify and segment critical characteristics, multivariate techniques can be used to classify and segregate the principal components influencing the process characteristics, SPC/SQC techniques to control critical process factors, and so on. In various chapters of this book, we have seen the extent of applications of statistical tools and their potential in improving the quality of a process, product, and services. Understanding SOP is another best way to balance the VA and NVA activities in process activity.

It is easy to explain the impact of VA and NVA activities through any administrative system. The majority of the time, the problem accumulates because of delays in executing a particular task. We know that delays generally do not add any value to the person concerned, and hence, it is a waste. It destroys the work atmosphere by influencing lethargy on others in many specific cases. The impact of bureaucracy, autocracy, and red-tapeism (a political term used for file movement and priority) is such that the entire administrative system can get collapsed. It can spoil the productivity and quality of the organization and the society at large if proper transparency is not maintained. Government administration at large is contributing a large amount of NVAs in their day-to-day activities. Look at the delays in processing an application (leave, transfer, appointment, admission, and so on), judging a request for correctness, accuracy, and approval in a government institution. Most of the time, there is a long waiting time and queue for final sanction and clearance, leaving no option for the person to get frustrated. If proper regulation is not made to make it in time, then there will be no use of these decisions and end up in only waste. Justice delayed is justice denied.

## 10.2.2 Cost of Quality/Cost of Poor Quality

The cost incurred to ensure that quality is maintained at an acceptable level is called the cost *of quality* (COQ). This type of cost is part of every process and cannot be visible every time. For instance, the inspection process is done routinely to see the performance of product and service quality; however, the cost associated with the inspection is generally not accounted for in figures or numbers. The cost incurred to correct processes that fail to perform as intended is called the *cost of poor quality* (COPQ). Almost all processes contain some rework, work-abounds, and unexplained redundancy, not to mention the malpractice and risk management. For instance, failure to get the right candidates in each recruitment process involves a considerable amount of COPQ. Still, as a part of acquiring eligible candidates, the process of recruitment should not be stopped.

Identifying NVA activities is essential to ascertain the COQ and COPQ associated with any process or organizational events. If any everyday activities provide no benefit to customers, then there will be uncontrolled amounts of wastes, rework, rejections, customer returns recalls, inspection costs, etc. This can lead to many chronic problems in the organization. The quality costs can be anything like: *internal* (scrap or rework, design changes, modifications, excess inventories, late deliveries, downtime, injuries, etc.), *external* (customer complaints, lawsuits, returns and allowances, product quality, etc.), *appraisal* (testing, inspection, audits, controls, safety and security checks, certifications, etc.), and *preventive* (planning, training, specifications, maintenance, capability, etc.) types (Butler, 2009; Rust et al., 1994). All the costs described above contribute to a high level of customer dissatisfaction. Mathematically, COPQ is expressed as the difference between the theoretical minimum cost and the actual cost. COPQ disappears if every task was done correctly the first time,

**Table 10.1** Sigma level versus quality costs

| Sigma level | DPMO | Quality yield (%) | COQ/COPQ Cost percent total |
|---|---|---|---|
| 2 | 308,537 | 69 | Non-competitive |
| 3 | 66,807 | 93.3 | 25–40% |
| 4 | 6210 | 99.4 | 15–25% |
| 5 | 233 | 99.98 | 5–15% |
| 6 | 3.4 | 99.9997 | World class |

every time. Lean Six Sigma relates quality costs with sigma level, as shown in Table 10.1. As quality increases, COQ and COPQ decrease (Butler et al., 2009; Pande et al., 2003).

Quality cost analyses have as their principal objective cost reduction through the identification of improvement opportunities. This is often done with Pareto analysis. The Pareto analysis consists of identifying quality costs by category, or by-product, or by type of defect or nonconformity (Montgomery, 2003). A typical COQ/COPQ for various sectors of industry according to multiple categories of costs in percentages of the total project cost is presented in Table 10.2. The comparison is based on lean, green, and clean initiatives, and therefore, the figures shown in the table is only a perception.

Some generally not estimable costs because of the nature of the occurrence are called "Hidden costs." They also occur in most of the sectors, as mentioned above (Gryna et al., 2007). Some of them are the costs associated with:

- Reprocessing
- Sorting and inspection
- Warranty expenses
- Customer returns
- Lost sales

**Table 10.2** Comparison of COQ/COPQ against industry sectors

| Cost types | Industrial Sectors | | | |
|---|---|---|---|---|
| | Manufacturing (%) | Production (%) | Automobile (%) | Service (%) |
| Internal quality failures | 30 | 30 | 20 | 40 |
| External quality failures | 25 | 25 | 20 | 15 |
| Appraisal | 18 | 20 | 15 | 25 |
| Cost of inefficient processes | 9 | 10 | 15 | 5 |
| Prevention | 8 | 8 | 10 | 5 |
| Hidden | 10 | 7 | 20 | 10 |
| **Total** | **100** | **100** | **100** | **100** |

- Overtime to correct errors
- Loss of goodwill
- Lost discounts
- Paperwork errors
- Damaged goods
- Premium freight charges
- Customer allowances
- Extra process capacity.

Reducing the cost of poor quality has a dramatic impact on company financial performance. So, in analyzing quality costs and in formulating plans for reducing the cost of quality, it is essential to monitor the role of prevention and appraisal costs. According to Montgomery (2003), many organizations spend a significant amount of their quality management budget on assessment and not enough on preventions. This is because appraisal costs are often budgeted items in quality assurance or manufacturing areas, whereas prevention costs are not usually budgeted. Generating the quality costs figures is not always easy because most quality cost categories are not directly reflected in the organization's accounting records.

Consequently, it may be challenging to obtain extremely accurate information on the costs incurred for the various categories. The reporting of quality costs is usually done on the basis that permits straightforward evaluation by management. Managers want quality costs expressed in an index that compares quality cost with the opportunity for quality cost.

The recovery of quality cost also depends on the improvement programs one initiates in the organization. Table 10.3 makes a comparison of various process improvement methods concerning the cost of quality recovery. The cost of quality recovery is assessed on a minimum level of the total cost of the project. So, the comparison done here is based on a particular transactional project carried out through various improvement methods and is not a stand-alone fact for reference and use.

## 10.3  Leadership and People Engagement

Inspired people are always an asset to any organization. They take positive steps in all-round improvement in quality and are greatly influenced by the nature of the work they perform, and they can influence others to follow. Good leadership can engage people for optimum use with their management, communication, problem-solving, and planning skills. A project/process leader can see the big picture, coordinate all the moving parts of the project, and decompose the project results into manageable components. The leadership can evolve from any section of the organization. Leadership can be of two types: white coat leadership and improvement leadership. The white coat leadership people are generally autocratic and impatient, whereas an improvement leadership people are usually facilitators, teachers, and students at the

**Table 10.3** Comparison of improvement methods concerning the cost of quality

| Program | Duration of the project | Improvement Potential | Cost of quality recovery (minimum of the total cost of the project) (%) | Focus and credentials |
|---|---|---|---|---|
| Company-wide | 1 month | 5–10% | > 5 | Simple, routine, flexible |
| TQM | 1–2 months | 10–20% | > 10 | Routine, flexible, focused on quality |
| PDCA | 1–2 months | 20–40% | > 20 | Specialized TQM, process-based |
| Lean | 2–3 months | 30–50% | > 30 | Simple, increase speed, flexibility, focused on waste elimination |
| Lean Six Sigma | 2–3 months | 40–80% | > 40 | Focused on waste elimination, problem-solving |
| LGC | 3–4 months | 70–90% | > 60 | Focused on waste elimination, problem-solving, sustainable practices |
| Six Sigma | 4–6 months | Very high | > 80 | Rigorous, focused on variation and processes, customer, problem-solving. Technical expertise required |

same time. They can very well communicate with their peers with all humility and perseverance.

Quality is not just the responsibility of one person or one department in the organization as it once was. Supervisors or process owners play a vital role in the success of any quality initiative. They serve the essential communications link between the management and the employee, as they understand both the workers' challenges and the expectations of top management. The quality philosophy is set into motion by the workers under the leadership and guidance of their supervisor. It is their responsibility to coordinate and schedule regular in-service training sessions for their progress and growth. Supervisors must demonstrate a positive attitude toward training by encouraging all employees to take part. Training to gain skills in quality control must be a top priority for the supervisor. Their approach and commitment to quality service exhibit then as the role model for all employees (Deming, 1986). Supervisors directly help management create a climate for innovation and continuous improvement.

An efficient leader can empower the workforce in any manageable way. They can encourage people to take initiatives and broader the project scope, thereby increasing the ownership and responsibility more meaningfully. With empowerment comes the need to redefine the fundamental role of upper management, middle management, and the workforce. *Top management* generally acts as the communicator and supports the organization's mission. *Middle management* should run its area of responsibility and work as a group to integrate all parts of the organization. Besides, it must support the workforce by eliminating obstacles to progress. The *workforce* is the primary producer of the output for customers.

Therefore, to establish complete transparency in the system, management must provide employees with the information, feedback, and means of regulating their work promptly. Transparent communication and data relevant information is the right of every citizen. For business or any organizational function to sustain, empowerment requires employees to have the capability, the authority, and the desire to act (Gryna et al. (2007). Despite all, Argyris (1998), opines that difficulties in achieving an empowerment environment are due to a failure by management to make an internal commitment from employees by jointly setting quality goals with employees rather than management unilaterally setting the goals. This arrangement can sometimes lead to a complete loss of communication between the employees and management.

When you are part of a project or steering a project team, you will encounter resistance at all stages. You will probably not get the same support, resistance, and opposition from the same people as you progress through the project. Still, you must continue to monitor and manage it. Quite often, project selection itself will demand the competence of its members. Sometimes, people with different specializations and conflicting interests have to cooperate to produce the result. So, resistance based on intelligence level and expertise level happens naturally and can have a positive and negative outcome. Creating an atmosphere of openness will be the sole responsibility of a team leader in those difficult times. According to Muir (2006), the common resistance types and their possible solutions are as follows:

- *Cognitive*—This kind of resistance is the result of overconfidence or misguidance. The people associated with the project believe that the diagnosis is wrong, and the proposed course of action is misguided or irrelevant.
- *Ideological*—This is because people believe that the proposed change violates fundamental values that have made the organization what it is.
- *Power-driven*—Here, people perceive that the proposed change will lead to a loss of power, autonomy, and self-control that will lead to reduced status and independence.
- *Psychological*—Here, people believe that they have to acquire new knowledge and learn new technologies. So psychologically, resistance will prevail.
- *Fear*—They are more generic and universal. New situations and unfamiliar processes can lead to a general fear of changing the ways things are done. It will eventually lead to one of the above types if not addressed in time.

Another attribute of good leadership is the attitude and integrity of the person. A person's attitude is determined by psychological factors like ideas, values, beliefs, perceptions, etc. They are fundamental determinants of our opinions and actions toward all aspects of our social environment. Even if one has excellent leadership qualities and high ethical character, the person can still not deliver if the environment is not favorable. More often, the negative attitude of the person or society brings all hindrances for development and progress. However, a person with high moral values, good discipline, integrity, and attitude can always influence people and manage any responsibility to higher esteems and returns. Therefore, the success of initiatives like lean, green, and clean philosophy begins with those kinds of people. Let us be a part of this contingent.

## 10.4   Moving Toward Sustainable Quality Life

In the previous chapters, we have articulated the importance, necessity, realization, and implementation of lean, green, and clean initiatives for a sustainable quality life. Most of the time, the discussions were organized on issues on the elimination of wastes and variations. This is the only way to improve quality. Both economic waste (lean waste) and environmental waste (green waste) contribute to a quality problem, and therefore, its elimination is a must. It is human nature that, in every aspect of life, we look for more comfortable, faster, better, and cheaper products and services. People, environment, and society are necessary for participatory human relations with the world at large. We know that lean is about tomorrow better than today (Zokaei et al., 2013); it is about creating a learning organization to pursue systematic eradication of all types of constraints to the organization's economic goals.

Leading a quality life is everyone's aspiration. Being modern and technologically sound is considered to be the objective of quality of life. Today's world is full of challenges and opportunities. This gives rise to competitions in all walks of life. Therefore, to resist all challenges and opportunities, one should present a high moral value, good character, positive attitude, and conduct, and above all, love and respect for all living beings. The best formula for a successful life is: material things, remain content, and be idealistic in spiritual matters. This is the only formula that gives one peace of mind (Maulana Wahiduddin Khan, The Speaking tree, October 9, 2018).

If happiness is the sole objective of living a quality life, then such a thing should come from the acceptance of society and the world around you. Being simple, humble, and flexible makes you understand your people and ecosystem more, and you become committed to sharing the tenets of joy with everyone around. It makes you more responsible, confident, and mentally balanced. If leading a long life is what quality of life means, staying healthy and fit (physical and mental) should be the main focus of life. This is possible through regular physical exercises, meditation, controlled diet, and positive thinking. Keeping the mind stable and active in an appropriate proportion is also necessary to stay calm and fit. Allow love and compassion to be part of life. Both bring the highest level of happiness in life. There is no investment or cost involved in sharing love and compassion.

From all the contexts and discussions did so far, one thing is clear that there is no objective definition for "quality of life." One can answer only with a subjective perception with descriptive vocabulary. The quality of life of a millionaire cannot be the same as that of a layman or a poor man. The opinion can be different from culture to culture, generation to generation, tradition to tradition, region to region, and religion to religion. It varies with the levels, status, time, and resources. There is science, art, and philosophy in living a quality life. We can define "quality of a product" or "quality of services" objectively, being the ability to satisfy the needs and desires in some amount. This can be measured in various dimensions like size, color, texture, volume, amount, efficiency, capability, reliability, accuracy, durability, flexibility, technology, etc. Having all comforts, resources, and material acquisitions are not, what describes "quality of life." Good health, people's wellness, universal love (transforming "I" to "we"), compassion, and one's contribution to their own country are the characteristics of a good quality of life. One should strive for the betterment of the people by being concerned is the best way of leading a quality life. These characteristics will automatically qualify for idealizing a good human being.

The universal account of the quality of life stands for the lives of oppressed or marginalized. Apart from the physical, mental, and intellectual dimensions, we also possess a spiritual dimension, which is equally important. This kind of universal account considers freedom from imperfection and suffering as the goal of human life and quality of life is to be attained in terms of catering to the legitimate needs of all these in a balanced way. (Mukherjee (2003), Amartya Sen, 1993). In pursuit of excellence, striving for betterment and attainment of quality of life has been a perennial human concern. Freedom from imperfections and consequent suffering has been the chief motivating factor for all cognitive enterprises and technological advances. Everyone should be able to contribute by the manifestation of one's capabilities through a dynamic discovery of one's potential being assisted in this process by the society and natural surroundings. The technological approach alone cannot support a viewpoint that coordinates work and welfare. Still, possession and enjoyment with a spirit of sacrifice, social progress and social justice, material well-being, and spiritual enhancement are also crucial for a strong foundation for a quality life.

Tagore viewed life in both particular and universal contexts and considered that the disparities and the inequalities among human groups might be brought to an optimum level for a better "quality of life." Through a careful intervention of education toward the development of a man's relationship with his "self" and "society" to the universal consciousness, it is possible to achieve. Man's primordial desires for a self-image and identity and to know the world is to be coordinated, synthesized, and universalized for the comprehensive appraisal of "quality of life" (Adhikari, 2003).

Swami Atmapriyananda articulates that the role of science and technology in enhancing quality of life should be in this way: Through the best use of the scientific application, one can maximize the quality of life under any constraints. Such an application engenders a scientific outlook and a temper that helps us meet the challenges coming in the way of quality maximization. While technology helps to enhance the quality of physical life—bodily comforts, a healthy environment, etc.—the scientific outlook of life gives us a different kind of productive mind and an enlightened intellect. Unless we achieve success in this mental and intellectual aspect of the quality of life, mere enhancement of the physical quality through technology is perhaps more of an evil and danger.

The Rio + 20 Summit, held in June 2012, stressed that every organization should strive for "continuous value enhancement for the whole of society." One should focus on those developments that meet the needs of the present without compromising future generations' ability to meet their own needs. This is the real sustainable development. According to the United Nations Development Program (UNDP), for sustainable development to be achieved, it is crucial to harmonize three core elements: economic growth, social inclusion, and environmental protection. These elements are interconnected, and all are crucial for the well-being of individuals and societies. With this objective, UNDP brought out the sustainable development goals (SDG) with a universal call to action to end poverty, protect the planet, and ensure that all people enjoy peace and prosperity. The SDGs came into effect in January 2016, and they will continue to guide UNDP policy and funding until 2030. UNDP is uniquely placed as the lead UN development agency to help implement the goals through their work in some 170 countries and territories. These 17 goals build on the successes of the Millennium Development Goals (MDG) while including new areas such as climate change, economic inequality, innovation, sustainable consumption, peace, and justice, among other priorities. The goals are often interconnected—the key to the success of one will involve tackling issues more commonly associated with another.

One of the significant developments of the plan is the strengthening of data production and better data in policymaking and monitoring, as they are becoming increasingly recognized as a fundamental means for growth. Like MDG, the SDG monitoring experience demonstrates that effective use of data can help galvanize development efforts, implement successful targeted interventions, track performance, and improve accountability. Thus, sustainable development demands a data revolution to improve the availability, quality, timeliness, and disaggregation of data to support the implementation of the new development agenda at all levels (https://www.undp.org/con tent/undp/en/home/sustainable-development-goals.html).

Achieving the SDGs requires the partnership of governments, private sector, civil society, and citizens alike to make sure we leave a better planet for future generations. For quality life to sustain, one should try to know how to live with minimal natural resources and demands. The use of anything in excess is a waste. As discussed in various chapters of this book, by eliminating unnecessary space, material handling, controlled water, and energy usage, one can achieve this. We need a tremendous attitudinal change to accept the changes and deviations for a better quality of life. When the attitude is wrong, motivation is of no use. When the approach is correct, there is no need for motivation. If we are concerned with the people–planet–profit policy, then we can create a win–win situation, and that model will be the best for sustaining a quality life.

Finally, my recommendations for sustainable quality life are as follows:

- Prepare a mission statement for your life.
- Be grateful to all natural resources and its blessings.
- Let the environment and society be a part of your life's priority list.
- Do not miss any opportunities and challenges for any improvement.
- Accept everything as a blessing.
- Respect time and value others.
- Use technology for better living and management of life.
- Adopt and follow a healthy database management system in life.
- Always think of lean, green, and clean opportunities for every aspect of life.
- Amend all quality improvements for better use and sustain.
- Use everything (!) to a minimum.

At the time of writing this chapter, the world is struggling to contain the novel coronavirus (COVID-19) and to find medicine for the virus. Almost all countries are in the grip of this pandemic. There is a complete lockdown and shutdown across the world. People are asked to stay put in their homes. There is an excellent display of empathy, honesty, integrity, and spirituality among the people all over. Service for humanity is on display in various ways. Social media is active in educating people about this deadly virus and its impact. All central and state government machinery is entirely on alert and concentrating all efforts to contain the disease. They brought an end to all ongoing projects and programs. All socioeconomic activities are on suspended animation. A role reversal of conventional norms and rules which man proposed is working against human existence. The environment is teaching a lesson to humanity. We need to understand the situation and rise to the occasion.

According to Vaidhyanathan (2020), the lesson from the viral infection will always remain the same—to move from fear to love, from selfishness to unselfishness, from taking to giving, and from hoarding to distribution and sharing. As and when we realize this, and we stop feeling scared, spreading panic and anxiety, and start becoming more robust, start trusting ourselves and our divine nature, put our faith in a higher power and surrender, let go, trust, accept, love, hope, forgive and be peaceful, our vibrations will rise to higher levels dramatically. The threat from the virus will recede automatically. We have to stop getting scared and realize that we have ignored those who suffer many more days due to poverty, hunger, extremism, illiteracy, outdated beliefs, accidents, and other diseases and turn our loving attention on them.

Quality excellence in every aspect of life will be the focus of this new century. The last century had seen the necessary seeds for technological development, industrial growth, and human intelligence. All these contributed to the explosive growth in science and technology that led to the rise of consumerism, intensified international competition in quality, and above all, threats to human safety, health, and the environment. So be careful and safe.V

*"Loka-samastha-sukhino-bhavanthu."*

# References

Adhikary, A. K. (2003). Tagore's view on Quality life, an edited volume of selected papers presented at the National seminar of Quality of life, Das and Bhattacharya (eds.), pp. 29–31.

Argyris, C. (1998). "Empowerment: The emperor's New Clothes," *Harward Business Review*, May–June, pp. 98–105.

Butler, G., Caldwell, C., & Poston, N. (2009). Lean six sigma for healthcare, ASQ, Quality Press.

Convey, S. (1994). Eliminating unproductive activities and processes. *CMA Magazine, 65*(9), 20–24.

Deming. (1986). Out of the crisis. MIT Centre for Advanced Engineering Study.

George, M. L. (2003). *Lean six sigma for service*. McGraw Hill.

George, M. L. (2002). Lean six sigma: Combining six sigma quality with lean speed. McGraw Hill.

Gryna, F. M., Chua, R. C. H., & Defeo, J. (2007). *Juran's quality planning and analysis for Enterprise Quality*. Tata McGraw-Hill.

Harry, M. J., & Mann, P. S. (2010). *Practitioner's guide to statistics and lean six sigma for process improvement*. Wiley.

Kaplan, R. S., & Cooper, R. (1998). Cost and effect: Using integrated cost system to drive profitability and performance. Havard business school press.

Koskela, L. (1992). Application of new production philosophy to construction. Technical Report, Stanford University.

Koskela, L. (2000). *An exploration towards a production theory and its application to construction (Doctoral dissertation)*. Helsinki University of Technology.

Montgomery, D. C. (2003). *Introduction to statistical quality control,* (4th ed.). Wiley.

Muir, A. (2006). *Lean six sigma way*. McGraw Hill.

Mukherjee, A. (2003). *Quality of life—Concept and its concern*, Quality of life, in B. Das, & G. Bhattacharya (eds.) Indian Association for Productivity, Quality and Reliability, pp. 59–65.

Muralidharan, K. (2015). Six sigma for organizational excellence: A statistical approach, Springer.

Pande, P. S. Newuman, R. P., & Cavanagh, R. R. (2003). The six sigma way. Tata McGraw Hill.

Rust, R., Zahorik, A., & Kewiningham, T. (1994). *Return on quality*. Probus Publishing.

Sen, A. (1993). Capability and well-being in quality of life, in M. Nussbaum & A. Sen, Oxford University Press.

Vaidhyanathan. (2020). The covid-19 challenge, The Speaking Tree, TOI dated 17.03.2020.

Womack, J. P., & Jones, D. T. (1996). Lean thinking: Banish waste and create wealth in your organization. New York: Simon and Schuster.

Womack, P., & Jones, D. T. (2003). *Lean thinking*. Simon & Schuster.

Zokaei, K., Lovins, H., Wood, A., & Hines, P. (2013). *Creating lean and green systems: Techniques for improving profits and sustainability*. CRC Press.

www.mmt-inst.com

www.un.org/sustainabledevelopment/development-agenda/

www.qualitygurus.com

https://www.undp.org/content/undp/en/home/sustainable-development-goals.html

# Appendix

See Tables A.1, A.2, A.3, A.4, A.5, A.6, A.7, and A.8.

**Table A.1** Tables of binomial probability sums

$$\Pr(x \le K) = \sum_{x=0}^{K} \binom{n}{x} p^x (1-p)^{n-x}$$

| K | \ | $p = 0.1$ | 0.2 | 0.3 | 0.4 | 0.5 | 0.6 | 0.7 | 0.8 | 0.9 |
|---|---|---|---|---|---|---|---|---|---|---|
| $n = 2$ | | | | | | | | | | |
| 0 | — | 0.81 | 0.64 | 0.49 | 0.36 | 0.25 | 0.16 | 0.09 | 0.04 | 0.01 |
| 1 | — | 0.99 | 0.96 | 0.91 | 0.84 | 0.75 | 0.64 | 0.51 | 0.36 | 0.19 |
| 2 | — | 1.00 | 1.00 | 1.00 | 1.00 | 1.00 | 1.00 | 1.00 | 1.00 | 1.00 |
| $n = 3$ | | | | | | | | | | |
| 0 | — | 0.729 | 0.512 | 0.343 | 0.216 | 0.125 | 0.064 | 0.027 | 0.008 | 0.001 |
| 1 | — | 0.972 | 0.896 | 0.784 | 0.648 | 0.500 | 0.352 | 0.216 | 0.104 | 0.028 |
| 2 | — | 0.999 | 0.992 | 0.973 | 0.936 | 0.875 | 0.784 | 0.657 | 0.488 | 0.271 |
| 3 | — | 1.000 | 1.000 | 1.000 | 1.000 | 1.000 | 1.000 | 1.000 | 1.000 | 1.000 |
| $n = 4$ | | | | | | | | | | |
| 0 | — | 0.6561 | 0.4096 | 0.2401 | 0.1296 | 0.0625 | 0.0256 | 0.0081 | 0.0016 | 0.0001 |
| 1 | — | 0.9477 | 0.8192 | 0.6517 | 0.4752 | 0.3125 | 0.1792 | 0.0837 | 0.0272 | 0.0037 |
| 2 | — | 0.9963 | 0.9728 | 0.9163 | 0.8208 | 0.6875 | 0.5248 | 0.3483 | 0.1808 | 0.0523 |
| 3 | — | 0.9999 | 0.9984 | 0.9919 | 0.9744 | 0.9375 | 0.8704 | 0.7599 | 0.5904 | 0.3439 |
| 4 | — | 1.0000 | 1.0000 | 1.0000 | 1.0000 | 1.0000 | 1.0000 | 1.0000 | 1.0000 | 1.0000 |
| $n = 5$ | | | | | | | | | | |
| 0 | — | 0.59049 | 0.32768 | 0.16807 | 0.07776 | 0.03125 | 0.01024 | 0.00243 | 0.00032 | 0.00001 |
| 1 | — | 0.91854 | 0.73728 | 0.52822 | 0.33696 | 0.18750 | 0.08704 | 0.03078 | 0.00672 | 0.00046 |
| 2 | — | 0.99144 | 0.94208 | 0.83692 | 0.68256 | 0.50000 | 0.31744 | 0.16308 | 0.05792 | 0.00856 |

(continued)

**Table A.1** (continued)

$$\Pr(x \le K) = \sum_{x=0}^{K} \binom{n}{x} p^x (1-p)^{n-x}$$

| K | | p = 0.1 | 0.2 | 0.3 | 0.4 | 0.5 | 0.6 | 0.7 | 0.8 | 0.9 |
|---|---|---|---|---|---|---|---|---|---|---|
| 3 | | 0.99954 | 0.99328 | 0.96922 | 0.91296 | 0.81250 | 0.66304 | 0.47178 | 0.26272 | 0.08146 |
| 4 | | 0.99999 | 0.99968 | 0.99757 | 0.98976 | 0.96875 | 0.92224 | 0.83193 | 0.67232 | 0.40951 |
| 5 | | 1.00000 | 1.00000 | 1.00000 | 1.00000 | 1.00000 | 1.00000 | 1.00000 | 1.00000 | 1.00000 |
| n = 6 | | | | | | | | | | |
| 0 | | 0.53144 | 0.26214 | 0.11765 | 0.04666 | 0.01562 | 0.00410 | 0.00073 | 0.00006 | 0.00000 |
| 1 | | 0.88574 | 0.65536 | 0.42018 | 0.23328 | 0.10938 | 0.04096 | 0.01094 | 0.00160 | 0.00006 |
| 2 | | 0.98415 | 0.90112 | 0.74431 | 0.54432 | 0.34375 | 0.17290 | 0.07047 | 0.01696 | 0.00127 |
| 3 | | 0.99873 | 0.98304 | 0.92953 | 0.82080 | 0.65625 | 0.45568 | 0.25569 | 0.09888 | 0.01585 |
| 4 | | 0.99994 | 0.99840 | 0.98906 | 0.95904 | 0.89062 | 0.76672 | 0.57982 | 0.34464 | 0.11426 |
| 5 | | 1.00000 | 0.99994 | 0.99927 | 0.99590 | 0.98438 | 0.95334 | 0.88235 | 0.73786 | 0.46856 |
| 6 | | 1.00000 | 1.00000 | 1.00000 | 1.00000 | 1.00000 | 1.00000 | 1.00000 | 1.00000 | 1.00000 |
| n = 7 | | | | | | | | | | |
| 0 | | 0.47830 | 0.20972 | 0.08235 | 0.02799 | 0.00781 | 0.00164 | 0.00022 | 0.00001 | 0.00000 |
| 1 | | 0.85031 | 0.57672 | 0.32942 | 0.15863 | 0.06250 | 0.01884 | 0.00379 | 0.00037 | 0.00001 |
| 2 | | 0.97431 | 0.85197 | 0.64707 | 0.41990 | 0.22656 | 0.09626 | 0.02880 | 0.00467 | 0.00018 |
| 3 | | 0.99727 | 0.96666 | 0.87396 | 0.71021 | 0.50000 | 0.28979 | 0.12604 | 0.03334 | 0.00273 |
| 4 | | 0.99982 | 0.99533 | 0.97120 | 0.90374 | 0.77344 | 0.58010 | 0.35293 | 0.14803 | 0.02569 |
| 5 | | 0.99999 | 0.99963 | 0.99621 | 0.98116 | 0.93750 | 0.84137 | 0.67058 | 0.42328 | 0.14969 |

(continued)

**Table A.1** (continued)

$$\Pr(x \le K) = \sum_{x=0}^{K} \binom{n}{x} p^x (1-p)^{n-x}$$

| K | | p = 0.1 | 0.2 | 0.3 | 0.4 | 0.5 | 0.6 | 0.7 | 0.8 | 0.9 |
|---|---|---|---|---|---|---|---|---|---|---|
| 6 | | 1.00000 | 0.99999 | 0.99978 | 0.99836 | 0.99219 | 0.97201 | 0.97165 | 0.79028 | 0.52170 |
| 7 | | 1.00000 | 1.00000 | 1.00000 | 1.00000 | 1.00000 | 1.00000 | 1.00000 | 1.00000 | 1.00000 |
| n = 8 | | | | | | | | | | |
| 0 | | 0.43047 | 0.16777 | 0.05765 | 0.01680 | 0.00391 | 0.00066 | 0.00007 | 0.00000 | 0.00000 |
| 1 | | 0.81310 | 0.50332 | 0.25530 | 0.10638 | 0.03516 | 0.00852 | 0.00129 | 0.00008 | 0.00000 |
| 2 | | 0.96191 | 0.79692 | 0.55177 | 0.31539 | 0.14453 | 0.04981 | 0.01129 | 0.00123 | 0.00002 |
| 3 | | 0.99498 | 0.94372 | 0.80590 | 0.59409 | 0.36328 | 0.17367 | 0.05797 | 0.01041 | 0.00043 |
| 4 | | 0.99957 | 0.98959 | 0.94203 | 0.82633 | 0.63672 | 0.40591 | 0.19410 | 0.05628 | 0.00502 |
| 5 | | 0.99998 | 0.99877 | 0.98871 | 0.95019 | 0.85547 | 0.68461 | 0.44823 | 0.20308 | 0.03809 |
| 6 | | 1.00000 | 0.99992 | 0.99871 | 0.99148 | 0.96484 | 0.89362 | 0.74470 | 0.49668 | 0.18690 |
| 7 | | 1.00000 | 1.00000 | 0.99993 | 0.99934 | 0.99609 | 0.98320 | 0.94235 | 0.83223 | 0.56953 |
| 8 | | 1.00000 | 1.00000 | 1.00000 | 1.00000 | 1.00000 | 1.00000 | 1.00000 | 1.00000 | 1.00000 |
| n = 9 | | | | | | | | | | |
| 0 | | 0.38742 | 0.13422 | 0.04035 | 0.01008 | 0.00195 | 0.00026 | 0.00002 | 0.00000 | 0.00000 |
| 1 | | 0.77484 | 0.43621 | 0.19600 | 0.07054 | 0.01953 | 0.00380 | 0.00043 | 0.00002 | 0.00000 |
| 2 | | 0.94703 | 0.73820 | 0.46283 | 0.23179 | 0.08984 | 0.02503 | 0.00429 | 0.00031 | 0.00000 |
| 3 | | 0.99167 | 0.91436 | 0.72966 | 0.48261 | 0.25391 | 0.09935 | 0.02529 | 0.00307 | 0.00006 |
| 4 | | 0.99911 | 0.98042 | 0.90119 | 0.73343 | 0.50000 | 0.26657 | 0.09881 | 0.01958 | 0.00089 |

(continued)

**Table A.1** (continued)

$$\Pr(x \le K) = \sum_{x=0}^{K} \binom{n}{x} p^x (1-p)^{n-x}$$

| K | | $p = 0.1$ | 0.2 | 0.3 | 0.4 | 0.5 | 0.6 | 0.7 | 0.8 | 0.9 |
|---|---|---|---|---|---|---|---|---|---|---|
| 5 | | 0.99994 | 0.99693 | 0.97471 | 0.90065 | 0.74609 | 0.51739 | 0.27034 | 0.08564 | 0.00833 |
| 6 | | 1.00000 | 0.99969 | 0.99571 | 0.97497 | 0.91016 | 0.76821 | 0.53717 | 0.26180 | 0.05297 |
| 7 | | 1.00000 | 0.99998 | 0.99957 | 0.99620 | 0.98047 | 0.92946 | 0.80400 | 0.56379 | 0.22516 |
| 8 | | 1.00000 | 1.00000 | 0.99998 | 0.99974 | 0.99805 | 0.98992 | 0.95965 | 0.86578 | 0.61258 |
| 9 | | 1.00000 | 1.00000 | 1.00000 | 1.00000 | 1.00000 | 1.00000 | 1.00000 | 1.00000 | 1.00000 |
| $n = 10$ | | | | | | | | | | |
| 0 | | 0.34868 | 0.10737 | 0.02825 | 0.00605 | 0.00098 | 0.00010 | 0.00001 | 0.00000 | 0.00000 |
| 1 | | 0.73610 | 0.37581 | 0.14931 | 0.04636 | 0.01074 | 0.00168 | 0.00014 | 0.00000 | 0.00000 |
| 2 | | 0.92981 | 0.67780 | 0.38278 | 0.16729 | 0.05469 | 0.01229 | 0.00159 | 0.00008 | 0.00000 |
| 3 | | 0.98720 | 0.87913 | 0.64691 | 0.38228 | 0.17188 | 0.05476 | 0.01059 | 0.00086 | 0.00001 |
| 4 | | 0.99837 | 0.96721 | 0.84973 | 0.63310 | 0.37695 | 0.16624 | 0.04735 | 0.00637 | 0.00015 |
| 5 | | 0.99985 | 0.99363 | 0.95265 | 0.83376 | 0.62305 | 0.36690 | 0.15027 | 0.03279 | 0.00163 |
| 6 | | 0.99999 | 0.99914 | 0.98941 | 0.94524 | 0.82812 | 0.61772 | 0.35039 | 0.12087 | 0.01280 |
| 7 | | 1.00000 | 0.99992 | 0.99841 | 0.98771 | 0.94531 | 0.83271 | 0.61722 | 0.32220 | 0.07019 |
| 8 | | 1.00000 | 1.00000 | 0.99986 | 0.99832 | 0.98926 | 0.95364 | 0.85069 | 0.62419 | 0.26390 |
| 9 | | 1.00000 | 1.00000 | 0.99999 | 0.99990 | 0.99902 | 0.99395 | 0.97175 | 0.89263 | 0.65132 |
| 10 | | 1.00000 | 1.00000 | 1.00000 | 1.00000 | 1.00000 | 1.00000 | 1.00000 | 1.00000 | 1.00000 |

(continued)

**Table A.1** (continued)

$$\Pr(x \le K) = \sum_{x=0}^{K} \binom{n}{x} p^x (1-p)^{n-x}$$

| K | \ | p = 0.1 | 0.2 | 0.3 | 0.4 | 0.5 | 0.6 | 0.7 | 0.8 | 0.9 |
|---|---|---|---|---|---|---|---|---|---|---|
| n = 11 | | | | | | | | | | |
| 0 | | 0.31381 | 0.08590 | 0.01977 | 0.00363 | 0.00049 | 0.00004 | 0.00000 | 0.00000 | 0.00000 |
| 1 | | 0.69736 | 0.32212 | 0.11299 | 0.03023 | 0.00586 | 0.00073 | 0.00005 | 0.00000 | 0.00000 |
| 2 | | 0.91044 | 0.61740 | 0.31274 | 0.11892 | 0.03271 | 0.00592 | 0.00058 | 0.00002 | 0.00000 |
| 3 | | 0.98147 | 0.83886 | 0.56956 | 0.29628 | 0.11328 | 0.02928 | 0.00429 | 0.00024 | 0.00000 |
| 4 | | 0.99725 | 0.94959 | 0.78790 | 0.53277 | 0.27441 | 0.09935 | 0.02162 | 0.00197 | 0.00002 |
| 5 | | 0.99970 | 0.98835 | 0.92178 | 0.75350 | 0.50000 | 0.24650 | 0.07822 | 0.01165 | 0.00030 |
| 6 | | 0.99998 | 0.99803 | 0.97838 | 0.90065 | 0.72259 | 0.46723 | 0.21030 | 0.05041 | 0.00275 |
| 7 | | 1.00000 | 0.99976 | 0.99571 | 0.97072 | 0.88672 | 0.70372 | 0.43044 | 0.16114 | 0.01853 |
| 8 | | 1.00000 | 0.99998 | 0.99942 | 0.99408 | 0.96729 | 0.88108 | 0.68726 | 0.38260 | 0.08956 |
| 9 | | 1.00000 | 1.00000 | 0.99995 | 0.99927 | 0.99414 | 0.96977 | 0.88701 | 0.67788 | 0.30264 |
| 10 | | 1.00000 | 1.00000 | 1.00000 | 0.99996 | 0.99951 | 0.99637 | 0.98023 | 0.91410 | 0.68619 |
| 11 | | 1.00000 | 1.00000 | 1.00000 | 1.00000 | 1.00000 | 1.00000 | 1.00000 | 1.00000 | 1.00000 |
| n = 12 | | | | | | | | | | |
| 0 | | 0.28243 | 0.06872 | 0.01384 | 0.00218 | 0.00024 | 0.00002 | 0.00000 | 0.00000 | 0.00000 |
| 1 | | 0.65900 | 0.27488 | 0.08503 | 0.01959 | 0.00317 | 0.00032 | 0.00002 | 0.00000 | 0.00000 |
| 2 | | 0.88913 | 0.55835 | 0.25282 | 0.08344 | 0.01929 | 0.00281 | 0.00021 | 0.00000 | 0.00000 |
| 3 | | 0.97436 | 0.79457 | 0.49252 | 0.22534 | 0.07300 | 0.01527 | 0.00169 | 0.00006 | 0.00000 |
| 4 | | 0.99567 | 0.92744 | 0.72366 | 0.43818 | 0.19385 | 0.05731 | 0.00949 | 0.00058 | 0.00000 |

(continued)

**Table A.1** (continued)

$$\Pr(x \le K) = \sum_{x=0}^{K} \binom{n}{x} p^x (1-p)^{n-x}$$

| K | \ | p = 0.1 | 0.2 | 0.3 | 0.4 | 0.5 | 0.6 | 0.7 | 0.8 | 0.9 |
|---|---|---------|-----|-----|-----|-----|-----|-----|-----|-----|
| 5 | \ | 0.99946 | 0.98059 | 0.88215 | 0.66521 | 0.38721 | 0.15821 | 0.03860 | 0.00390 | 0.00005 |
| 6 | \ | 0.99995 | 0.99610 | 0.96140 | 0.84179 | 0.61279 | 0.33479 | 0.11785 | 0.01941 | 0.00054 |
| 7 | \ | 1.00000 | 0.99942 | 0.99051 | 0.94269 | 0.80615 | 0.56182 | 0.27634 | 0.07256 | 0.00433 |
| 8 | \ | 1.00000 | 0.99994 | 0.99831 | 0.98473 | 0.92700 | 0.77466 | 0.50748 | 0.20543 | 0.02564 |
| 9 | \ | 1.00000 | 1.00000 | 0.99979 | 0.99719 | 0.98071 | 0.91656 | 0.74718 | 0.44165 | 0.11087 |
| 10 | \ | 1.00000 | 1.00000 | 0.99998 | 0.99968 | 0.99683 | 0.98041 | 0.91497 | 0.72512 | 0.34100 |
| 11 | \ | 1.00000 | 1.00000 | 1.00000 | 0.99998 | 0.99976 | 0.99782 | 0.98616 | 0.93128 | 0.71757 |
| 12 | \ | 1.00000 | 1.00000 | 1.00000 | 1.00000 | 1.00000 | 1.00000 | 1.00000 | 1.00000 | 1.00000 |
| n = 13 | | | | | | | | | | |
| 0 | \ | 0.25419 | 0.05498 | 0.00969 | 0.00131 | 0.00012 | 0.00001 | 0.00000 | 0.00000 | 0.00000 |
| 1 | \ | 0.62134 | 0.23365 | 0.06367 | 0.01263 | 0.00171 | 0.00014 | 0.00000 | 0.00000 | 0.00000 |
| 2 | \ | 0.86612 | 0.50165 | 0.20248 | 0.05790 | 0.01123 | 0.00132 | 0.00007 | 0.00000 | 0.00000 |
| 3 | \ | 0.96584 | 0.74732 | 0.42061 | 0.16858 | 0.04614 | 0.00779 | 0.00065 | 0.00002 | 0.00000 |
| 4 | \ | 0.99354 | 0.90087 | 0.65431 | 0.35304 | 0.13342 | 0.03208 | 0.00403 | 0.00017 | 0.00000 |
| 5 | \ | 0.99908 | 0.96996 | 0.83460 | 0.57440 | 0.29053 | 0.09767 | 0.01822 | 0.00125 | 0.00001 |
| 6 | \ | 0.99990 | 0.99300 | 0.93762 | 0.77116 | 0.50000 | 0.22884 | 0.06238 | 0.00700 | 0.00010 |
| 7 | \ | 0.99999 | 0.99875 | 0.98178 | 0.90233 | 0.70947 | 0.42560 | 0.16540 | 0.03004 | 0.00092 |
| 8 | \ | 1.00000 | 0.99983 | 0.99597 | 0.96792 | 0.86658 | 0.64696 | 0.34569 | 0.09913 | 0.00646 |

(continued)

**Table A.1** (continued)

$$\Pr(x \le K) = \sum_{x=0}^{K} \binom{n}{x} p^x (1-p)^{n-x}$$

| K | \ | p = 0.1 | 0.2 | 0.3 | 0.4 | 0.5 | 0.6 | 0.7 | 0.8 | 0.9 |
|---|---|---|---|---|---|---|---|---|---|---|
| 9 | | 1.00000 | 0.99998 | 0.99935 | 0.99221 | 0.95386 | 0.83142 | 0.57939 | 0.25268 | 0.03416 |
| 10 | | 1.00000 | 1.00000 | 0.99993 | 0.99868 | 0.98877 | 0.94210 | 0.79752 | 0.49835 | 0.13388 |
| 11 | | 1.00000 | 1.00000 | 1.00000 | 0.99986 | 0.99829 | 0.98737 | 0.93633 | 0.76635 | 0.37866 |
| 12 | | 1.00000 | 1.00000 | 1.00000 | 0.99999 | 0.99988 | 0.99869 | 0.99031 | 0.94502 | 0.74581 |
| 13 | | 1.00000 | 1.00000 | 1.00000 | 1.00000 | 1.00000 | 1.00000 | 1.00000 | 1.00000 | 1.00000 |
| n = 14 | | | | | | | | | | |
| 0 | | 0.22877 | 0.04398 | 0.00678 | 0.00078 | 0.00006 | 0.00000 | 0.00000 | 0.00000 | 0.00000 |
| 1 | | 0.58463 | 0.19791 | 0.04748 | 0.00810 | 0.00092 | 0.00006 | 0.00000 | 0.00000 | 0.00000 |
| 2 | | 0.84164 | 0.44805 | 0.16084 | 0.03979 | 0.00647 | 0.00061 | 0.00003 | 0.00000 | 0.00000 |
| 3 | | 0.95587 | 0.69819 | 0.35517 | 0.12431 | 0.02869 | 0.00391 | 0.00025 | 0.00000 | 0.00000 |
| 4 | | 0.99077 | 0.87016 | 0.58420 | 0.27926 | 0.08978 | 0.01751 | 0.00167 | 0.00005 | 0.00000 |
| 5 | | 0.99853 | 0.95615 | 0.78052 | 0.48585 | 0.21198 | 0.05832 | 0.00829 | 0.00038 | 0.00000 |
| 6 | | 0.99982 | 0.98839 | 0.90672 | 0.69245 | 0.39526 | 0.15014 | 0.03147 | 0.00240 | 0.00002 |
| 7 | | 0.99998 | 0.99760 | 0.96853 | 0.84986 | 0.60474 | 0.30755 | 0.09328 | 0.01161 | 0.00018 |
| 8 | | 1.00000 | 0.99962 | 0.99171 | 0.94168 | 0.78802 | 0.51415 | 0.21948 | 0.04385 | 0.00147 |
| 9 | | 1.00000 | 0.99995 | 0.99833 | 0.98249 | 0.91022 | 0.72074 | 0.41580 | 0.12984 | 0.00923 |
| 10 | | 1.00000 | 1.00000 | 0.99975 | 0.99609 | 0.97131 | 0.87569 | 0.64483 | 0.30181 | 0.04413 |
| 11 | | 1.00000 | 1.00000 | 0.99997 | 0.99939 | 0.99353 | 0.96021 | 0.83916 | 0.55195 | 0.15836 |

(continued)

**Table A.1** (continued)

$$\Pr(x \le K) = \sum_{x=0}^{K} \binom{n}{x} p^x (1-p)^{n-x}$$

| K | \ | p = 0.1 | 0.2 | 0.3 | 0.4 | 0.5 | 0.6 | 0.7 | 0.8 | 0.9 |
|---|---|---------|-----|-----|-----|-----|-----|-----|-----|-----|
| 12 | | 1.00000 | 1.00000 | 1.00000 | 0.99994 | 0.99908 | 0.99190 | 0.95252 | 0.80209 | 0.41537 |
| 13 | | 1.00000 | 1.00000 | 1.00000 | 1.00000 | 0.99994 | 0.99922 | 0.99322 | 0.95602 | 0.77123 |
| 14 | | 1.00000 | 1.00000 | 1.00000 | 1.00000 | 1.00000 | 1.00000 | 1.00000 | 1.00000 | 1.00000 |
| *n* = 15 | | | | | | | | | | |
| 0 | | 0.20589 | 0.03518 | 0.00475 | 0.00047 | 0.00003 | 0.00000 | 0.00000 | 0.00000 | 0.00000 |
| 1 | | 0.54904 | 0.16713 | 0.03527 | 0.00517 | 0.00049 | 0.00003 | 0.00000 | 0.00000 | 0.00000 |
| 2 | | 0.81594 | 0.39802 | 0.12683 | 0.02711 | 0.00369 | 0.00028 | 0.00001 | 0.00000 | 0.00000 |
| 3 | | 0.94444 | 0.64816 | 0.29687 | 0.09050 | 0.01758 | 0.00193 | 0.00009 | 0.00000 | 0.00000 |
| 4 | | 0.98728 | 0.83577 | 0.51549 | 0.21728 | 0.05923 | 0.00935 | 0.00067 | 0.00001 | 0.00000 |
| 5 | | 0.99775 | 0.93895 | 0.72162 | 0.40322 | 0.15088 | 0.03383 | 0.00365 | 0.00011 | 0.00000 |
| 6 | | 0.99969 | 0.98194 | 0.86886 | 0.60981 | 0.30362 | 0.09505 | 0.01524 | 0.00078 | 0.00000 |
| 7 | | 0.99997 | 0.99576 | 0.94999 | 0.78690 | 0.50000 | 0.21310 | 0.05001 | 0.00424 | 0.00003 |
| 8 | | 1.00000 | 0.99922 | 0.98476 | 0.90495 | 0.69638 | 0.39019 | 0.13114 | 0.01806 | 0.00031 |
| 9 | | 1.00000 | 0.99989 | 0.99635 | 0.96617 | 0.84912 | 0.59678 | 0.27838 | 0.06105 | 0.00225 |
| 10 | | 1.00000 | 0.99999 | 0.99933 | 0.99065 | 0.94077 | 0.78272 | 0.48451 | 0.16423 | 0.01272 |
| 11 | | 1.00000 | 1.00000 | 0.99991 | 0.99807 | 0.98242 | 0.90950 | 0.70313 | 0.35317 | 0.05556 |
| 12 | | 1.00000 | 1.00000 | 0.99999 | 0.99972 | 0.99631 | 0.97289 | 0.87317 | 0.60198 | 0.18406 |
| 13 | | 1.00000 | 1.00000 | 1.00000 | 0.99997 | 0.99951 | 0.99483 | 0.96473 | 0.83287 | 0.45096 |

(continued)

**Table A.1** (continued)

$$\Pr(x \leq K) = \sum_{x=0}^{K} \binom{n}{x} p^x (1-p)^{n-x}$$

| K | \ | p = 0.1 | 0.2 | 0.3 | 0.4 | 0.5 | 0.6 | 0.7 | 0.8 | 0.9 |
|---|---|---------|-----|-----|-----|-----|-----|-----|-----|-----|
| 14 | | 1.00000 | 1.00000 | 1.00000 | 1.00000 | 0.99997 | 0.99953 | 0.99525 | 0.96482 | 0.79411 |
| 15 | | 1.00000 | 1.00000 | 1.00000 | 1.00000 | 1.00000 | 1.00000 | 1.00000 | 1.00000 | 1.00000 |
| *n* = 16 | | | | | | | | | | |
| 0 | | 0.18530 | 0.02815 | 0.00332 | 0.00028 | 0.00002 | 0.00000 | 0.00000 | 0.00000 | 0.00000 |
| 1 | | 0.51473 | 0.14074 | 0.02611 | 0.00328 | 0.00026 | 0.00001 | 0.00000 | 0.00000 | 0.00000 |
| 2 | | 0.78925 | 0.35184 | 0.09936 | 0.01834 | 0.00209 | 0.00013 | 0.00000 | 0.00000 | 0.00000 |
| 3 | | 0.93159 | 0.59813 | 0.24586 | 0.06515 | 0.01064 | 0.00094 | 0.00003 | 0.00000 | 0.00000 |
| 4 | | 0.98300 | 0.79825 | 0.44990 | 0.16657 | 0.03841 | 0.00490 | 0.00027 | 0.00000 | 0.00000 |
| 5 | | 0.99670 | 0.91831 | 0.65978 | 0.32884 | 0.10506 | 0.01914 | 0.00157 | 0.00003 | 0.00000 |
| 6 | | 0.99950 | 0.97334 | 0.82469 | 0.52717 | 0.22725 | 0.05832 | 0.00713 | 0.00025 | 0.00000 |
| 7 | | 0.99994 | 0.99300 | 0.92565 | 0.71606 | 0.40181 | 0.14227 | 0.02567 | 0.00148 | 0.00001 |
| 8 | | 0.99999 | 0.99852 | 0.97433 | 0.85773 | 0.59819 | 0.28394 | 0.07435 | 0.00700 | 0.00006 |
| 9 | | 1.00000 | 0.99975 | 0.99287 | 0.94168 | 0.77275 | 0.47283 | 0.17531 | 0.02666 | 0.00050 |
| 10 | | 1.00000 | 0.99997 | 0.99843 | 0.98086 | 0.89494 | 0.67116 | 0.34022 | 0.08169 | 0.00300 |
| 11 | | 1.00000 | 1.00000 | 0.99973 | 0.99510 | 0.96159 | 0.83343 | 0.55010 | 0.20175 | 0.01700 |
| 12 | | 1.00000 | 1.00000 | 0.99997 | 0.99906 | 0.98936 | 0.93485 | 0.75414 | 0.40187 | 0.06841 |
| 13 | | 1.00000 | 1.00000 | 1.00000 | 0.99987 | 0.99791 | 0.98166 | 0.90064 | 0.64816 | 0.21075 |
| 14 | | 1.00000 | 1.00000 | 1.00000 | 0.99999 | 0.99974 | 0.99671 | 0.97389 | 0.85926 | 0.48527 |

(continued)

**Table A.1** (continued)

$$\Pr(x \leq K) = \sum_{x=0}^{K} \binom{n}{x} p^x (1-p)^{n-x}$$

| K | \ | p = 0.1 | 0.2 | 0.3 | 0.4 | 0.5 | 0.6 | 0.7 | 0.8 | 0.9 |
|---|---|---------|-----|-----|-----|-----|-----|-----|-----|-----|
| 15 | — | 1.00000 | 1.00000 | 1.00000 | 1.00000 | 0.99998 | 0.99972 | 0.99668 | 0.97185 | 0.81470 |
| 16 | — | 1.00000 | 1.00000 | 1.00000 | 1.00000 | 1.00000 | 1.00000 | 1.00000 | 1.00000 | 1.00000 |
| *n* = 17 | | | | | | | | | | |
| 0 | — | 0.16677 | 0.02252 | 0.00233 | 0.00017 | 0.00001 | 0.00000 | 0.00000 | 0.00000 | 0.00000 |
| 1 | — | 0.48179 | 0.11822 | 0.01928 | 0.00209 | 0.00014 | 0.00000 | 0.00000 | 0.00000 | 0.00000 |
| 2 | — | 0.76180 | 0.30962 | 0.07739 | 0.01232 | 0.00117 | 0.00006 | 0.00000 | 0.00000 | 0.00000 |
| 3 | — | 0.91736 | 0.54888 | 0.20191 | 0.04642 | 0.00636 | 0.00045 | 0.00001 | 0.00000 | 0.00000 |
| 4 | — | 0.97786 | 0.75822 | 0.38869 | 0.12600 | 0.02452 | 0.00252 | 0.00010 | 0.00000 | 0.00000 |
| 5 | — | 0.99533 | 0.89430 | 0.59682 | 0.26393 | 0.07173 | 0.01059 | 0.00066 | 0.00001 | 0.00000 |
| 6 | — | 0.99922 | 0.96234 | 0.77522 | 0.44784 | 0.16615 | 0.03481 | 0.00324 | 0.00008 | 0.00000 |
| 7 | — | 0.99989 | 0.98907 | 0.89536 | 0.64051 | 0.31453 | 0.09190 | 0.01269 | 0.00049 | 0.00000 |
| 8 | — | 0.99999 | 0.99742 | 0.95972 | 0.80106 | 0.50000 | 0.19894 | 0.04028 | 0.00258 | 0.00001 |
| 9 | — | 1.00000 | 0.99951 | 0.98731 | 0.90810 | 0.68547 | 0.35949 | 0.10464 | 0.01093 | 0.00011 |
| 10 | — | 1.00000 | 0.99992 | 0.99676 | 0.96519 | 0.83385 | 0.55216 | 0.22478 | 0.03766 | 0.00078 |
| 11 | — | 1.00000 | 0.99999 | 0.99934 | 0.98941 | 0.92827 | 0.73607 | 0.40318 | 0.10570 | 0.00467 |
| 12 | — | 1.00000 | 1.00000 | 0.99990 | 0.99748 | 0.97548 | 0.87400 | 0.61131 | 0.24178 | 0.02214 |
| 13 | — | 1.00000 | 1.00000 | 0.99999 | 0.99955 | 0.99364 | 0.95358 | 0.79809 | 0.45112 | 0.08264 |
| 14 | — | 1.00000 | 1.00000 | 1.00000 | 0.99994 | 0.99883 | 0.98768 | 0.92261 | 0.69038 | 0.23820 |

(continued)

**Table A.1** (continued)

$$\Pr(x \le K) = \sum_{x=0}^{K} \binom{n}{x} p^x (1-p)^{n-x}$$

| K | \ | p = 0.1 | 0.2 | 0.3 | 0.4 | 0.5 | 0.6 | 0.7 | 0.8 | 0.9 |
|---|---|---|---|---|---|---|---|---|---|---|
| 15 | | 1.00000 | 1.00000 | 1.00000 | 1.00000 | 0.99986 | 0.99791 | 0.98072 | 0.88178 | 0.51821 |
| 16 | | 1.00000 | 1.00000 | 1.00000 | 1.00000 | 0.99999 | 0.99983 | 0.99767 | 0.97748 | 0.83323 |
| 17 | | 1.00000 | 1.00000 | 1.00000 | 1.00000 | 1.00000 | 1.00000 | 1.00000 | 1.00000 | 1.00000 |
| n = 18 | | | | | | | | | | |
| 0 | | 0.15009 | 0.01801 | 0.00163 | 0.00010 | 0.00000 | 0.00000 | 0.00000 | 0.00000 | 0.00000 |
| 1 | | 0.45028 | 0.09908 | 0.01419 | 0.00132 | 0.00007 | 0.00000 | 0.00000 | 0.00000 | 0.00000 |
| 2 | | 0.73380 | 0.27134 | 0.05995 | 0.00823 | 0.00066 | 0.00003 | 0.00000 | 0.00000 | 0.00000 |
| 3 | | 0.90180 | 0.50103 | 0.16455 | 0.03278 | 0.00377 | 0.00021 | 0.00000 | 0.00000 | 0.00000 |
| 4 | | 0.97181 | 0.71635 | 0.33265 | 0.09417 | 0.01544 | 0.00128 | 0.00004 | 0.00000 | 0.00000 |
| 5 | | 0.99358 | 0.86708 | 0.53438 | 0.20876 | 0.04813 | 0.00575 | 0.00027 | 0.00000 | 0.00000 |
| 6 | | 0.99883 | 0.94873 | 0.72170 | 0.37428 | 0.11894 | 0.02028 | 0.00143 | 0.00002 | 0.00000 |
| 7 | | 0.99983 | 0.98372 | 0.85932 | 0.56344 | 0.24034 | 0.05765 | 0.00607 | 0.00016 | 0.00000 |
| 8 | | 0.99998 | 0.99575 | 0.94041 | 0.73684 | 0.40726 | 0.13471 | 0.02097 | 0.00091 | 0.00000 |
| 9 | | 1.00000 | 0.99909 | 0.97903 | 0.86529 | 0.59274 | 0.26316 | 0.05959 | 0.00425 | 0.00002 |
| 10 | | 1.00000 | 0.99984 | 0.99393 | 0.94235 | 0.75966 | 0.43656 | 0.14068 | 0.01628 | 0.00017 |
| 11 | | 1.00000 | 0.99998 | 0.99857 | 0.97972 | 0.88106 | 0.62572 | 0.27830 | 0.05127 | 0.00117 |
| 12 | | 1.00000 | 1.00000 | 0.99973 | 0.99425 | 0.95187 | 0.79124 | 0.46562 | 0.13292 | 0.00642 |
| 13 | | 1.00000 | 1.00000 | 0.99996 | 0.99872 | 0.98456 | 0.90583 | 0.66735 | 0.28365 | 0.02819 |

(continued)

**Table A.1** (continued)

$$\Pr(x \le K) = \sum_{x=0}^{K} \binom{n}{x} p^x (1-p)^{n-x}$$

| K | \ | p = 0.1 | 0.2 | 0.3 | 0.4 | 0.5 | 0.6 | 0.7 | 0.8 | 0.9 |
|---|---|---------|-----|-----|-----|-----|-----|-----|-----|-----|
| 14 | — | 1.00000 | 1.00000 | 1.00000 | 0.99979 | 0.99623 | 0.96722 | 0.83545 | 0.49897 | 0.09820 |
| 15 | — | 1.00000 | 1.00000 | 1.00000 | 0.99997 | 0.99934 | 0.99177 | 0.94005 | 0.72866 | 0.26620 |
| 16 | — | 1.00000 | 1.00000 | 1.00000 | 1.00000 | 0.99993 | 0.99868 | 0.98581 | 0.90092 | 0.54972 |
| 17 | — | 1.00000 | 1.00000 | 1.00000 | 1.00000 | 1.00000 | 0.99990 | 0.99837 | 0.98199 | 0.84991 |
| 18 | — | 1.00000 | 1.00000 | 1.00000 | 1.00000 | 1.00000 | 1.00000 | 1.00000 | 1.00000 | 1.00000 |
| n = 19 | | | | | | | | | | |
| 0 | — | 0.13509 | 0.01441 | 0.00114 | 0.00006 | 0.00000 | 0.00000 | 0.00000 | 0.00000 | 0.00000 |
| 1 | — | 0.42026 | 0.08287 | 0.01042 | 0.00083 | 0.00004 | 0.00000 | 0.00000 | 0.00000 | 0.00000 |
| 2 | — | 0.70544 | 0.23689 | 0.04622 | 0.00546 | 0.00036 | 0.00001 | 0.00000 | 0.00000 | 0.00000 |
| 3 | — | 0.88500 | 0.45509 | 0.13317 | 0.02296 | 0.00221 | 0.00010 | 0.00000 | 0.00000 | 0.00000 |
| 4 | — | 0.96481 | 0.67329 | 0.28222 | 0.06961 | 0.00961 | 0.00064 | 0.00001 | 0.00000 | 0.00000 |
| 5 | — | 0.99141 | 0.83694 | 0.47386 | 0.16292 | 0.03178 | 0.00307 | 0.00011 | 0.00000 | 0.00000 |
| 6 | — | 0.99830 | 0.93240 | 0.66550 | 0.30807 | 0.08353 | 0.01156 | 0.00062 | 0.00001 | 0.00000 |
| 7 | — | 0.99973 | 0.97672 | 0.81083 | 0.48778 | 0.17964 | 0.03523 | 0.00282 | 0.00005 | 0.00000 |
| 8 | — | 0.99996 | 0.99334 | 0.91608 | 0.66748 | 0.32380 | 0.08847 | 0.01054 | 0.00031 | 0.00000 |
| 9 | — | 1.00000 | 0.99842 | 0.96745 | 0.81391 | 0.50000 | 0.18609 | 0.03255 | 0.00158 | 0.00000 |
| 10 | — | 1.00000 | 0.99969 | 0.98946 | 0.91153 | 0.67620 | 0.33252 | 0.08392 | 0.00666 | 0.00004 |
| 11 | — | 1.00000 | 0.99995 | 0.99718 | 0.96477 | 0.82036 | 0.51222 | 0.18197 | 0.02328 | 0.00027 |

(continued)

**Table A.1**  (continued)

$$\Pr(x \leq K) = \sum_{x=0}^{K} \binom{n}{x} p^x (1-p)^{n-x}$$

| K | \ | p = 0.1 | 0.2 | 0.3 | 0.4 | 0.5 | 0.6 | 0.7 | 0.8 | 0.9 |
|---|---|---------|-----|-----|-----|-----|-----|-----|-----|-----|
| 12 |  | 1.00000 | 0.99999 | 0.99938 | 0.98844 | 0.91647 | 0.69193 | 0.33450 | 0.06760 | 0.00170 |
| 13 |  | 1.00000 | 1.00000 | 0.99989 | 0.99693 | 0.96822 | 0.83708 | 0.52614 | 0.16306 | 0.00859 |
| 14 |  | 1.00000 | 1.00000 | 0.99999 | 0.99936 | 0.99039 | 0.93039 | 0.71778 | 0.32671 | 0.03519 |
| 15 |  | 1.00000 | 1.00000 | 1.00000 | 0.99990 | 0.99779 | 0.97704 | 0.86683 | 0.54491 | 0.11500 |
| 16 |  | 1.00000 | 1.00000 | 1.00000 | 0.99999 | 0.99964 | 0.99454 | 0.95378 | 0.76311 | 0.29456 |
| 17 |  | 1.00000 | 1.00000 | 1.00000 | 1.00000 | 0.99996 | 0.99917 | 0.98958 | 0.91713 | 0.57974 |
| 18 |  | 1.00000 | 1.00000 | 1.00000 | 1.00000 | 1.00000 | 0.99994 | 0.99886 | 0.98559 | 0.86491 |
| 19 |  | 1.00000 | 1.00000 | 1.00000 | 1.00000 | 1.00000 | 1.00000 | 1.00000 | 1.00000 | 1.00000 |
| n = 20 |  |  |  |  |  |  |  |  |  |  |
| 0 |  | 0.12158 | 0.01153 | 0.00080 | 0.00004 | 0.00000 | 0.00000 | 0.00000 | 0.00000 | 0.00000 |
| 1 |  | 0.39175 | 0.06918 | 0.00764 | 0.00052 | 0.00002 | 0.00000 | 0.00000 | 0.00000 | 0.00000 |
| 2 |  | 0.67693 | 0.20608 | 0.03548 | 0.00361 | 0.00020 | 0.00001 | 0.00000 | 0.00000 | 0.00000 |
| 3 |  | 0.86705 | 0.41145 | 0.10709 | 0.01596 | 0.00129 | 0.00005 | 0.00000 | 0.00000 | 0.00000 |
| 4 |  | 0.95683 | 0.62965 | 0.23751 | 0.05095 | 0.00591 | 0.00032 | 0.00001 | 0.00000 | 0.00000 |
| 5 |  | 0.98875 | 0.80421 | 0.41637 | 0.12560 | 0.02069 | 0.00161 | 0.00004 | 0.00000 | 0.00000 |
| 6 |  | 0.99761 | 0.91331 | 0.60801 | 0.25001 | 0.05766 | 0.00647 | 0.00026 | 0.00000 | 0.00000 |
| 7 |  | 0.99958 | 0.96786 | 0.77227 | 0.41589 | 0.13159 | 0.02103 | 0.00128 | 0.00002 | 0.00000 |
| 8 |  | 0.99994 | 0.99002 | 0.88667 | 0.59560 | 0.25172 | 0.05653 | 0.00514 | 0.00010 | 0.00000 |

(continued)

**Table A.1** (continued)

$$\Pr(x \le K) = \sum_{x=0}^{K} \binom{n}{x} p^x (1-p)^{n-x}$$

| K | \ | p = 0.1 | 0.2 | 0.3 | 0.4 | 0.5 | 0.6 | 0.7 | 0.8 | 0.9 |
|---|---|---------|-----|-----|-----|-----|-----|-----|-----|-----|
| 9  | | 0.99999 | 0.99741 | 0.95204 | 0.75534 | 0.41190 | 0.12752 | 0.01714 | 0.00056 | 0.00000 |
| 10 | | 1.00000 | 0.99944 | 0.98286 | 0.87248 | 0.58810 | 0.24466 | 0.04796 | 0.00259 | 0.00001 |
| 11 | | 1.00000 | 0.99990 | 0.99486 | 0.94347 | 0.74828 | 0.40440 | 0.11333 | 0.00998 | 0.00006 |
| 12 | | 1.00000 | 0.99998 | 0.99872 | 0.97897 | 0.86841 | 0.58411 | 0.22773 | 0.03214 | 0.00042 |
| 13 | | 1.00000 | 1.00000 | 0.99974 | 0.99353 | 0.94234 | 0.74999 | 0.39199 | 0.08669 | 0.00239 |
| 14 | | 1.00000 | 1.00000 | 0.99996 | 0.99839 | 0.97931 | 0.87440 | 0.58363 | 0.19579 | 0.01125 |
| 15 | | 1.00000 | 1.00000 | 0.99999 | 0.99968 | 0.99409 | 0.94905 | 0.76249 | 0.37035 | 0.04317 |
| 16 | | 1.00000 | 1.00000 | 1.00000 | 0.99995 | 0.99871 | 0.98404 | 0.89291 | 0.58855 | 0.13295 |
| 17 | | 1.00000 | 1.00000 | 1.00000 | 0.99999 | 0.99980 | 0.99639 | 0.96452 | 0.79392 | 0.32307 |
| 18 | | 1.00000 | 1.00000 | 1.00000 | 1.00000 | 0.99998 | 0.99948 | 0.99236 | 0.93082 | 0.60825 |
| 19 | | 1.00000 | 1.00000 | 1.00000 | 1.00000 | 1.00000 | 0.99996 | 0.99920 | 0.98847 | 0.87842 |
| 20 | | 1.00000 | 1.00000 | 1.00000 | 1.00000 | 1.00000 | 1.00000 | 1.00000 | 1.00000 | 1.00000 |

**Table A.2** Cumulative Poisson distribution tables

$\lambda$ = mean

| X | 0.01 | 0.05 | 0.1 | 0.2 | 0.3 | 0.4 | 0.5 | 0.6 | 0.7 | 0.8 | 0.9 |
|---|---|---|---|---|---|---|---|---|---|---|---|
| 0 | 0.990 | 0.951 | 0.905 | 0.819 | 0.741 | 0.670 | 0.607 | 0.549 | 0.497 | 0.449 | 0.407 |
| 1 | 1.000 | 0.999 | 0.995 | 0.982 | 0.963 | 0.938 | 0.910 | 0.878 | 0.844 | 0.809 | 0.772 |
| 2 | | 1.000 | 1.000 | 0.999 | 0.996 | 0.992 | 0.986 | 0.977 | 0.966 | 0.953 | 0.937 |
| 3 | | | | 1.000 | 1.000 | 0.999 | 0.998 | 0.997 | 0.994 | 0.991 | 0.987 |
| 4 | | | | | | 1.000 | 1.000 | 1.000 | 0.999 | 0.999 | 0.998 |
| 5 | | | | | | | | | 1.000 | 1.000 | 1.000 |

| X | 1.0 | 1.1 | 1.2 | 1.3 | 1.4 | 1.5 | 1.6 | 1.7 | 1.8 | 1.9 | 2.0 |
|---|---|---|---|---|---|---|---|---|---|---|---|
| 0 | 0.368 | 0.333 | 0.301 | 0.273 | 0.247 | 0.223 | 0.202 | 0.183 | 0.165 | 0.150 | 0.135 |
| 1 | 0.736 | 0.699 | 0.663 | 0.627 | 0.592 | 0.558 | 0.525 | 0.493 | 0.463 | 0.434 | 0.406 |
| 2 | 0.920 | 0.900 | 0.879 | 0.857 | 0.833 | 0.809 | 0.783 | 0.757 | 0.731 | 0.704 | 0.677 |
| 3 | 0.981 | 0.974 | 0.966 | 0.957 | 0.946 | 0.934 | 0.921 | 0.907 | 0.891 | 0.875 | 0.857 |
| 4 | 0.996 | 0.995 | 0.992 | 0.989 | 0.986 | 0.981 | 0.976 | 0.970 | 0.964 | 0.956 | 0.947 |
| 5 | 0.999 | 0.999 | 0.998 | 0.998 | 0.997 | 0.996 | 0.994 | 0.992 | 0.990 | 0.987 | 0.983 |
| 6 | 1.000 | 1.000 | 1.000 | 1.000 | 0.999 | 0.999 | 0.999 | 0.998 | 0.997 | 0.997 | 0.995 |
| 7 | | | | 1.000 | 1.000 | 1.000 | 1.000 | 0.999 | 0.999 | 0.999 | |
| 8 | | | | | | | | | 1.000 | 1.000 | 1.000 |

| X | 2.2 | 2.4 | 2.6 | 2.8 | 3.0 | 3.5 | 4.0 | 4.5 | 5.0 | 5.5 | 6.0 |
|---|---|---|---|---|---|---|---|---|---|---|---|
| 0 | 0.111 | 0.091 | 0.074 | 0.061 | 0.050 | 0.030 | 0.018 | 0.011 | 0.007 | 0.004 | 0.002 |
| 1 | 0.355 | 0.308 | 0.267 | 0.231 | 0.199 | 0.136 | 0.092 | 0.061 | 0.040 | 0.027 | 0.017 |
| 2 | 0.623 | 0.570 | 0.518 | 0.469 | 0.423 | 0.321 | 0.238 | 0.174 | 0.125 | 0.088 | 0.062 |
| 3 | 0.819 | 0.779 | 0.736 | 0.692 | 0.647 | 0.537 | 0.433 | 0.342 | 0.265 | 0.202 | 0.151 |
| 4 | 0.928 | 0.904 | 0.877 | 0.848 | 0.815 | 0.725 | 0.629 | 0.532 | 0.440 | 0.358 | 0.285 |
| 5 | 0.975 | 0.964 | 0.951 | 0.935 | 0.916 | 0.858 | 0.785 | 0.703 | 0.616 | 0.529 | 0.446 |
| 6 | 0.993 | 0.988 | 0.983 | 0.976 | 0.966 | 0.935 | 0.889 | 0.831 | 0.762 | 0.686 | 0.606 |
| 7 | 0.998 | 0.997 | 0.995 | 0.992 | 0.988 | 0.973 | 0.949 | 0.913 | 0.867 | 0.809 | 0.744 |
| 8 | 1.000 | 0.999 | 0.999 | 0.998 | 0.996 | 0.990 | 0.979 | 0.960 | 0.932 | 0.894 | 0.847 |
| 9 | | 1.000 | 1.000 | 0.999 | 0.999 | 0.997 | 0.992 | 0.983 | 0.968 | 0.946 | 0.916 |
| 10 | | | | 1.000 | 1.000 | 0.999 | 0.997 | 0.993 | 0.986 | 0.975 | 0.957 |
| 11 | | | | | | 1.000 | 0.999 | 0.998 | 0.995 | 0.989 | 0.980 |
| 12 | | | | | | | 1.000 | 0.999 | 0.998 | 0.996 | 0.991 |
| 13 | | | | | | | | 1.000 | 0.999 | 0.998 | 0.996 |
| 14 | | | | | | | | | 1.000 | 0.999 | 0.999 |
| 15 | | | | | | | | | | 1.000 | 0.999 |
| 16 | | | | | | | | | | | 1.000 |

(continued)

**Table A.2** (continued)

| X | 6.5 | 7.0 | 7.5 | 8.0 | 9.0 | 10.0 | 12.0 | 14.0 | 16.0 | 18.0 | 20.0 |
|---|-----|-----|-----|-----|-----|------|------|------|------|------|------|
| 0 | 0.002 | 0.001 | 0.001 | 0.000 | | | | | | | |
| 1 | 0.011 | 0.007 | 0.005 | 0.003 | 0.001 | | | | | | |
| 2 | 0.043 | 0.030 | 0.020 | 0.014 | 0.006 | 0.003 | 0.001 | | | | |
| 3 | 0.112 | 0.082 | 0.059 | 0.042 | 0.021 | 0.010 | 0.002 | | | | |
| 4 | 0.224 | 0.173 | 0.132 | 0.100 | 0.055 | 0.029 | 0.008 | 0.002 | | | |
| 5 | 0.369 | 0.301 | 0.241 | 0.191 | 0.116 | 0.067 | 0.020 | 0.006 | 0.001 | | |
| 6 | 0.527 | 0.450 | 0.378 | 0.313 | 0.207 | 0.130 | 0.046 | 0.014 | 0.004 | 0.001 | |
| 7 | 0.673 | 0.599 | 0.525 | 0.453 | 0.324 | 0.220 | 0.090 | 0.032 | 0.010 | 0.003 | 0.001 |
| 8 | 0.792 | 0.729 | 0.662 | 0.593 | 0.456 | 0.333 | 0.155 | 0.062 | 0.022 | 0.007 | 0.002 |
| 9 | 0.877 | 0.830 | 0.776 | 0.717 | 0.587 | 0.458 | 0.242 | 0.109 | 0.043 | 0.015 | 0.005 |
| 10 | 0.933 | 0.901 | 0.862 | 0.816 | 0.706 | 0.583 | 0.347 | 0.176 | 0.077 | 0.030 | 0.011 |
| 11 | 0.966 | 0.947 | 0.921 | 0.888 | 0.803 | 0.697 | 0.462 | 0.260 | 0.127 | 0.055 | 0.021 |
| 12 | 0.984 | 0.973 | 0.957 | 0.936 | 0.876 | 0.792 | 0.576 | 0.358 | 0.193 | 0.092 | 0.039 |
| 13 | 0.993 | 0.987 | 0.978 | 0.966 | 0.926 | 0.864 | 0.682 | 0.464 | 0.275 | 0.143 | 0.066 |
| 14 | 0.997 | 0.994 | 0.990 | 0.983 | 0.959 | 0.917 | 0.772 | 0.570 | 0.368 | 0.208 | 0.105 |
| 15 | 0.999 | 0.998 | 0.995 | 0.992 | 0.978 | 0.951 | 0.844 | 0.669 | 0.467 | 0.287 | 0.157 |
| 16 | 1.000 | 0.999 | 0.998 | 0.996 | 0.989 | 0.973 | 0.899 | 0.756 | 0.566 | 0.375 | 0.221 |
| 17 | | 1.000 | 0.999 | 0.998 | 0.995 | 0.986 | 0.937 | 0.827 | 0.659 | 0.469 | 0.297 |
| 18 | | | 1.000 | 0.999 | 0.998 | 0.993 | 0.963 | 0.883 | 0.742 | 0.562 | 0.381 |
| 19 | | | | 1.000 | 0.999 | 0.997 | 0.979 | 0.923 | 0.812 | 0.651 | 0.470 |
| 20 | | | | | 1.000 | 0.998 | 0.988 | 0.952 | 0.868 | 0.731 | 0.559 |
| 21 | | | | | | 0.999 | 0.994 | 0.971 | 0.911 | 0.799 | 0.644 |
| 22 | | | | | | 1.000 | 0.997 | 0.983 | 0.942 | 0.855 | 0.721 |
| 23 | | | | | | | 0.999 | 0.991 | 0.963 | 0.899 | 0.787 |
| 24 | | | | | | | 0.999 | 0.995 | 0.978 | 0.932 | 0.843 |
| 25 | | | | | | | 1.000 | 0.997 | 0.987 | 0.955 | 0.888 |
| 26 | | | | | | | | 0.999 | 0.993 | 0.972 | 0.922 |
| 27 | | | | | | | | 0.999 | 0.996 | 0.983 | 0.948 |
| 28 | | | | | | | | 1.000 | 0.998 | 0.990 | 0.966 |
| 29 | | | | | | | | | 0.999 | 0.994 | 0.978 |
| 30 | | | | | | | | | 0.999 | 0.997 | 0.987 |
| 31 | | | | | | | | | 1.000 | 0.998 | 0.992 |
| 32 | | | | | | | | | | 0.999 | 0.995 |
| 33 | | | | | | | | | | 1.000 | 0.997 |
| 34 | | | | | | | | | | | 0.999 |
| 35 | | | | | | | | | | | 0.999 |
| 36 | | | | | | | | | | | 1.000 |

**Table A.3** Cumulative standard normal distribution

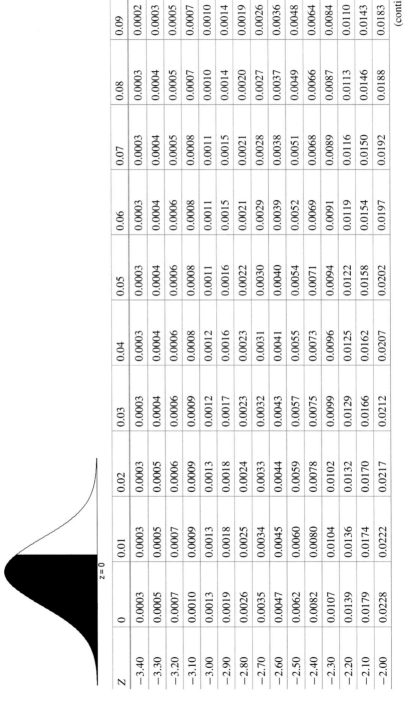

| Z | 0 | 0.01 | 0.02 | 0.03 | 0.04 | 0.05 | 0.06 | 0.07 | 0.08 | 0.09 |
|---|---|---|---|---|---|---|---|---|---|---|
| −3.40 | 0.0003 | 0.0003 | 0.0003 | 0.0003 | 0.0003 | 0.0003 | 0.0003 | 0.0003 | 0.0003 | 0.0002 |
| −3.30 | 0.0005 | 0.0005 | 0.0005 | 0.0004 | 0.0004 | 0.0004 | 0.0004 | 0.0004 | 0.0004 | 0.0003 |
| −3.20 | 0.0007 | 0.0007 | 0.0006 | 0.0006 | 0.0006 | 0.0006 | 0.0006 | 0.0005 | 0.0005 | 0.0005 |
| −3.10 | 0.0010 | 0.0009 | 0.0009 | 0.0009 | 0.0008 | 0.0008 | 0.0008 | 0.0008 | 0.0007 | 0.0007 |
| −3.00 | 0.0013 | 0.0013 | 0.0013 | 0.0012 | 0.0012 | 0.0011 | 0.0011 | 0.0011 | 0.0010 | 0.0010 |
| −2.90 | 0.0019 | 0.0018 | 0.0018 | 0.0017 | 0.0016 | 0.0016 | 0.0015 | 0.0015 | 0.0014 | 0.0014 |
| −2.80 | 0.0026 | 0.0025 | 0.0024 | 0.0023 | 0.0023 | 0.0022 | 0.0021 | 0.0021 | 0.0020 | 0.0019 |
| −2.70 | 0.0035 | 0.0034 | 0.0033 | 0.0032 | 0.0031 | 0.0030 | 0.0029 | 0.0028 | 0.0027 | 0.0026 |
| −2.60 | 0.0047 | 0.0045 | 0.0044 | 0.0043 | 0.0041 | 0.0040 | 0.0039 | 0.0038 | 0.0037 | 0.0036 |
| −2.50 | 0.0062 | 0.0060 | 0.0059 | 0.0057 | 0.0055 | 0.0054 | 0.0052 | 0.0051 | 0.0049 | 0.0048 |
| −2.40 | 0.0082 | 0.0080 | 0.0078 | 0.0075 | 0.0073 | 0.0071 | 0.0069 | 0.0068 | 0.0066 | 0.0064 |
| −2.30 | 0.0107 | 0.0104 | 0.0102 | 0.0099 | 0.0096 | 0.0094 | 0.0091 | 0.0089 | 0.0087 | 0.0084 |
| −2.20 | 0.0139 | 0.0136 | 0.0132 | 0.0129 | 0.0125 | 0.0122 | 0.0119 | 0.0116 | 0.0113 | 0.0110 |
| −2.10 | 0.0179 | 0.0174 | 0.0170 | 0.0166 | 0.0162 | 0.0158 | 0.0154 | 0.0150 | 0.0146 | 0.0143 |
| −2.00 | 0.0228 | 0.0222 | 0.0217 | 0.0212 | 0.0207 | 0.0202 | 0.0197 | 0.0192 | 0.0188 | 0.0183 |

(continued)

**Table A.3** (continued)

z = 0

| Z | 0 | 0.01 | 0.02 | 0.03 | 0.04 | 0.05 | 0.06 | 0.07 | 0.08 | 0.09 |
|---|---|------|------|------|------|------|------|------|------|------|
| −1.90 | 0.0287 | 0.0281 | 0.0274 | 0.0268 | 0.0262 | 0.0256 | 0.0250 | 0.0244 | 0.0239 | 0.0233 |
| −1.80 | 0.0359 | 0.0351 | 0.0344 | 0.0336 | 0.0329 | 0.0322 | 0.0314 | 0.0307 | 0.0301 | 0.0294 |
| −1.70 | 0.0446 | 0.0436 | 0.0427 | 0.0418 | 0.0409 | 0.0401 | 0.0392 | 0.0384 | 0.0375 | 0.0367 |
| −1.60 | 0.0548 | 0.0537 | 0.0526 | 0.0516 | 0.0505 | 0.0495 | 0.0485 | 0.0475 | 0.0465 | 0.0455 |
| −1.50 | 0.0668 | 0.0655 | 0.0643 | 0.0630 | 0.0618 | 0.0606 | 0.0594 | 0.0582 | 0.0571 | 0.0559 |
| −1.40 | 0.0808 | 0.0793 | 0.0778 | 0.0764 | 0.0749 | 0.0735 | 0.0721 | 0.0708 | 0.0694 | 0.0681 |
| −1.30 | 0.0968 | 0.0951 | 0.0934 | 0.0918 | 0.0901 | 0.0885 | 0.0869 | 0.0853 | 0.0838 | 0.0823 |
| −1.20 | 0.1151 | 0.1131 | 0.1112 | 0.1093 | 0.1075 | 0.1056 | 0.1038 | 0.1020 | 0.1003 | 0.0985 |
| −1.10 | 0.1357 | 0.1335 | 0.1314 | 0.1292 | 0.1271 | 0.1251 | 0.1230 | 0.1210 | 0.1190 | 0.1170 |
| −1.00 | 0.1587 | 0.1562 | 0.1539 | 0.1515 | 0.1492 | 0.1469 | 0.1446 | 0.1423 | 0.1401 | 0.1379 |
| −0.90 | 0.1841 | 0.1814 | 0.1788 | 0.1762 | 0.1736 | 0.1711 | 0.1685 | 0.1660 | 0.1635 | 0.1611 |
| −0.80 | 0.2119 | 0.2090 | 0.2061 | 0.2033 | 0.2005 | 0.1977 | 0.1949 | 0.1922 | 0.1894 | 0.1867 |
| −0.70 | 0.2420 | 0.2389 | 0.2358 | 0.2327 | 0.2296 | 0.2266 | 0.2236 | 0.2206 | 0.2177 | 0.2148 |
| −0.60 | 0.2743 | 0.2709 | 0.2676 | 0.2643 | 0.2611 | 0.2578 | 0.2546 | 0.2514 | 0.2483 | 0.2451 |

(continued)

**Table A.3** (continued)

z = 0

| Z | 0 | 0.01 | 0.02 | 0.03 | 0.04 | 0.05 | 0.06 | 0.07 | 0.08 | 0.09 |
|------|--------|--------|--------|--------|--------|--------|--------|--------|--------|--------|
| −0.50 | 0.3085 | 0.3050 | 0.3015 | 0.2981 | 0.2946 | 0.2912 | 0.2877 | 0.2843 | 0.2810 | 0.2776 |
| −0.40 | 0.3446 | 0.3409 | 0.3372 | 0.3336 | 0.3300 | 0.3264 | 0.3228 | 0.3192 | 0.3156 | 0.3121 |
| −0.30 | 0.3821 | 0.3783 | 0.3745 | 0.3707 | 0.3669 | 0.3632 | 0.3594 | 0.3557 | 0.3520 | 0.3483 |
| −0.20 | 0.4207 | 0.4168 | 0.4129 | 0.4090 | 0.4052 | 0.4013 | 0.3974 | 0.3936 | 0.3897 | 0.3859 |
| −0.10 | 0.4602 | 0.4562 | 0.4522 | 0.4483 | 0.4443 | 0.4404 | 0.4364 | 0.4325 | 0.4286 | 0.4247 |
| 0.00 | 0.5000 | 0.4960 | 0.4920 | 0.4880 | 0.4840 | 0.4801 | 0.4761 | 0.4721 | 0.4681 | 0.4641 |
| 0.00 | 0.5000 | 0.5040 | 0.5080 | 0.5120 | 0.5160 | 0.5199 | 0.5239 | 0.5279 | 0.5319 | 0.5359 |
| 0.10 | 0.5398 | 0.5438 | 0.5478 | 0.5517 | 0.5557 | 0.5596 | 0.5636 | 0.5675 | 0.5714 | 0.5753 |
| 0.20 | 0.5793 | 0.5832 | 0.5871 | 0.5910 | 0.5948 | 0.5987 | 0.6026 | 0.6064 | 0.6103 | 0.6141 |
| 0.30 | 0.6179 | 0.6217 | 0.6255 | 0.6293 | 0.6331 | 0.6368 | 0.6406 | 0.6443 | 0.6480 | 0.6517 |
| 0.40 | 0.6554 | 0.6591 | 0.6628 | 0.6664 | 0.6700 | 0.6736 | 0.6772 | 0.6808 | 0.6844 | 0.6879 |
| 0.50 | 0.6915 | 0.6950 | 0.6985 | 0.7019 | 0.7054 | 0.7088 | 0.7123 | 0.7157 | 0.7190 | 0.7224 |
| 0.60 | 0.7257 | 0.7291 | 0.7324 | 0.7357 | 0.7389 | 0.7422 | 0.7454 | 0.7486 | 0.7517 | 0.7549 |
| 0.70 | 0.7580 | 0.7611 | 0.7642 | 0.7673 | 0.7704 | 0.7734 | 0.7764 | 0.7794 | 0.7823 | 0.7852 |

(continued)

**Table A.3** (continued)

z = 0

| Z | 0 | 0.01 | 0.02 | 0.03 | 0.04 | 0.05 | 0.06 | 0.07 | 0.08 | 0.09 |
|------|--------|--------|--------|--------|--------|--------|--------|--------|--------|--------|
| 0.80 | 0.7881 | 0.7910 | 0.7939 | 0.7967 | 0.7995 | 0.8023 | 0.8051 | 0.8078 | 0.8106 | 0.8133 |
| 0.90 | 0.8159 | 0.8186 | 0.8212 | 0.8238 | 0.8264 | 0.8289 | 0.8315 | 0.8340 | 0.8365 | 0.8389 |
| 1.00 | 0.8413 | 0.8438 | 0.8461 | 0.8485 | 0.8508 | 0.8531 | 0.8554 | 0.8577 | 0.8599 | 0.8621 |
| 1.10 | 0.8643 | 0.8665 | 0.8686 | 0.8708 | 0.8729 | 0.8749 | 0.8770 | 0.8790 | 0.8810 | 0.8830 |
| 1.20 | 0.8849 | 0.8869 | 0.8888 | 0.8907 | 0.8925 | 0.8944 | 0.8962 | 0.8980 | 0.8997 | 0.9015 |
| 1.30 | 0.9032 | 0.9049 | 0.9066 | 0.9082 | 0.9099 | 0.9115 | 0.9131 | 0.9147 | 0.9162 | 0.9177 |
| 1.40 | 0.9192 | 0.9207 | 0.9222 | 0.9236 | 0.9251 | 0.9265 | 0.9279 | 0.9292 | 0.9306 | 0.9319 |
| 1.50 | 0.9332 | 0.9345 | 0.9357 | 0.9370 | 0.9382 | 0.9394 | 0.9406 | 0.9418 | 0.9429 | 0.9441 |
| 1.60 | 0.9452 | 0.9463 | 0.9474 | 0.9484 | 0.9495 | 0.9505 | 0.9515 | 0.9525 | 0.9535 | 0.9545 |
| 1.70 | 0.9554 | 0.9564 | 0.9573 | 0.9582 | 0.9591 | 0.9599 | 0.9608 | 0.9616 | 0.9625 | 0.9633 |
| 1.80 | 0.9641 | 0.9649 | 0.9656 | 0.9664 | 0.9671 | 0.9678 | 0.9686 | 0.9693 | 0.9699 | 0.9706 |
| 1.90 | 0.9713 | 0.9719 | 0.9726 | 0.9732 | 0.9738 | 0.9744 | 0.9750 | 0.9756 | 0.9761 | 0.9767 |
| 2.00 | 0.9772 | 0.9778 | 0.9783 | 0.9788 | 0.9793 | 0.9798 | 0.9803 | 0.9808 | 0.9812 | 0.9817 |
| 2.10 | 0.9821 | 0.9826 | 0.9830 | 0.9834 | 0.9838 | 0.9842 | 0.9846 | 0.9850 | 0.9854 | 0.9857 |

(continued)

**Table A.3** (continued)

z = 0

| Z | 0 | 0.01 | 0.02 | 0.03 | 0.04 | 0.05 | 0.06 | 0.07 | 0.08 | 0.09 |
|---|---|---|---|---|---|---|---|---|---|---|
| 2.20 | 0.9861 | 0.9864 | 0.9868 | 0.9871 | 0.9875 | 0.9878 | 0.9881 | 0.9884 | 0.9887 | 0.9890 |
| 2.30 | 0.9893 | 0.9896 | 0.9898 | 0.9901 | 0.9904 | 0.9906 | 0.9909 | 0.9911 | 0.9913 | 0.9916 |
| 2.40 | 0.9918 | 0.9920 | 0.9922 | 0.9925 | 0.9927 | 0.9929 | 0.9931 | 0.9932 | 0.9934 | 0.9936 |
| 2.50 | 0.9938 | 0.9940 | 0.9941 | 0.9943 | 0.9945 | 0.9946 | 0.9948 | 0.9949 | 0.9951 | 0.9952 |
| 2.60 | 0.9953 | 0.9955 | 0.9956 | 0.9957 | 0.9959 | 0.9960 | 0.9961 | 0.9962 | 0.9963 | 0.9964 |
| 2.70 | 0.9965 | 0.9966 | 0.9967 | 0.9968 | 0.9969 | 0.9970 | 0.9971 | 0.9972 | 0.9973 | 0.9974 |
| 2.80 | 0.9974 | 0.9975 | 0.9976 | 0.9977 | 0.9977 | 0.9978 | 0.9979 | 0.9979 | 0.9980 | 0.9981 |
| 2.90 | 0.9981 | 0.9982 | 0.9982 | 0.9983 | 0.9984 | 0.9984 | 0.9985 | 0.9985 | 0.9986 | 0.9986 |
| 3.00 | 0.9987 | 0.9987 | 0.9987 | 0.9988 | 0.9988 | 0.9989 | 0.9989 | 0.9989 | 0.9990 | 0.9990 |
| 3.10 | 0.9990 | 0.9991 | 0.9991 | 0.9991 | 0.9992 | 0.9992 | 0.9992 | 0.9992 | 0.9993 | 0.9993 |
| 3.20 | 0.9993 | 0.9993 | 0.9994 | 0.9994 | 0.9994 | 0.9994 | 0.9994 | 0.9995 | 0.9995 | 0.9995 |
| 3.30 | 0.9995 | 0.9995 | 0.9995 | 0.9996 | 0.9996 | 0.9996 | 0.9996 | 0.9996 | 0.9996 | 0.9997 |
| 3.40 | 0.9997 | 0.9997 | 0.9997 | 0.9997 | 0.9997 | 0.9997 | 0.9997 | 0.9997 | 0.9997 | 0.9998 |

**Table A.4** Sigma value versus DPMO

| Sigma value | DPMO | Sigma value | DPMO | Sigma value | DPMO |
|---|---|---|---|---|---|
| 0.00 | 933,193 | 4.05 | 5386 | 5.05 | 193 |
| 0.50 | 841,345 | 4.10 | 4661 | 5.10 | 159 |
| 0.75 | 773,373 | 4.15 | 4024 | 5.15 | 131 |
| 1.00 | 691,462 | 4.20 | 3467 | 5.20 | 108 |
| 1.25 | 598,706 | 4.25 | 2980 | 5.25 | 89 |
| 1.50 | 500,000 | 4.30 | 2555 | 5.30 | 72 |
| 1.75 | 401,294 | 4.35 | 2186 | 5.35 | 59 |
| 2.00 | 308,537 | 4.40 | 1866 | 5.40 | 48 |
| 2.25 | 226,627 | 4.45 | 1589 | 5.45 | 39 |
| 2.50 | 158,655 | 4.50 | 1350 | 5.50 | 32 |
| 2.75 | 105,650 | 4.55 | 1144 | 5.55 | 25.6 |
| 3.00 | 66,807 | 4.60 | 968 | 5.60 | 20.7 |
| 3.25 | 40,059 | 4.65 | 816 | 5.65 | 16.6 |
| 3.50 | 22,750 | 4.70 | 687 | 5.70 | 13.4 |
| 3.60 | 17,865 | 4.75 | 577 | 5.75 | 10.7 |
| 3.70 | 13,904 | 4.80 | 483 | 5.80 | 8.5 |
| 3.75 | 12,225 | 4.85 | 404 | 5.85 | 6.8 |
| 3.80 | 10,724 | 4.90 | 337 | 5.90 | 5.4 |
| 3.90 | 8198 | 4.95 | 280 | 5.95 | 4.3 |
| 4.00 | 6210 | 5.00 | 233 | 6.00 | 3.4 |

*Note* This table includes a 1.5 Sigma shift for all listed values of Z

**Table A.5** Values of $t$ for a specified right tail area

$t_{\alpha,\nu}$

| | Level of Significance $\alpha$ | | | | | | | | | |
|---|---|---|---|---|---|---|---|---|---|---|
| $\nu$ | 0.4 | 0.25 | 0.1 | 0.05 | 0.025 | 0.01 | 0.005 | 0.0025 | 0.001 | 0.0005 |
| 1 | 0.325 | 1.000 | 3.078 | 6.314 | 12.706 | 31.821 | 63.656 | 127.321 | 318.289 | 636.578 |
| 2 | 0.289 | 0.816 | 1.886 | 2.920 | 4.303 | 6.965 | 9.925 | 14.089 | 22.328 | 31.600 |
| 3 | 0.277 | 0.765 | 1.638 | 2.353 | 3.182 | 4.541 | 5.841 | 7.453 | 10.214 | 12.924 |
| 4 | 0.271 | 0.741 | 1.533 | 2.132 | 2.776 | 3.747 | 4.604 | 5.598 | 7.173 | 8.610 |
| 5 | 0.267 | 0.727 | 1.476 | 2.015 | 2.571 | 3.365 | 4.032 | 4.773 | 5.894 | 6.869 |
| 6 | 0.265 | 0.718 | 1.440 | 1.943 | 2.447 | 3.143 | 3.707 | 4.317 | 5.208 | 5.959 |
| 7 | 0.263 | 0.711 | 1.415 | 1.895 | 2.365 | 2.998 | 3.499 | 4.029 | 4.785 | 5.408 |
| 8 | 0.262 | 0.706 | 1.397 | 1.860 | 2.306 | 2.896 | 3.355 | 3.833 | 4.501 | 5.041 |
| 9 | 0.261 | 0.703 | 1.383 | 1.833 | 2.262 | 2.821 | 3.250 | 3.690 | 4.297 | 4.781 |
| 10 | 0.260 | 0.700 | 1.372 | 1.812 | 2.228 | 2.764 | 3.169 | 3.581 | 4.144 | 4.587 |
| 11 | 0.260 | 0.697 | 1.363 | 1.796 | 2.201 | 2.718 | 3.106 | 3.497 | 4.025 | 4.437 |
| 12 | 0.259 | 0.695 | 1.356 | 1.782 | 2.179 | 2.681 | 3.055 | 3.428 | 3.930 | 4.318 |
| 13 | 0.259 | 0.694 | 1.350 | 1.771 | 2.160 | 2.650 | 3.012 | 3.372 | 3.852 | 4.221 |

(continued)

**Table A.5** (continued)

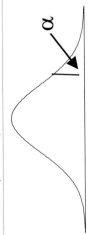

$t_{\alpha,\nu}$

| | Level of Significance $\alpha$ | | | | | | | | | | |
|---|---|---|---|---|---|---|---|---|---|---|---|
| $\nu$ | 0.4 | 0.25 | 0.1 | 0.05 | 0.025 | 0.01 | 0.005 | 0.0025 | 0.001 | 0.0005 |
| 14 | 0.258 | 0.692 | 1.345 | 1.761 | 2.145 | 2.624 | 2.977 | 3.326 | 3.787 | 4.140 |
| 15 | 0.258 | 0.691 | 1.341 | 1.753 | 2.131 | 2.602 | 2.947 | 3.286 | 3.733 | 4.073 |
| 16 | 0.258 | 0.690 | 1.337 | 1.746 | 2.120 | 2.583 | 2.921 | 3.252 | 3.686 | 4.015 |
| 17 | 0.257 | 0.689 | 1.333 | 1.740 | 2.110 | 2.567 | 2.898 | 3.222 | 3.646 | 3.965 |
| 18 | 0.257 | 0.688 | 1.330 | 1.734 | 2.101 | 2.552 | 2.878 | 3.197 | 3.610 | 3.922 |
| 19 | 0.257 | 0.688 | 1.328 | 1.729 | 2.093 | 2.539 | 2.861 | 3.174 | 3.579 | 3.883 |
| 20 | 0.257 | 0.687 | 1.325 | 1.725 | 2.086 | 2.528 | 2.845 | 3.153 | 3.552 | 3.850 |
| 21 | 0.257 | 0.686 | 1.323 | 1.721 | 2.080 | 2.518 | 2.831 | 3.135 | 3.527 | 3.819 |
| 22 | 0.256 | 0.686 | 1.321 | 1.717 | 2.074 | 2.508 | 2.819 | 3.119 | 3.505 | 3.792 |
| 23 | 0.256 | 0.685 | 1.319 | 1.714 | 2.069 | 2.500 | 2.807 | 3.104 | 3.485 | 3.768 |
| 24 | 0.256 | 0.685 | 1.318 | 1.711 | 2.064 | 2.492 | 2.797 | 3.091 | 3.467 | 3.745 |

(continued)

**Table A.5** (continued)

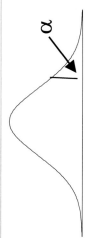

$t_{\alpha,\nu}$

| $\nu$ | Level of Significance $\alpha$ | | | | | | | | | |
|---|---|---|---|---|---|---|---|---|---|---|
| | 0.4 | 0.25 | 0.1 | 0.05 | 0.025 | 0.01 | 0.005 | 0.0025 | 0.001 | 0.0005 |
| 25 | 0.256 | 0.684 | 1.316 | 1.708 | 2.060 | 2.485 | 2.787 | 3.078 | 3.450 | 3.725 |
| 26 | 0.256 | 0.684 | 1.315 | 1.706 | 2.056 | 2.479 | 2.779 | 3.067 | 3.435 | 3.707 |
| 27 | 0.256 | 0.684 | 1.314 | 1.703 | 2.052 | 2.473 | 2.771 | 3.057 | 3.421 | 3.689 |
| 28 | 0.256 | 0.683 | 1.313 | 1.701 | 2.048 | 2.467 | 2.763 | 3.047 | 3.408 | 3.674 |
| 29 | 0.256 | 0.683 | 1.311 | 1.699 | 2.045 | 2.462 | 2.756 | 3.038 | 3.396 | 3.660 |
| 30 | 0.256 | 0.683 | 1.310 | 1.697 | 2.042 | 2.457 | 2.750 | 3.030 | 3.385 | 3.646 |
| 40 | 0.255 | 0.681 | 1.303 | 1.684 | 2.021 | 2.423 | 2.704 | 2.971 | 3.307 | 3.551 |
| 60 | 0.254 | 0.679 | 1.296 | 1.671 | 2.000 | 2.390 | 2.660 | 2.915 | 3.232 | 3.460 |
| 120 | 0.254 | 0.677 | 1.289 | 1.658 | 1.980 | 2.358 | 2.617 | 2.860 | 3.160 | 3.373 |
| $\infty$ | 0.253 | 0.674 | 1.282 | 1.645 | 1.960 | 2.326 | 2.576 | 2.807 | 3.090 | 3.291 |

**Table A.6** Chi-squared values for a specified right tail area

$\chi^2_{\alpha,\nu}$

| $\nu/\alpha$ | 0.999 | 0.995 | 0.99 | 0.975 | 0.95 | 0.9 | 0.1 | 0.05 | 0.025 | 0.01 | 0.005 | 0.001 |
|---|---|---|---|---|---|---|---|---|---|---|---|---|
| 1 | 0.00 | 0.00 | 0.00 | 0.00 | 0.00 | 0.02 | 2.71 | 3.84 | 5.02 | 6.63 | 7.88 | 10.83 |
| 2 | 0.00 | 0.01 | 0.02 | 0.05 | 0.10 | 0.21 | 4.61 | 5.99 | 7.38 | 9.21 | 10.60 | 13.82 |
| 3 | 0.02 | 0.07 | 0.11 | 0.22 | 0.35 | 0.58 | 6.25 | 7.81 | 9.35 | 11.34 | 12.84 | 16.27 |
| 4 | 0.09 | 0.21 | 0.30 | 0.48 | 0.71 | 1.06 | 7.78 | 9.49 | 11.14 | 13.28 | 14.86 | 18.47 |
| 5 | 0.21 | 0.41 | 0.55 | 0.83 | 1.15 | 1.61 | 9.24 | 11.07 | 12.83 | 15.09 | 16.75 | 20.51 |
| 6 | 0.38 | 0.68 | 0.87 | 1.24 | 1.64 | 2.20 | 10.64 | 12.59 | 14.45 | 16.81 | 18.55 | 22.46 |
| 7 | 0.60 | 0.99 | 1.24 | 1.69 | 2.17 | 2.83 | 12.02 | 14.07 | 16.01 | 18.48 | 20.28 | 24.32 |
| 8 | 0.86 | 1.34 | 1.65 | 2.18 | 2.73 | 3.49 | 13.36 | 15.51 | 17.53 | 20.09 | 21.95 | 26.12 |
| 9 | 1.15 | 1.73 | 2.09 | 2.70 | 3.33 | 4.17 | 14.68 | 16.92 | 19.02 | 21.67 | 23.59 | 27.88 |
| 10 | 1.48 | 2.16 | 2.56 | 3.25 | 3.94 | 4.87 | 15.99 | 18.31 | 20.48 | 23.21 | 25.19 | 29.59 |
| 11 | 1.83 | 2.60 | 3.05 | 3.82 | 4.57 | 5.58 | 17.28 | 19.68 | 21.92 | 24.73 | 26.76 | 31.26 |
| 12 | 2.21 | 3.07 | 3.57 | 4.40 | 5.23 | 6.30 | 18.55 | 21.03 | 23.34 | 26.22 | 28.30 | 32.91 |
| 13 | 2.62 | 3.57 | 4.11 | 5.01 | 5.89 | 7.04 | 19.81 | 22.36 | 24.74 | 27.69 | 29.82 | 34.53 |

(continued)

**Table A.6** (continued)

$$\chi^2_{\alpha,\nu}$$

| $\nu/\alpha$ | 0.999 | 0.995 | 0.99 | 0.975 | 0.95 | 0.9 | 0.1 | 0.05 | 0.025 | 0.01 | 0.005 | 0.001 |
|---|---|---|---|---|---|---|---|---|---|---|---|---|
| 14 | 3.04 | 4.07 | 4.66 | 5.63 | 6.57 | 7.79 | 21.06 | 23.68 | 26.12 | 29.14 | 31.32 | 36.12 |
| 15 | 3.48 | 4.60 | 5.23 | 6.26 | 7.26 | 8.55 | 22.31 | 25.00 | 27.49 | 30.58 | 32.80 | 37.70 |
| 16 | 3.94 | 5.14 | 5.81 | 6.91 | 7.96 | 9.31 | 23.54 | 26.30 | 28.85 | 32.00 | 34.27 | 39.25 |
| 17 | 4.42 | 5.70 | 6.41 | 7.56 | 8.67 | 10.09 | 24.77 | 27.59 | 30.19 | 33.41 | 35.72 | 40.79 |
| 18 | 4.90 | 6.26 | 7.01 | 8.23 | 9.39 | 10.86 | 25.99 | 28.87 | 31.53 | 34.81 | 37.16 | 42.31 |
| 19 | 5.41 | 6.84 | 7.63 | 8.91 | 10.12 | 11.65 | 27.20 | 30.14 | 32.85 | 36.19 | 38.58 | 43.82 |
| 20 | 5.92 | 7.43 | 8.26 | 9.59 | 10.85 | 12.44 | 28.41 | 31.41 | 34.17 | 37.57 | 40.00 | 45.31 |
| 21 | 6.45 | 8.03 | 8.90 | 10.28 | 11.59 | 13.24 | 29.62 | 32.67 | 35.48 | 38.93 | 41.40 | 46.80 |
| 22 | 6.98 | 8.64 | 9.54 | 10.98 | 12.34 | 14.04 | 30.81 | 33.92 | 36.78 | 40.29 | 42.80 | 48.27 |
| 23 | 7.53 | 9.26 | 10.20 | 11.69 | 13.09 | 14.85 | 32.01 | 35.17 | 38.08 | 41.64 | 44.18 | 49.73 |
| 24 | 8.08 | 9.89 | 10.86 | 12.40 | 13.85 | 15.66 | 33.20 | 36.42 | 39.36 | 42.98 | 45.56 | 51.18 |
| 25 | 8.65 | 10.52 | 11.52 | 13.12 | 14.61 | 16.47 | 34.38 | 37.65 | 40.65 | 44.31 | 46.93 | 52.62 |

(continued)

**Table A.6** (continued)

$$\chi^2_{\alpha,\nu}$$

| ν/α | 0.999 | 0.995 | 0.99 | 0.975 | 0.95 | 0.9 | 0.1 | 0.05 | 0.025 | 0.01 | 0.005 | 0.001 |
|---|---|---|---|---|---|---|---|---|---|---|---|---|
| 26 | 9.22 | 11.16 | 12.20 | 13.84 | 15.38 | 17.29 | 35.56 | 38.89 | 41.92 | 45.64 | 48.29 | 54.05 |
| 27 | 9.80 | 11.81 | 12.88 | 14.57 | 16.15 | 18.11 | 36.74 | 40.11 | 43.19 | 46.96 | 49.65 | 55.48 |
| 28 | 10.39 | 12.46 | 13.56 | 15.31 | 16.93 | 18.94 | 37.92 | 41.34 | 44.46 | 48.28 | 50.99 | 56.89 |
| 29 | 10.99 | 13.12 | 14.26 | 16.05 | 17.71 | 19.77 | 39.09 | 42.56 | 45.72 | 49.59 | 52.34 | 58.30 |
| 30 | 11.59 | 13.79 | 14.95 | 16.79 | 18.49 | 20.60 | 40.26 | 43.77 | 46.98 | 50.89 | 53.67 | 59.70 |
| 32 | 12.81 | 15.13 | 16.36 | 18.29 | 20.07 | 22.27 | 42.58 | 46.19 | 49.48 | 53.49 | 56.33 | 62.49 |
| 34 | 14.06 | 16.50 | 17.79 | 19.81 | 21.66 | 23.95 | 44.90 | 48.60 | 51.97 | 56.06 | 58.96 | 65.25 |
| 36 | 15.32 | 17.89 | 19.23 | 21.34 | 23.27 | 25.64 | 47.21 | 51.00 | 54.44 | 58.62 | 61.58 | 67.98 |
| 38 | 16.61 | 19.29 | 20.69 | 22.88 | 24.88 | 27.34 | 49.51 | 53.38 | 56.90 | 61.16 | 64.18 | 70.70 |
| 40 | 17.92 | 20.71 | 22.16 | 24.43 | 26.51 | 29.05 | 51.81 | 55.76 | 59.34 | 63.69 | 66.77 | 73.40 |
| 42 | 19.24 | 22.14 | 23.65 | 26.00 | 28.14 | 30.77 | 54.09 | 58.12 | 61.78 | 66.21 | 69.34 | 76.08 |
| 44 | 20.58 | 23.58 | 25.15 | 27.57 | 29.79 | 32.49 | 56.37 | 60.48 | 64.20 | 68.71 | 71.89 | 78.75 |

(continued)

**Table A.6** (continued)

$$\chi^2_{\alpha,\nu}$$

| $\nu/\alpha$ | 0.999 | 0.995 | 0.99 | 0.975 | 0.95 | 0.9 | 0.1 | 0.05 | 0.025 | 0.01 | 0.005 | 0.001 |
|---|---|---|---|---|---|---|---|---|---|---|---|---|
| 46 | 21.93 | 25.04 | 26.66 | 29.16 | 31.44 | 34.22 | 58.64 | 62.83 | 66.62 | 71.20 | 74.44 | 81.40 |
| 48 | 23.29 | 26.51 | 28.18 | 30.75 | 33.10 | 35.95 | 60.91 | 65.17 | 69.02 | 73.68 | 76.97 | 84.04 |
| 50 | 24.67 | 27.99 | 29.71 | 32.36 | 34.76 | 37.69 | 63.17 | 67.50 | 71.42 | 76.15 | 79.49 | 86.66 |
| 55 | 28.17 | 31.73 | 33.57 | 36.40 | 38.96 | 42.06 | 68.80 | 73.31 | 77.38 | 82.29 | 85.75 | 93.17 |
| 60 | 31.74 | 35.53 | 37.48 | 40.48 | 43.19 | 46.46 | 74.40 | 79.08 | 83.30 | 88.38 | 91.95 | 99.61 |
| 65 | 35.36 | 39.38 | 41.44 | 44.60 | 47.45 | 50.88 | 79.97 | 84.82 | 89.18 | 94.42 | 98.10 | 105.99 |
| 70 | 39.04 | 43.28 | 45.44 | 48.76 | 51.74 | 55.33 | 85.53 | 90.53 | 95.02 | 100.43 | 104.21 | 112.32 |
| 75 | 42.76 | 47.21 | 49.48 | 52.94 | 56.05 | 59.79 | 91.06 | 96.22 | 100.84 | 106.39 | 110.29 | 118.60 |
| 80 | 46.52 | 51.17 | 53.54 | 57.15 | 60.39 | 64.28 | 96.58 | 101.88 | 106.63 | 112.33 | 116.32 | 124.84 |
| 85 | 50.32 | 55.17 | 57.63 | 61.39 | 64.75 | 68.78 | 102.08 | 107.52 | 112.39 | 118.24 | 122.32 | 131.04 |
| 90 | 54.16 | 59.20 | 61.75 | 65.65 | 69.13 | 73.29 | 107.57 | 113.15 | 118.14 | 124.12 | 128.30 | 137.21 |
| 95 | 58.02 | 63.25 | 65.90 | 69.92 | 73.52 | 77.82 | 113.04 | 118.75 | 123.86 | 129.97 | 134.25 | 143.34 |
| 100 | 61.92 | 67.33 | 70.06 | 74.22 | 77.93 | 82.36 | 118.50 | 124.34 | 129.56 | 135.81 | 140.17 | 149.45 |

**Table A.7** Values of F for a specified right tail area $F_{0.01(v_1, v_2)}$

| $v_2$ \ $v_1$ | 1 | 2 | 3 | 4 | 5 | 6 | 7 | 8 | 9 | 10 | 12 | 15 | 20 | 24 | 30 | 40 | 60 | 120 | ∞ |
|---|---|---|---|---|---|---|---|---|---|---|---|---|---|---|---|---|---|---|---|
| 1 | 4052 | 4999 | 5404 | 5624 | 5764 | 5859 | 5928 | 5981 | 6022 | 6056 | 6107 | 6157 | 6209 | 6234 | 6260 | 6286 | 6313 | 6340 | 6366 |
| 2 | 98.5 | 99.0 | 99.2 | 99.3 | 99.3 | 99.3 | 99.4 | 99.4 | 99.4 | 99.4 | 99.4 | 99.4 | 99.4 | 99.5 | 99.5 | 99.5 | 99.5 | 99.5 | 99.5 |
| 3 | 34.1 | 30.8 | 29.5 | 28.7 | 28.2 | 27.9 | 27.7 | 27.5 | 27.3 | 27.2 | 27.1 | 26.9 | 26.7 | 26.6 | 26.5 | 26.4 | 26.3 | 26.2 | 26.1 |
| 4 | 21.2 | 18.0 | 16.7 | 16.0 | 15.5 | 15.2 | 15.0 | 14.8 | 14.7 | 14.5 | 14.4 | 14.2 | 14.0 | 13.9 | 13.8 | 13.7 | 13.7 | 13.6 | 13.5 |
| 5 | 16.3 | 13.3 | 12.1 | 11.4 | 11.0 | 10.7 | 10.5 | 10.3 | 10.2 | 10.1 | 9.89 | 9.72 | 9.55 | 9.47 | 9.38 | 9.29 | 9.20 | 9.11 | 9.02 |
| 6 | 13.7 | 10.9 | 9.78 | 9.15 | 8.75 | 8.47 | 8.26 | 8.10 | 7.98 | 7.87 | 7.72 | 7.56 | 7.40 | 7.31 | 7.23 | 7.14 | 7.06 | 6.97 | 6.88 |
| 7 | 12.2 | 9.55 | 8.45 | 7.85 | 7.46 | 7.19 | 6.99 | 6.84 | 6.72 | 6.62 | 6.47 | 6.31 | 6.16 | 6.07 | 5.99 | 5.91 | 5.82 | 5.74 | 5.65 |
| 8 | 11.3 | 8.65 | 7.59 | 7.01 | 6.63 | 6.37 | 6.18 | 6.03 | 5.91 | 5.81 | 5.67 | 5.52 | 5.36 | 5.28 | 5.20 | 5.12 | 5.03 | 4.95 | 4.86 |
| 9 | 10.6 | 8.02 | 6.99 | 6.42 | 6.06 | 5.80 | 5.61 | 5.47 | 5.35 | 5.26 | 5.11 | 4.96 | 4.81 | 4.73 | 4.65 | 4.57 | 4.48 | 4.40 | 4.31 |
| 10 | 10.0 | 7.56 | 6.55 | 5.99 | 5.64 | 5.39 | 5.20 | 5.06 | 4.94 | 4.85 | 4.71 | 4.56 | 4.41 | 4.33 | 4.25 | 4.17 | 4.08 | 4.00 | 3.91 |
| 11 | 9.65 | 7.21 | 6.22 | 5.67 | 5.32 | 5.07 | 4.89 | 4.74 | 4.63 | 4.54 | 4.40 | 4.25 | 4.10 | 4.02 | 3.94 | 3.86 | 3.78 | 3.69 | 3.60 |
| 12 | 9.33 | 6.93 | 5.95 | 5.41 | 5.06 | 4.82 | 4.64 | 4.50 | 4.39 | 4.30 | 4.16 | 4.01 | 3.86 | 3.78 | 3.70 | 3.62 | 3.54 | 3.45 | 3.36 |
| 13 | 9.07 | 6.70 | 5.74 | 5.21 | 4.86 | 4.62 | 4.44 | 4.30 | 4.19 | 4.10 | 3.96 | 3.82 | 3.66 | 3.59 | 3.51 | 3.43 | 3.34 | 3.25 | 3.17 |
| 14 | 8.86 | 6.51 | 5.56 | 5.04 | 4.69 | 4.46 | 4.28 | 4.14 | 4.03 | 3.94 | 3.80 | 3.66 | 3.51 | 3.43 | 3.35 | 3.27 | 3.18 | 3.09 | 3.00 |
| 15 | 8.68 | 6.36 | 5.42 | 4.89 | 4.56 | 4.32 | 4.14 | 4.00 | 3.89 | 3.80 | 3.67 | 3.52 | 3.37 | 3.29 | 3.21 | 3.13 | 3.05 | 2.96 | 2.87 |
| 16 | 8.53 | 6.23 | 5.29 | 4.77 | 4.44 | 4.20 | 4.03 | 3.89 | 3.78 | 3.69 | 3.55 | 3.41 | 3.26 | 3.18 | 3.10 | 3.02 | 2.93 | 2.84 | 2.75 |
| 17 | 8.40 | 6.11 | 5.19 | 4.67 | 4.34 | 4.10 | 3.93 | 3.79 | 3.68 | 3.59 | 3.46 | 3.31 | 3.16 | 3.08 | 3.00 | 2.92 | 2.83 | 2.75 | 2.65 |
| 18 | 8.29 | 6.01 | 5.09 | 4.58 | 4.25 | 4.01 | 3.84 | 3.71 | 3.60 | 3.51 | 3.37 | 3.23 | 3.08 | 3.00 | 2.92 | 2.84 | 2.75 | 2.66 | 2.57 |
| 19 | 8.18 | 5.93 | 5.01 | 4.50 | 4.17 | 3.94 | 3.77 | 3.63 | 3.52 | 3.43 | 3.30 | 3.15 | 3.00 | 2.92 | 2.84 | 2.76 | 2.67 | 2.58 | 2.49 |
| 20 | 8.10 | 5.85 | 4.94 | 4.43 | 4.10 | 3.87 | 3.70 | 3.56 | 3.46 | 3.37 | 3.23 | 3.09 | 2.94 | 2.86 | 2.78 | 2.69 | 2.61 | 2.52 | 2.42 |

(continued)

**Table A.7** (continued)

| df $v2$ | \ $v1$ 1 | 2 | 3 | 4 | 5 | 6 | 7 | 8 | 9 | 10 | 12 | 15 | 20 | 24 | 30 | 40 | 60 | 120 | ∞ |
|---|---|---|---|---|---|---|---|---|---|---|---|---|---|---|---|---|---|---|---|
| 21 | 8.02 | 5.78 | 4.87 | 4.37 | 4.04 | 3.81 | 3.64 | 3.51 | 3.40 | 3.31 | 3.17 | 3.03 | 2.88 | 2.80 | 2.72 | 2.64 | 2.55 | 2.46 | 2.36 |
| 22 | 7.95 | 5.72 | 4.82 | 4.31 | 3.99 | 3.76 | 3.59 | 3.45 | 3.35 | 3.26 | 3.12 | 2.98 | 2.83 | 2.75 | 2.67 | 2.58 | 2.50 | 2.40 | 2.31 |
| 23 | 7.88 | 5.66 | 4.76 | 4.26 | 3.94 | 3.71 | 3.54 | 3.41 | 3.30 | 3.21 | 3.07 | 2.93 | 2.78 | 2.70 | 2.62 | 2.54 | 2.45 | 2.35 | 2.26 |
| 24 | 7.82 | 5.61 | 4.72 | 4.22 | 3.90 | 3.67 | 3.50 | 3.36 | 3.26 | 3.17 | 3.03 | 2.89 | 2.74 | 2.66 | 2.58 | 2.49 | 2.40 | 2.31 | 2.21 |
| 25 | 7.77 | 5.57 | 4.68 | 4.18 | 3.85 | 3.63 | 3.46 | 3.32 | 3.22 | 3.13 | 2.99 | 2.85 | 2.70 | 2.62 | 2.54 | 2.45 | 2.36 | 2.27 | 2.17 |
| 26 | 7.72 | 5.53 | 4.64 | 4.14 | 3.82 | 3.59 | 3.42 | 3.29 | 3.18 | 3.09 | 2.96 | 2.81 | 2.66 | 2.58 | 2.50 | 2.42 | 2.33 | 2.23 | 2.13 |
| 27 | 7.68 | 5.49 | 4.60 | 4.11 | 3.78 | 3.56 | 3.39 | 3.26 | 3.15 | 3.06 | 2.93 | 2.78 | 2.63 | 2.55 | 2.47 | 2.38 | 2.29 | 2.20 | 2.10 |
| 28 | 7.64 | 5.45 | 4.57 | 4.07 | 3.75 | 3.53 | 3.36 | 3.23 | 3.12 | 3.03 | 2.90 | 2.75 | 2.60 | 2.52 | 2.44 | 2.35 | 2.26 | 2.17 | 2.06 |
| 29 | 7.60 | 5.42 | 4.54 | 4.04 | 3.73 | 3.50 | 3.33 | 3.20 | 3.09 | 3.00 | 2.87 | 2.73 | 2.57 | 2.49 | 2.41 | 2.33 | 2.23 | 2.14 | 2.03 |
| 30 | 7.56 | 5.39 | 4.51 | 4.02 | 3.70 | 3.47 | 3.30 | 3.17 | 3.07 | 2.98 | 2.84 | 2.70 | 2.55 | 2.47 | 2.39 | 2.30 | 2.21 | 2.11 | 2.01 |
| 40 | 7.31 | 5.18 | 4.31 | 3.83 | 3.51 | 3.29 | 3.12 | 2.99 | 2.89 | 2.80 | 2.66 | 2.52 | 2.37 | 2.29 | 2.20 | 2.11 | 2.02 | 1.92 | 1.80 |
| 60 | 7.08 | 4.98 | 4.13 | 3.65 | 3.34 | 3.12 | 2.95 | 2.82 | 2.72 | 2.63 | 2.50 | 2.35 | 2.20 | 2.12 | 2.03 | 1.94 | 1.84 | 1.73 | 1.60 |
| 120 | 6.85 | 4.79 | 3.95 | 3.48 | 3.17 | 2.96 | 2.79 | 2.66 | 2.56 | 2.47 | 2.34 | 2.19 | 2.03 | 1.95 | 1.86 | 1.76 | 1.66 | 1.53 | 1.38 |
| ∞ | 6.64 | 4.61 | 3.78 | 3.32 | 3.02 | 2.80 | 2.64 | 2.51 | 2.41 | 2.32 | 2.18 | 2.04 | 1.88 | 1.79 | 1.70 | 1.59 | 1.47 | 1.32 | 1.00 |

Values of $F$ for a specified right tail area $F_{0.025(v_1, v_2)}$

| $v2$ | 1 | 2 | 3 | 4 | 5 | 6 | 7 | 8 | 9 | 10 | 12 | 15 | 20 | 24 | 30 | 40 | 60 | 120 | ∞ |
|---|---|---|---|---|---|---|---|---|---|---|---|---|---|---|---|---|---|---|---|
| 1 | 648 | 799 | 864 | 900 | 922 | 937 | 948 | 957 | 963 | 969 | 977 | 985 | 993 | 997 | 1001 | 1006 | 1010 | 1014 | 1018 |
| 2 | 38.5 | 39.0 | 39.2 | 39.2 | 39.3 | 39.3 | 39.4 | 39.4 | 39.4 | 39.4 | 39.4 | 39.4 | 39.4 | 39.5 | 39.5 | 39.5 | 39.5 | 39.5 | 39.5 |
| 3 | 17.4 | 16.0 | 15.4 | 15.1 | 14.9 | 14.7 | 14.6 | 14.5 | 14.5 | 14.4 | 14.3 | 14.3 | 14.2 | 14.1 | 14.1 | 14.0 | 14.0 | 13.9 | 13.9 |
| 4 | 12.2 | 10.6 | 9.98 | 9.60 | 9.36 | 9.20 | 9.07 | 8.98 | 8.90 | 8.84 | 8.75 | 8.66 | 8.56 | 8.51 | 8.46 | 8.41 | 8.36 | 8.31 | 8.26 |

(continued)

**Table A.7** (continued)

| df | v1 | | | | | | | | | | | | | | | | | | |
|---|---|---|---|---|---|---|---|---|---|---|---|---|---|---|---|---|---|---|---|
| v2 | 1 | 2 | 3 | 4 | 5 | 6 | 7 | 8 | 9 | 10 | 12 | 15 | 20 | 24 | 30 | 40 | 60 | 120 | ∞ |
| 5 | 10.0 | 8.43 | 7.76 | 7.39 | 7.15 | 6.98 | 6.85 | 6.76 | 6.68 | 6.62 | 6.52 | 6.43 | 6.33 | 6.28 | 6.23 | 6.18 | 6.12 | 6.07 | 6.02 |
| 6 | 8.81 | 7.26 | 6.60 | 6.23 | 5.99 | 5.82 | 5.70 | 5.60 | 5.52 | 5.46 | 5.37 | 5.27 | 5.17 | 5.12 | 5.07 | 5.01 | 4.96 | 4.90 | 4.85 |
| 7 | 8.07 | 6.54 | 5.89 | 5.52 | 5.29 | 5.12 | 4.99 | 4.90 | 4.82 | 4.76 | 4.67 | 4.57 | 4.47 | 4.41 | 4.36 | 4.31 | 4.25 | 4.20 | 4.14 |
| 8 | 7.57 | 6.06 | 5.42 | 5.05 | 4.82 | 4.65 | 4.53 | 4.43 | 4.36 | 4.30 | 4.20 | 4.10 | 4.00 | 3.95 | 3.89 | 3.84 | 3.78 | 3.73 | 3.67 |
| 9 | 7.21 | 5.71 | 5.08 | 4.72 | 4.48 | 4.32 | 4.20 | 4.10 | 4.03 | 3.96 | 3.87 | 3.77 | 3.67 | 3.61 | 3.56 | 3.51 | 3.45 | 3.39 | 3.33 |
| 10 | 6.94 | 5.46 | 4.83 | 4.47 | 4.24 | 4.07 | 3.95 | 3.85 | 3.78 | 3.72 | 3.62 | 3.52 | 3.42 | 3.37 | 3.31 | 3.26 | 3.20 | 3.14 | 3.08 |
| 11 | 6.72 | 5.26 | 4.63 | 4.28 | 4.04 | 3.88 | 3.76 | 3.66 | 3.59 | 3.53 | 3.43 | 3.33 | 3.23 | 3.17 | 3.12 | 3.06 | 3.00 | 2.94 | 2.88 |
| 12 | 6.55 | 5.10 | 4.47 | 4.12 | 3.89 | 3.73 | 3.61 | 3.51 | 3.44 | 3.37 | 3.28 | 3.18 | 3.07 | 3.02 | 2.96 | 2.91 | 2.85 | 2.79 | 2.73 |
| 13 | 6.41 | 4.97 | 4.35 | 4.00 | 3.77 | 3.60 | 3.48 | 3.39 | 3.31 | 3.25 | 3.15 | 3.05 | 2.95 | 2.89 | 2.84 | 2.78 | 2.72 | 2.66 | 2.60 |
| 14 | 6.30 | 4.86 | 4.24 | 3.89 | 3.66 | 3.50 | 3.38 | 3.29 | 3.21 | 3.15 | 3.05 | 2.95 | 2.84 | 2.79 | 2.73 | 2.67 | 2.61 | 2.55 | 2.49 |
| 15 | 6.20 | 4.77 | 4.15 | 3.80 | 3.58 | 3.41 | 3.29 | 3.20 | 3.12 | 3.06 | 2.96 | 2.86 | 2.76 | 2.70 | 2.64 | 2.59 | 2.52 | 2.46 | 2.40 |
| 16 | 6.12 | 4.69 | 4.08 | 3.73 | 3.50 | 3.34 | 3.22 | 3.12 | 3.05 | 2.99 | 2.89 | 2.79 | 2.68 | 2.63 | 2.57 | 2.51 | 2.45 | 2.38 | 2.32 |
| 17 | 6.04 | 4.62 | 4.01 | 3.66 | 3.44 | 3.28 | 3.16 | 3.06 | 2.98 | 2.92 | 2.82 | 2.72 | 2.62 | 2.56 | 2.50 | 2.44 | 2.38 | 2.32 | 2.25 |
| 18 | 5.98 | 4.56 | 3.95 | 3.61 | 3.38 | 3.22 | 3.10 | 3.01 | 2.93 | 2.87 | 2.77 | 2.67 | 2.56 | 2.50 | 2.44 | 2.38 | 2.32 | 2.26 | 2.19 |
| 19 | 5.92 | 4.51 | 3.90 | 3.56 | 3.33 | 3.17 | 3.05 | 2.96 | 2.88 | 2.82 | 2.72 | 2.62 | 2.51 | 2.45 | 2.39 | 2.33 | 2.27 | 2.20 | 2.13 |
| 20 | 5.87 | 4.46 | 3.86 | 3.51 | 3.29 | 3.13 | 3.01 | 2.91 | 2.84 | 2.77 | 2.68 | 2.57 | 2.46 | 2.41 | 2.35 | 2.29 | 2.22 | 2.16 | 2.09 |
| 21 | 5.83 | 4.42 | 3.82 | 3.48 | 3.25 | 3.09 | 2.97 | 2.87 | 2.80 | 2.73 | 2.64 | 2.53 | 2.42 | 2.37 | 2.31 | 2.25 | 2.18 | 2.11 | 2.04 |
| 22 | 5.79 | 4.38 | 3.78 | 3.44 | 3.22 | 3.05 | 2.93 | 2.84 | 2.76 | 2.70 | 2.60 | 2.50 | 2.39 | 2.33 | 2.27 | 2.21 | 2.14 | 2.08 | 2.00 |
| 23 | 5.75 | 4.35 | 3.75 | 3.41 | 3.18 | 3.02 | 2.90 | 2.81 | 2.73 | 2.67 | 2.57 | 2.47 | 2.36 | 2.30 | 2.24 | 2.18 | 2.11 | 2.04 | 1.97 |

(continued)

**Table A.7** (continued)

| $df$ $\nu_1$ | | | | | | | | | | | | | | | | | | | |
|---|---|---|---|---|---|---|---|---|---|---|---|---|---|---|---|---|---|---|---|
| $\nu_2$ | 1 | 2 | 3 | 4 | 5 | 6 | 7 | 8 | 9 | 10 | 12 | 15 | 20 | 24 | 30 | 40 | 60 | 120 | ∞ |
| 24 | 5.72 | 4.32 | 3.72 | 3.38 | 3.15 | 2.99 | 2.87 | 2.78 | 2.70 | 2.64 | 2.54 | 2.44 | 2.33 | 2.27 | 2.21 | 2.15 | 2.08 | 2.01 | 1.94 |
| 25 | 5.69 | 4.29 | 3.69 | 3.35 | 3.13 | 2.97 | 2.85 | 2.75 | 2.68 | 2.61 | 2.51 | 2.41 | 2.30 | 2.24 | 2.18 | 2.12 | 2.05 | 1.98 | 1.91 |
| 26 | 5.66 | 4.27 | 3.67 | 3.33 | 3.10 | 2.94 | 2.82 | 2.73 | 2.65 | 2.59 | 2.49 | 2.39 | 2.28 | 2.22 | 2.16 | 2.09 | 2.03 | 1.95 | 1.88 |
| 27 | 5.63 | 4.24 | 3.65 | 3.31 | 3.08 | 2.92 | 2.80 | 2.71 | 2.63 | 2.57 | 2.47 | 2.36 | 2.25 | 2.19 | 2.13 | 2.07 | 2.00 | 1.93 | 1.85 |
| 28 | 5.61 | 4.22 | 3.63 | 3.29 | 3.06 | 2.90 | 2.78 | 2.69 | 2.61 | 2.55 | 2.45 | 2.34 | 2.23 | 2.17 | 2.11 | 2.05 | 1.98 | 1.91 | 1.83 |
| 29 | 5.59 | 4.20 | 3.61 | 3.27 | 3.04 | 2.88 | 2.76 | 2.67 | 2.59 | 2.53 | 2.43 | 2.32 | 2.21 | 2.15 | 2.09 | 2.03 | 1.96 | 1.89 | 1.81 |
| 30 | 5.57 | 4.18 | 3.59 | 3.25 | 3.03 | 2.87 | 2.75 | 2.65 | 2.57 | 2.51 | 2.41 | 2.31 | 2.20 | 2.14 | 2.07 | 2.01 | 1.94 | 1.87 | 1.79 |
| 40 | 5.42 | 4.05 | 3.46 | 3.13 | 2.90 | 2.74 | 2.62 | 2.53 | 2.45 | 2.39 | 2.29 | 2.18 | 2.07 | 2.01 | 1.94 | 1.88 | 1.80 | 1.72 | 1.64 |
| 60 | 5.29 | 3.93 | 3.34 | 3.01 | 2.79 | 2.63 | 2.51 | 2.41 | 2.33 | 2.27 | 2.17 | 2.06 | 1.94 | 1.88 | 1.82 | 1.74 | 1.67 | 1.58 | 1.48 |
| 120 | 5.15 | 3.80 | 3.23 | 2.89 | 2.67 | 2.52 | 2.39 | 2.30 | 2.22 | 2.16 | 2.05 | 1.94 | 1.82 | 1.76 | 1.69 | 1.61 | 1.53 | 1.43 | 1.31 |
| ∞ | 5.02 | 3.69 | 3.12 | 2.79 | 2.57 | 2.41 | 2.29 | 2.19 | 2.11 | 2.05 | 1.94 | 1.83 | 1.71 | 1.64 | 1.57 | 1.48 | 1.39 | 1.27 | 1.00 |

Values of $F$ for a specified right tail area $F_{0.05(\nu_1, \nu_2)}$

| | 1 | 2 | 3 | 4 | 5 | 6 | 7 | 8 | 9 | 10 | 12 | 15 | 20 | 24 | 30 | 40 | 60 | 120 | ∞ |
|---|---|---|---|---|---|---|---|---|---|---|---|---|---|---|---|---|---|---|---|
| 1 | 161 | 199 | 216 | 225 | 230 | 234 | 237 | 239 | 241 | 242 | 244 | 246 | 248 | 249 | 250 | 251 | 252 | 253 | 254 |
| 2 | 18.5 | 19.0 | 19.2 | 19.2 | 19.3 | 19.3 | 19.4 | 19.4 | 19.4 | 19.4 | 19.4 | 19.4 | 19.4 | 19.5 | 19.5 | 19.5 | 19.5 | 19.5 | 19.5 |
| 3 | 10.1 | 9.55 | 9.28 | 9.12 | 9.01 | 8.94 | 8.89 | 8.85 | 8.81 | 8.79 | 8.74 | 8.70 | 8.66 | 8.64 | 8.62 | 8.59 | 8.57 | 8.55 | 8.53 |
| 4 | 7.71 | 6.94 | 6.59 | 6.39 | 6.26 | 6.16 | 6.09 | 6.04 | 6.00 | 5.96 | 5.91 | 5.86 | 5.80 | 5.77 | 5.75 | 5.72 | 5.69 | 5.66 | 5.63 |
| 5 | 6.61 | 5.79 | 5.41 | 5.19 | 5.05 | 4.95 | 4.88 | 4.82 | 4.77 | 4.74 | 4.68 | 4.62 | 4.56 | 4.53 | 4.50 | 4.46 | 4.43 | 4.40 | 4.37 |
| 6 | 5.99 | 5.14 | 4.76 | 4.53 | 4.39 | 4.28 | 4.21 | 4.15 | 4.10 | 4.06 | 4.00 | 3.94 | 3.87 | 3.84 | 3.81 | 3.77 | 3.74 | 3.70 | 3.67 |
| 7 | 5.59 | 4.74 | 4.35 | 4.12 | 3.97 | 3.87 | 3.79 | 3.73 | 3.68 | 3.64 | 3.57 | 3.51 | 3.44 | 3.41 | 3.38 | 3.34 | 3.30 | 3.27 | 3.23 |

(continued)

**Table A.7** (continued)

| df v2 \ v1 | 1 | 2 | 3 | 4 | 5 | 6 | 7 | 8 | 9 | 10 | 12 | 15 | 20 | 24 | 30 | 40 | 60 | 120 | ∞ |
|---|---|---|---|---|---|---|---|---|---|---|---|---|---|---|---|---|---|---|---|
| 8 | 5.32 | 4.46 | 4.07 | 3.84 | 3.69 | 3.58 | 3.50 | 3.44 | 3.39 | 3.35 | 3.28 | 3.22 | 3.15 | 3.12 | 3.08 | 3.04 | 3.01 | 2.97 | 2.93 |
| 9 | 5.12 | 4.26 | 3.86 | 3.63 | 3.48 | 3.37 | 3.29 | 3.23 | 3.18 | 3.14 | 3.07 | 3.01 | 2.94 | 2.90 | 2.86 | 2.83 | 2.79 | 2.75 | 2.71 |
| 10 | 4.96 | 4.10 | 3.71 | 3.48 | 3.33 | 3.22 | 3.14 | 3.07 | 3.02 | 2.98 | 2.91 | 2.85 | 2.77 | 2.74 | 2.70 | 2.66 | 2.62 | 2.58 | 2.54 |
| 11 | 4.84 | 3.98 | 3.59 | 3.36 | 3.20 | 3.09 | 3.01 | 2.95 | 2.90 | 2.85 | 2.79 | 2.72 | 2.65 | 2.61 | 2.57 | 2.53 | 2.49 | 2.45 | 2.40 |
| 12 | 4.75 | 3.89 | 3.49 | 3.26 | 3.11 | 3.00 | 2.91 | 2.85 | 2.80 | 2.75 | 2.69 | 2.62 | 2.54 | 2.51 | 2.47 | 2.43 | 2.38 | 2.34 | 2.30 |
| 13 | 4.67 | 3.81 | 3.41 | 3.18 | 3.03 | 2.92 | 2.83 | 2.77 | 2.71 | 2.67 | 2.60 | 2.53 | 2.46 | 2.42 | 2.38 | 2.34 | 2.30 | 2.25 | 2.13 |
| 14 | 4.60 | 3.74 | 3.34 | 3.11 | 2.96 | 2.85 | 2.76 | 2.70 | 2.65 | 2.60 | 2.53 | 2.46 | 2.39 | 2.35 | 2.31 | 2.27 | 2.22 | 2.18 | 2.13 |
| 15 | 4.54 | 3.68 | 3.29 | 3.06 | 2.90 | 2.79 | 2.71 | 2.64 | 2.59 | 2.54 | 2.48 | 2.40 | 2.33 | 2.29 | 2.25 | 2.20 | 2.16 | 2.11 | 2.07 |
| 16 | 4.49 | 3.63 | 3.24 | 3.01 | 2.85 | 2.74 | 2.66 | 2.59 | 2.54 | 2.49 | 2.42 | 2.35 | 2.28 | 2.24 | 2.19 | 2.15 | 2.11 | 2.06 | 2.01 |
| 17 | 4.45 | 3.59 | 3.20 | 2.96 | 2.81 | 2.70 | 2.61 | 2.55 | 2.49 | 2.45 | 2.38 | 2.31 | 2.23 | 2.19 | 2.15 | 2.10 | 2.06 | 2.01 | 1.96 |
| 18 | 4.41 | 3.55 | 3.16 | 2.93 | 2.77 | 2.66 | 2.58 | 2.51 | 2.46 | 2.41 | 2.34 | 2.27 | 2.19 | 2.15 | 2.11 | 2.06 | 2.02 | 1.97 | 1.92 |
| 19 | 4.38 | 3.52 | 3.13 | 2.90 | 2.74 | 2.63 | 2.54 | 2.48 | 2.42 | 2.38 | 2.31 | 2.23 | 2.16 | 2.11 | 2.07 | 2.03 | 1.98 | 1.93 | 1.88 |
| 20 | 4.35 | 3.49 | 3.10 | 2.87 | 2.71 | 2.60 | 2.51 | 2.45 | 2.39 | 2.35 | 2.28 | 2.20 | 2.12 | 2.08 | 2.04 | 1.99 | 1.95 | 1.90 | 1.84 |
| 21 | 4.32 | 3.47 | 3.07 | 2.84 | 2.68 | 2.57 | 2.49 | 2.42 | 2.37 | 2.32 | 2.25 | 2.18 | 2.10 | 2.05 | 2.01 | 1.96 | 1.92 | 1.87 | 1.81 |
| 22 | 4.30 | 3.44 | 3.05 | 2.82 | 2.66 | 2.55 | 2.46 | 2.40 | 2.34 | 2.30 | 2.23 | 2.15 | 2.07 | 2.03 | 1.98 | 1.94 | 1.89 | 1.84 | 1.78 |
| 23 | 4.28 | 3.42 | 3.03 | 2.80 | 2.64 | 2.53 | 2.44 | 2.37 | 2.32 | 2.27 | 2.20 | 2.13 | 2.05 | 2.01 | 1.96 | 1.91 | 1.86 | 1.81 | 1.76 |
| 24 | 4.26 | 3.40 | 3.01 | 2.78 | 2.62 | 2.51 | 2.42 | 2.36 | 2.30 | 2.25 | 2.18 | 2.11 | 2.03 | 1.98 | 1.94 | 1.89 | 1.84 | 1.79 | 1.73 |
| 25 | 4.24 | 3.39 | 2.99 | 2.76 | 2.60 | 2.49 | 2.40 | 2.34 | 2.28 | 2.24 | 2.16 | 2.09 | 2.01 | 1.96 | 1.92 | 1.87 | 1.82 | 1.77 | 1.71 |
| 26 | 4.23 | 3.37 | 2.98 | 2.74 | 2.59 | 2.47 | 2.39 | 2.32 | 2.27 | 2.22 | 2.15 | 2.07 | 1.99 | 1.95 | 1.90 | 1.85 | 1.80 | 1.75 | 1.69 |

(continued)

**Table A.7** (continued)

| $df$ $v_2$ | $v_1$ 1 | 2 | 3 | 4 | 5 | 6 | 7 | 8 | 9 | 10 | 12 | 15 | 20 | 24 | 30 | 40 | 60 | 120 | ∞ |
|---|---|---|---|---|---|---|---|---|---|---|---|---|---|---|---|---|---|---|---|
| 27 | 4.21 | 3.35 | 2.96 | 2.73 | 2.57 | 2.46 | 2.37 | 2.31 | 2.25 | 2.20 | 2.13 | 2.06 | 1.97 | 1.93 | 1.88 | 1.84 | 1.79 | 1.73 | 1.67 |
| 28 | 4.20 | 3.34 | 2.95 | 2.71 | 2.56 | 2.45 | 2.36 | 2.29 | 2.24 | 2.19 | 2.12 | 2.04 | 1.96 | 1.91 | 1.87 | 1.82 | 1.77 | 1.71 | 1.65 |
| 29 | 4.18 | 3.33 | 2.93 | 2.70 | 2.55 | 2.43 | 2.35 | 2.28 | 2.22 | 2.18 | 2.10 | 2.03 | 1.94 | 1.90 | 1.85 | 1.81 | 1.75 | 1.70 | 1.64 |
| 30 | 4.17 | 3.32 | 2.92 | 2.69 | 2.53 | 2.42 | 2.33 | 2.27 | 2.21 | 2.16 | 2.09 | 2.01 | 1.93 | 1.89 | 1.84 | 1.79 | 1.74 | 1.68 | 1.62 |
| 40 | 4.08 | 3.23 | 2.84 | 2.61 | 2.45 | 2.34 | 2.25 | 2.18 | 2.12 | 2.08 | 2.00 | 1.92 | 1.84 | 1.79 | 1.74 | 1.69 | 1.64 | 1.58 | 1.51 |
| 60 | 4.00 | 3.15 | 2.76 | 2.53 | 2.37 | 2.25 | 2.17 | 2.10 | 2.04 | 1.99 | 1.92 | 1.84 | 1.75 | 1.70 | 1.65 | 1.59 | 1.53 | 1.47 | 1.39 |
| 120 | 3.92 | 3.07 | 2.68 | 2.45 | 2.29 | 2.18 | 2.09 | 2.02 | 1.96 | 1.91 | 1.83 | 1.75 | 1.66 | 1.61 | 1.55 | 1.50 | 1.43 | 1.35 | 1.25 |
| ∞ | 3.84 | 3.00 | 2.60 | 2.37 | 2.21 | 2.10 | 2.01 | 1.94 | 1.88 | 1.83 | 1.75 | 1.67 | 1.57 | 1.52 | 1.46 | 1.39 | 1.32 | 1.22 | 1.00 |

Values of $F$ for a specified right tail area $F_{0.10(v_1,v_2)}$

| $v_2$ | 1 | 2 | 3 | 4 | 5 | 6 | 7 | 8 | 9 | 10 | 12 | 15 | 20 | 24 | 30 | 40 | 60 | 120 | ∞ |
|---|---|---|---|---|---|---|---|---|---|---|---|---|---|---|---|---|---|---|---|
| 1 | 39.9 | 49.5 | 53.6 | 55.8 | 57.2 | 58.2 | 58.9 | 59.4 | 59.9 | 60.2 | 60.7 | 61.2 | 61.7 | 62.0 | 62.3 | 62.5 | 62.8 | 63.1 | 63.3 |
| 2 | 8.53 | 9.00 | 9.16 | 9.24 | 9.29 | 9.33 | 9.35 | 9.37 | 9.38 | 9.39 | 9.41 | 9.42 | 9.44 | 9.45 | 9.46 | 9.47 | 9.47 | 9.48 | 9.49 |
| 3 | 5.54 | 5.46 | 5.39 | 5.34 | 5.31 | 5.28 | 5.27 | 5.25 | 5.24 | 5.23 | 5.22 | 5.20 | 5.18 | 5.18 | 5.17 | 5.16 | 5.15 | 5.14 | 5.13 |
| 4 | 4.54 | 4.32 | 4.19 | 4.11 | 4.05 | 4.01 | 3.98 | 3.95 | 3.94 | 3.92 | 3.90 | 3.87 | 3.84 | 3.83 | 3.82 | 3.80 | 3.79 | 3.78 | 3.76 |
| 5 | 4.06 | 3.78 | 3.62 | 3.52 | 3.45 | 3.40 | 3.37 | 3.34 | 3.32 | 3.30 | 3.27 | 3.24 | 3.21 | 3.19 | 3.17 | 3.16 | 3.14 | 3.12 | 3.11 |
| 6 | 3.78 | 3.46 | 3.29 | 3.18 | 3.11 | 3.05 | 3.01 | 2.98 | 2.96 | 2.94 | 2.90 | 2.87 | 2.84 | 2.82 | 2.80 | 2.78 | 2.76 | 2.74 | 2.72 |
| 7 | 3.59 | 3.26 | 3.07 | 2.96 | 2.88 | 2.83 | 2.78 | 2.75 | 2.72 | 2.70 | 2.67 | 2.63 | 2.59 | 2.58 | 2.56 | 2.54 | 2.51 | 2.49 | 2.47 |
| 8 | 3.46 | 3.11 | 2.92 | 2.81 | 2.73 | 2.67 | 2.62 | 2.59 | 2.56 | 2.54 | 2.50 | 2.46 | 2.42 | 2.40 | 2.38 | 2.36 | 2.34 | 2.32 | 2.29 |
| 9 | 3.36 | 3.01 | 2.81 | 2.69 | 2.61 | 2.55 | 2.51 | 2.47 | 2.44 | 2.42 | 2.38 | 2.34 | 2.30 | 2.28 | 2.25 | 2.23 | 2.21 | 2.18 | 2.16 |
| 10 | 3.29 | 2.92 | 2.73 | 2.61 | 2.52 | 2.46 | 2.41 | 2.38 | 2.35 | 2.32 | 2.28 | 2.24 | 2.20 | 2.18 | 2.16 | 2.13 | 2.11 | 2.08 | 2.06 |

(continued)

**Table A.7** (continued)

| df | v1 | | | | | | | | | | | | | | | | | | |
|---|---|---|---|---|---|---|---|---|---|---|---|---|---|---|---|---|---|---|---|
| v2 | 1 | 2 | 3 | 4 | 5 | 6 | 7 | 8 | 9 | 10 | 12 | 15 | 20 | 24 | 30 | 40 | 60 | 120 | ∞ |
| 11 | 3.23 | 2.86 | 2.66 | 2.54 | 2.45 | 2.39 | 2.34 | 2.30 | 2.27 | 2.25 | 2.21 | 2.17 | 2.12 | 2.10 | 2.08 | 2.05 | 2.03 | 2.00 | 1.97 |
| 12 | 3.18 | 2.81 | 2.61 | 2.48 | 2.39 | 2.33 | 2.28 | 2.24 | 2.21 | 2.19 | 2.15 | 2.10 | 2.06 | 2.04 | 2.01 | 1.99 | 1.96 | 1.93 | 1.90 |
| 13 | 3.14 | 2.76 | 2.56 | 2.43 | 2.35 | 2.28 | 2.23 | 2.20 | 2.16 | 2.14 | 2.10 | 2.05 | 2.01 | 1.98 | 1.96 | 1.93 | 1.90 | 1.88 | 1.85 |
| 14 | 3.10 | 2.73 | 2.52 | 2.39 | 2.31 | 2.24 | 2.19 | 2.15 | 2.12 | 2.10 | 2.05 | 2.01 | 1.96 | 1.94 | 1.91 | 1.89 | 1.86 | 1.83 | 1.80 |
| 15 | 3.07 | 2.70 | 2.49 | 2.36 | 2.27 | 2.21 | 2.16 | 2.12 | 2.09 | 2.06 | 2.02 | 1.97 | 1.92 | 1.90 | 1.87 | 1.85 | 1.82 | 1.79 | 1.76 |
| 16 | 3.05 | 2.67 | 2.46 | 2.33 | 2.24 | 2.18 | 2.13 | 2.09 | 2.06 | 2.03 | 1.99 | 1.94 | 1.89 | 1.87 | 1.84 | 1.81 | 1.78 | 1.75 | 1.72 |
| 17 | 3.03 | 2.64 | 2.44 | 2.31 | 2.22 | 2.15 | 2.10 | 2.06 | 2.03 | 2.00 | 1.96 | 1.91 | 1.86 | 1.84 | 1.81 | 1.78 | 1.75 | 1.72 | 1.69 |
| 18 | 3.01 | 2.62 | 2.42 | 2.29 | 2.20 | 2.13 | 2.08 | 2.04 | 2.00 | 1.98 | 1.93 | 1.89 | 1.84 | 1.81 | 1.78 | 1.75 | 1.72 | 1.69 | 1.66 |
| 19 | 2.99 | 2.61 | 2.40 | 2.27 | 2.18 | 2.11 | 2.06 | 2.02 | 1.98 | 1.96 | 1.91 | 1.86 | 1.81 | 1.79 | 1.76 | 1.73 | 1.70 | 1.67 | 1.63 |
| 20 | 2.97 | 2.59 | 2.38 | 2.25 | 2.16 | 2.09 | 2.04 | 2.00 | 1.96 | 1.94 | 1.89 | 1.84 | 1.79 | 1.77 | 1.74 | 1.71 | 1.68 | 1.64 | 1.61 |
| 21 | 2.96 | 2.57 | 2.36 | 2.23 | 2.14 | 2.08 | 2.02 | 1.98 | 1.95 | 1.92 | 1.87 | 1.83 | 1.78 | 1.75 | 1.72 | 1.69 | 1.66 | 1.62 | 1.59 |
| 22 | 2.95 | 2.56 | 2.35 | 2.22 | 2.13 | 2.06 | 2.01 | 1.97 | 1.93 | 1.90 | 1.86 | 1.81 | 1.76 | 1.73 | 1.70 | 1.67 | 1.64 | 1.60 | 1.57 |
| 23 | 2.94 | 2.55 | 2.34 | 2.21 | 2.11 | 2.05 | 1.99 | 1.95 | 1.92 | 1.89 | 1.84 | 1.80 | 1.74 | 1.72 | 1.69 | 1.66 | 1.62 | 1.59 | 1.55 |
| 24 | 2.93 | 2.54 | 2.33 | 2.19 | 2.10 | 2.04 | 1.98 | 1.94 | 1.91 | 1.88 | 1.83 | 1.78 | 1.73 | 1.70 | 1.67 | 1.64 | 1.61 | 1.57 | 1.53 |
| 25 | 2.92 | 2.53 | 2.32 | 2.18 | 2.09 | 2.02 | 1.97 | 1.93 | 1.89 | 1.87 | 1.82 | 1.77 | 1.72 | 1.69 | 1.66 | 1.63 | 1.59 | 1.56 | 1.52 |
| 26 | 2.91 | 2.52 | 2.31 | 2.17 | 2.08 | 2.01 | 1.96 | 1.92 | 1.88 | 1.86 | 1.81 | 1.76 | 1.71 | 1.68 | 1.65 | 1.61 | 1.58 | 1.54 | 1.50 |
| 27 | 2.90 | 2.51 | 2.30 | 2.17 | 2.07 | 2.00 | 1.95 | 1.91 | 1.87 | 1.85 | 1.80 | 1.75 | 1.70 | 1.67 | 1.64 | 1.60 | 1.57 | 1.53 | 1.49 |
| 28 | 2.89 | 2.50 | 2.29 | 2.16 | 2.06 | 2.00 | 1.94 | 1.90 | 1.87 | 1.84 | 1.79 | 1.74 | 1.69 | 1.66 | 1.63 | 1.59 | 1.56 | 1.52 | 1.48 |
| 29 | 2.89 | 2.50 | 2.28 | 2.15 | 2.06 | 1.99 | 1.93 | 1.89 | 1.86 | 1.83 | 1.78 | 1.73 | 1.68 | 1.65 | 1.62 | 1.58 | 1.55 | 1.51 | 1.47 |

(continued)

**Table A.7** (continued)

| $df$ $v2$ \ $v1$ | 1 | 2 | 3 | 4 | 5 | 6 | 7 | 8 | 9 | 10 | 12 | 15 | 20 | 24 | 30 | 40 | 60 | 120 | ∞ |
|---|---|---|---|---|---|---|---|---|---|---|---|---|---|---|---|---|---|---|---|
| 30 | 2.88 | 2.49 | 2.28 | 2.14 | 2.05 | 1.98 | 1.93 | 1.88 | 1.85 | 1.82 | 1.77 | 1.72 | 1.67 | 1.64 | 1.61 | 1.57 | 1.54 | 1.50 | 1.46 |
| 40 | 2.84 | 2.44 | 2.23 | 2.09 | 2.00 | 1.93 | 1.87 | 1.83 | 1.79 | 1.76 | 1.71 | 1.66 | 1.61 | 1.57 | 1.54 | 1.51 | 1.47 | 1.42 | 1.38 |
| 60 | 2.79 | 2.39 | 2.18 | 2.04 | 1.95 | 1.87 | 1.82 | 1.77 | 1.74 | 1.71 | 1.66 | 1.60 | 1.54 | 1.51 | 1.48 | 1.44 | 1.40 | 1.35 | 1.29 |
| 120 | 2.75 | 2.35 | 2.13 | 1.99 | 1.90 | 1.82 | 1.77 | 1.72 | 1.68 | 1.65 | 1.60 | 1.55 | 1.48 | 1.45 | 1.41 | 1.37 | 1.32 | 1.26 | 1.19 |
| ∞ | 2.71 | 2.30 | 2.08 | 1.94 | 1.85 | 1.77 | 1.72 | 1.67 | 1.63 | 1.60 | 1.55 | 1.49 | 1.42 | 1.38 | 1.34 | 1.30 | 1.24 | 1.17 | 1.00 |

Values of $F$ for a specified right tail area $F_{0.25(v_1,v_2)}$

| $df$ $v2$ \ $v1$ | 1 | 2 | 3 | 4 | 5 | 6 | 7 | 8 | 9 | 10 | 12 | 15 | 20 | 24 | 30 | 40 | 60 | 120 | ∞ |
|---|---|---|---|---|---|---|---|---|---|---|---|---|---|---|---|---|---|---|---|
| 1 | 5.83 | 7.50 | 8.20 | 8.58 | 8.82 | 8.98 | 9.10 | 9.19 | 9.26 | 9.32 | 9.41 | 9.49 | 9.58 | 9.63 | 9.67 | 9.71 | 9.76 | 9.80 | 9.85 |
| 2 | 2.57 | 3.00 | 3.15 | 3.23 | 3.28 | 3.31 | 3.34 | 3.35 | 3.37 | 3.38 | 3.39 | 3.41 | 3.43 | 3.43 | 3.44 | 3.45 | 3.46 | 3.47 | 3.48 |
| 3 | 2.02 | 2.28 | 2.36 | 2.39 | 2.41 | 2.42 | 2.43 | 2.44 | 2.44 | 2.44 | 2.45 | 2.46 | 2.46 | 2.46 | 2.47 | 2.47 | 2.47 | 2.47 | 2.47 |
| 4 | 1.81 | 2.00 | 2.05 | 2.06 | 2.07 | 2.08 | 2.08 | 2.08 | 2.08 | 2.08 | 2.08 | 2.08 | 2.08 | 2.08 | 2.08 | 2.08 | 2.08 | 2.08 | 2.08 |
| 5 | 1.69 | 1.85 | 1.88 | 1.89 | 1.89 | 1.89 | 1.89 | 1.89 | 1.89 | 1.89 | 1.89 | 1.89 | 1.88 | 1.88 | 1.88 | 1.88 | 1.87 | 1.87 | 1.87 |
| 6 | 1.62 | 1.76 | 1.78 | 1.79 | 1.79 | 1.78 | 1.78 | 1.78 | 1.77 | 1.77 | 1.77 | 1.76 | 1.76 | 1.75 | 1.75 | 1.75 | 1.74 | 1.74 | 1.74 |
| 7 | 1.57 | 1.70 | 1.72 | 1.72 | 1.71 | 1.71 | 1.70 | 1.70 | 1.69 | 1.69 | 1.68 | 1.68 | 1.67 | 1.67 | 1.66 | 1.66 | 1.65 | 1.65 | 1.65 |
| 8 | 1.54 | 1.66 | 1.67 | 1.66 | 1.66 | 1.65 | 1.64 | 1.64 | 1.63 | 1.63 | 1.62 | 1.62 | 1.61 | 1.60 | 1.60 | 1.59 | 1.59 | 1.58 | 1.58 |
| 9 | 1.51 | 1.62 | 1.63 | 1.63 | 1.62 | 1.61 | 1.60 | 1.60 | 1.59 | 1.59 | 1.58 | 1.57 | 1.56 | 1.56 | 1.55 | 1.54 | 1.54 | 1.53 | 1.53 |
| 10 | 1.49 | 1.60 | 1.60 | 1.59 | 1.59 | 1.58 | 1.57 | 1.56 | 1.56 | 1.55 | 1.54 | 1.53 | 1.52 | 1.52 | 1.51 | 1.51 | 1.50 | 1.49 | 1.48 |
| 11 | 1.47 | 1.58 | 1.58 | 1.57 | 1.56 | 1.55 | 1.54 | 1.53 | 1.53 | 1.52 | 1.51 | 1.50 | 1.49 | 1.49 | 1.48 | 1.47 | 1.47 | 1.46 | 1.45 |
| 12 | 1.46 | 1.56 | 1.56 | 1.55 | 1.54 | 1.53 | 1.52 | 1.51 | 1.51 | 1.50 | 1.49 | 1.48 | 1.47 | 1.46 | 1.45 | 1.45 | 1.44 | 1.43 | 1.42 |
| 13 | 1.45 | 1.55 | 1.55 | 1.53 | 1.52 | 1.51 | 1.50 | 1.49 | 1.49 | 1.48 | 1.47 | 1.46 | 1.45 | 1.44 | 1.43 | 1.42 | 1.42 | 1.41 | 1.40 |

(continued)

**Table A.7** (continued)

| $df$ | $v1$ | | | | | | | | | | | | | | | | | | |
|---|---|---|---|---|---|---|---|---|---|---|---|---|---|---|---|---|---|---|---|
| $v2$ | 1 | 2 | 3 | 4 | 5 | 6 | 7 | 8 | 9 | 10 | 12 | 15 | 20 | 24 | 30 | 40 | 60 | 120 | ∞ |
| 14 | 1.44 | 1.53 | 1.53 | 1.52 | 1.51 | 1.50 | 1.49 | 1.48 | 1.47 | 1.46 | 1.45 | 1.44 | 1.43 | 1.42 | 1.41 | 1.41 | 1.40 | 1.39 | 1.38 |
| 15 | 1.43 | 1.52 | 1.52 | 1.51 | 1.49 | 1.48 | 1.47 | 1.46 | 1.46 | 1.45 | 1.44 | 1.43 | 1.41 | 1.41 | 1.40 | 1.39 | 1.38 | 1.37 | 1.36 |
| 16 | 1.42 | 1.51 | 1.51 | 1.50 | 1.48 | 1.47 | 1.46 | 1.45 | 1.44 | 1.44 | 1.43 | 1.41 | 1.40 | 1.39 | 1.38 | 1.37 | 1.36 | 1.35 | 1.34 |
| 17 | 1.42 | 1.51 | 1.50 | 1.49 | 1.47 | 1.46 | 1.45 | 1.44 | 1.43 | 1.43 | 1.41 | 1.40 | 1.39 | 1.38 | 1.37 | 1.36 | 1.35 | 1.34 | 1.33 |
| 18 | 1.41 | 1.50 | 1.49 | 1.48 | 1.46 | 1.45 | 1.44 | 1.43 | 1.42 | 1.42 | 1.40 | 1.39 | 1.38 | 1.37 | 1.36 | 1.35 | 1.34 | 1.33 | 1.32 |
| 19 | 1.41 | 1.49 | 1.49 | 1.47 | 1.46 | 1.44 | 1.43 | 1.42 | 1.41 | 1.41 | 1.40 | 1.38 | 1.37 | 1.36 | 1.35 | 1.34 | 1.33 | 1.32 | 1.30 |
| 20 | 1.40 | 1.49 | 1.48 | 1.47 | 1.45 | 1.44 | 1.43 | 1.42 | 1.41 | 1.40 | 1.39 | 1.37 | 1.36 | 1.35 | 1.34 | 1.33 | 1.32 | 1.31 | 1.29 |
| 21 | 1.40 | 1.48 | 1.48 | 1.46 | 1.44 | 1.43 | 1.42 | 1.41 | 1.40 | 1.39 | 1.38 | 1.37 | 1.35 | 1.34 | 1.33 | 1.32 | 1.31 | 1.30 | 1.28 |
| 22 | 1.40 | 1.48 | 1.47 | 1.45 | 1.44 | 1.42 | 1.41 | 1.40 | 1.39 | 1.39 | 1.37 | 1.36 | 1.34 | 1.33 | 1.32 | 1.31 | 1.30 | 1.29 | 1.28 |
| 23 | 1.39 | 1.47 | 1.47 | 1.45 | 1.43 | 1.42 | 1.41 | 1.40 | 1.39 | 1.38 | 1.37 | 1.35 | 1.34 | 1.33 | 1.32 | 1.31 | 1.30 | 1.28 | 1.27 |
| 24 | 1.39 | 1.47 | 1.46 | 1.44 | 1.43 | 1.41 | 1.40 | 1.39 | 1.38 | 1.38 | 1.36 | 1.35 | 1.33 | 1.32 | 1.31 | 1.30 | 1.29 | 1.28 | 1.26 |
| 25 | 1.39 | 1.47 | 1.46 | 1.44 | 1.42 | 1.41 | 1.40 | 1.39 | 1.38 | 1.37 | 1.36 | 1.34 | 1.33 | 1.32 | 1.31 | 1.29 | 1.28 | 1.27 | 1.25 |
| 26 | 1.38 | 1.46 | 1.45 | 1.44 | 1.42 | 1.41 | 1.39 | 1.38 | 1.37 | 1.37 | 1.35 | 1.34 | 1.32 | 1.31 | 1.30 | 1.29 | 1.28 | 1.26 | 1.25 |
| 27 | 1.38 | 1.46 | 1.45 | 1.43 | 1.42 | 1.40 | 1.39 | 1.38 | 1.37 | 1.36 | 1.35 | 1.33 | 1.32 | 1.31 | 1.30 | 1.28 | 1.27 | 1.26 | 1.24 |
| 28 | 1.38 | 1.46 | 1.45 | 1.43 | 1.41 | 1.40 | 1.39 | 1.38 | 1.37 | 1.36 | 1.34 | 1.33 | 1.31 | 1.30 | 1.29 | 1.28 | 1.27 | 1.25 | 1.24 |
| 29 | 1.38 | 1.45 | 1.45 | 1.43 | 1.41 | 1.40 | 1.38 | 1.37 | 1.36 | 1.35 | 1.34 | 1.32 | 1.31 | 1.30 | 1.29 | 1.27 | 1.26 | 1.25 | 1.23 |
| 30 | 1.38 | 1.45 | 1.44 | 1.42 | 1.41 | 1.39 | 1.38 | 1.37 | 1.36 | 1.35 | 1.34 | 1.32 | 1.30 | 1.29 | 1.28 | 1.27 | 1.26 | 1.24 | 1.23 |
| 40 | 1.36 | 1.44 | 1.42 | 1.40 | 1.39 | 1.37 | 1.36 | 1.35 | 1.34 | 1.33 | 1.31 | 1.30 | 1.28 | 1.26 | 1.25 | 1.24 | 1.22 | 1.21 | 1.19 |
| 60 | 1.35 | 1.42 | 1.41 | 1.38 | 1.37 | 1.35 | 1.33 | 1.32 | 1.31 | 1.30 | 1.29 | 1.27 | 1.25 | 1.24 | 1.22 | 1.21 | 1.19 | 1.17 | 1.15 |
| 120 | 1.34 | 1.40 | 1.39 | 1.37 | 1.35 | 1.33 | 1.31 | 1.30 | 1.29 | 1.28 | 1.26 | 1.24 | 1.22 | 1.21 | 1.19 | 1.18 | 1.16 | 1.13 | 1.10 |

(continued)

**Table A.7** (continued)

| df | $v1$ | | | | | | | | | | | | | | | | | | |
|----|------|---|---|---|---|---|---|---|---|---|---|---|---|---|---|---|---|---|---|
| $v2$ | 1 | 2 | 3 | 4 | 5 | 6 | 7 | 8 | 9 | 10 | 12 | 15 | 20 | 24 | 30 | 40 | 60 | 120 | $\infty$ |
| $\infty$ | 1.32 | 1.39 | 1.37 | 1.35 | 1.33 | 1.31 | 1.29 | 1.28 | 1.27 | 1.25 | 1.24 | 1.22 | 1.19 | 1.18 | 1.16 | 1.14 | 1.12 | 1.08 | 1.00 |

**Table A.8** Control chart coefficients for variables

| Subgroup size n | Control limits for $\overline{X}$-chart | | | Control limits for $\sigma$-chart | | | | | | | $d_2$ | $d_3$ | Control limits for $\overline{R}$-chart | | | |
|---|---|---|---|---|---|---|---|---|---|---|---|---|---|---|---|---|
| | $A$ | $A_2$ | $A_3$ | $c_4$ | $B_3$ | $B_4$ | $B_5$ | $B_6$ | | | $d_2$ | $d_3$ | $D_1$ | $D_2$ | $D_3$ | $D_4$ |
| 2 | 2.121 | 1.880 | 2.659 | 0.7979 | 0 | 3.267 | 0 | 2.606 | | | 1.128 | 0.853 | 0 | 3.686 | 0 | 3.267 |
| 3 | 1.732 | 1.023 | 1.954 | 0.8862 | 0 | 2.568 | 0 | 2.276 | | | 1.693 | 0.888 | 0 | 4.358 | 0 | 2.575 |
| 4 | 1.500 | 0.729 | 1.628 | 0.9213 | 0 | 2.266 | 0 | 2.088 | | | 2.059 | 0.880 | 0 | 4.698 | 0 | 2.282 |
| 5 | 1.342 | 0.577 | 1.427 | 0.9400 | 0 | 2.089 | 0 | 1.964 | | | 2.326 | 0.864 | 0 | 4.918 | 0 | 2.115 |
| 6 | 1.225 | 0.483 | 1.287 | 0.9515 | 0.030 | 1.970 | 0.029 | 1.874 | | | 2.534 | 0.848 | 0 | 5.078 | 0 | 2.004 |
| 7 | 1.134 | 0.419 | 1.182 | 0.9594 | 0.118 | 1.882 | 0.113 | 1.806 | | | 2.704 | 0.833 | 0.204 | 5.204 | 0.076 | 1.924 |
| 8 | 1.061 | 0.373 | 1.099 | 0.9650 | 0.185 | 1.815 | 0.179 | 1.751 | | | 2.847 | 0.820 | 0.388 | 5.306 | 0.136 | 1.864 |
| 9 | 1.000 | 0.337 | 1.032 | 0.9693 | 0.239 | 1.761 | 0.232 | 1.707 | | | 2.970 | 0.808 | 0.547 | 5.393 | 0.184 | 1.816 |
| 10 | 0.949 | 0.308 | 0.975 | 0.9727 | 0.284 | 1.716 | 0.276 | 1.669 | | | 3.078 | 0.797 | 0.687 | 5.469 | 0.223 | 1.777 |
| 11 | 0.905 | 0.285 | 0.927 | 0.9754 | 0.321 | 1.679 | 0.313 | 1.637 | | | 3.173 | 0.787 | 0.811 | 5.355 | 0.256 | 1.744 |
| 12 | 0.866 | 0.266 | 0.886 | 0.9776 | 0.354 | 1.646 | 0.346 | 1.610 | | | 3.258 | 0.778 | 0.922 | 5.594 | 0.283 | 1.717 |
| 13 | 0.832 | 0.249 | 0.850 | 0.9794 | 0.382 | 1.618 | 0.374 | 1.585 | | | 3.336 | 0.770 | 1.025 | 5.647 | 0.307 | 1.693 |
| 14 | 0.802 | 0.235 | 0.817 | 0.9810 | 0.406 | 1.594 | 0.399 | 1.563 | | | 3.407 | 0.763 | 1.118 | 5.696 | 0.328 | 1.672 |
| 15 | 0.775 | 0.223 | 0.789 | 0.9823 | 0.428 | 1.572 | 0.421 | 1.544 | | | 3.472 | 0.756 | 1.203 | 5.741 | 0.347 | 1.653 |
| 20 | 0.671 | 0.18 | 0.680 | 0.9869 | 0.510 | 1.490 | 0.504 | 1.470 | | | 3.735 | 0.729 | 1.549 | 5.921 | 0.415 | 1.585 |

(continued)

**Table A.8** (continued)

| Subgroup size $n$ | Control limits for $\overline{X}$-chart | | | Control limits for $\sigma$-chart | | | | | $d_2$ | Control limits for $\overline{R}$-chart | | | | |
|---|---|---|---|---|---|---|---|---|---|---|---|---|---|---|
| | $A$ | $A_2$ | $A_3$ | $c_4$ | $B_3$ | $B_4$ | $B_5$ | $B_6$ | | $d_3$ | $D_1$ | $D_2$ | $D_3$ | $D_4$ |
| 25 | 0.600 | 0.153 | 0.606 | 0.9896 | 0.565 | 1.435 | 0.559 | 1.420 | 3.931 | 0.708 | 1.806 | 6.056 | 0.459 | 1.541 |

For $n > 25$: The quantities are approximated as

$A = \dfrac{3}{\sqrt{n}}$, $A_3 = \dfrac{3}{c_4\sqrt{n}}$, $c_4 = \dfrac{4(n-1)}{4n-3}$.

$B_3 = 1 - \dfrac{3}{c_4\sqrt{2(n-1)}}$, $B_4 = 1 + \dfrac{3}{c_4\sqrt{2(n-1)}}$.

$B_5 = c_4 - \dfrac{3}{\sqrt{2(n-1)}}$, $B_6 = c_4 + \dfrac{3}{\sqrt{2(n-1)}}$.

# Glossary of Terms

**Accuracy** The measured value has a little deviation from the actual value.

**Affinity diagrams** The tools used to organize information and help achieve order out of the chaos that can develop in a brainstorming session.

**Analysis of variance** A statistical method of identifying variations associated with various components.

**Arithmetic means** The most commonly used accuracy measure in the Six Sigma project. It is defined as the ratio of the sum of all observations to the total number of observations.

**Artificial intelligence** Artificial intelligence, also called machine intelligence, is intelligence demonstrated by machines, in contrast to the natural intelligence displayed by humans and other animals.

**Balanced scorecard** The measurement tool used to assess the degree of accomplishment of strategic results and the effectiveness of projects.

**Benchmarking** One of the most cost-effective ways to understand and to predict what your competition is going to do by your competitive products.

**Big data** It is all about tapping into diverse data sets, finding, and monetizing unknown relationships, and, therefore, ultimately a data-driven process technique.

**Blockchain** A methodology to de-tangling all the data/documents/communication exchanges happening within the supply chain ecosystem.

**Brainstorming** A useful and effective technique for idea generation, by the company's think tank, generally done in a group. These ideas are then scrutinized for their viability of execution and implementation thereafter.

**Capability maturity model integration** A framework that organizes the project components used in generating models, training materials, and appraisal methods. This concept was first introduced and popularized by the Software Engineering Institute (SEI) of Carnegie Mellon University, Pittsburg.

**Carbon profile** Refers to the total greenhouse gas emissions caused directly and indirectly by a person, organization, event or product, gases that include non-carbon compounds.

**Cause and effect diagram** Also called the *Ishikawa diagram* or *Fishbone diagram*, is to generate in a structured manner the maximum number of ideas regarding possible causes for a problem by using a brainstorming technique.

**Cleaner production** It is the continuous application of an integrated, preventive environmental strategy to processes and products to reduce risks to humans and the environment.

**Climate engineering** It is all about the reduction of $CO_2$ in air–carbon sequestration.

**Coefficient of variation** The ratio of the standard deviation to the mean used as a measure of consistency.

**Cloud computing** Cloud computing is a new way of looking at technology and services. It makes data and applications available through the Internet.

**Common cause variations** The variation as the sum of the multitude of effects of a complex interaction of random or common causes. It is common to all occasions and places and always present to some degree.

**Control charts** A quality tool to monitor the stability of the process, determine when the special cause is present, and when to take appropriate action.

**Corporate social responsibility** An approach that creates long-term stakeholder value by implementing a business strategy that considers every dimension of how a business operates in the ethical, social, environmental, cultural, and economic spheres.

**Correlation** The extent of association between any two variables.

**Cost of quality** The cost incurred in a process to assure that the quality standard is met. These costs do not add value to the customer of the process but assure that high quality is achieved. Types of COQ include inspection and prevention costs. COQ should be kept as low as possible while maintaining a high level of quality.

**Cost of poor quality** The cost incurred when processes fail. These costs include the cost of repeating the process to get it right, costs to explain to the patient, MD, family what happened, and how you will correct the problem, excessive malpractice costs, complaint management, risk management, legal costs, etc. These costs are categorized as internal failures and external failures.

**Critical to quality** They describe the requirements of quality in general but lack the specificity to be measurable.

**Critical to cost** They are similar to critical to quality but deal exclusively with the impact of cost on the customer.

**Critical to process** They are typically the key process input variables (or independent variables).

**Critical to safety** They are stated customer needs regarding the safety of the product or process. Though identical to the CTQ and CTC, it is identified by the customer preference to quality.

**Customer** The stakeholder who wants to use, influence, or is affected by the outcomes.

**Cycle time** The time required to complete one cycle of operation in a process, which includes the actual work time and wait time.

**Dashboard** An easy to read, often single page, real-time user interface, showing a graphical presentation of the current status (snapshot) and historical trends of an organization's key performance indicators to enable instantaneous and informed decisions to be made at a glance.

**Defect** Any characteristic of the product that fails to meet customer requirements.

**Defectives** The total number of units containing different types of defects.

**Designed for Six Sigma** A Six Sigma application of tools to product development and process design efforts with the goal of "designing in" Six Sigma performance capability. DFSS is a methodology by which new products and services can be designed and implemented.

**DMADV** The five-step approach (D—define, M—measure, A—analyze, D—design, V—validate) for redesign to improve a process.

**DMAIC** The five steps (D—define, M—measure, A—analyze, I—improve, C—control), Six Sigma methodology to improve a process by reducing variation. The hallmarks of this Six Sigma methodology are statistical methods and a strong commitment of executives and project managers.

**DMEDI** A five-step (D—define, M—measure, E—explore, D—develop, I—implement) Lean Six Sigma approach to redesign a process to ensure efficiency or speed.

**DPU** Defect per unit.

**DPMO** Defects per million opportunities.

**Environmental management system** EMS is a structured framework for managing an organization's significant impact on the environment. These impacts can include waste, emissions, energy use, transport and consumption of materials, and increasingly, climate change factors.

**Eco-management and audit scheme** A voluntary European-wide standard introduced by the European Union and applied to all European countries. The aim of EMAS is "to recognize and reward those organizations that go beyond minimum legal compliance and continuously improve their environmental performance".

**Estimate** The observed value of the estimator.

**Environmental stress screening** A technique with universal application to improve product reliability.

**Estimator** Any statistic used to estimate the value of an unknown parameter.

**Exhaustive events** If two or more events together define the total sample space, the events are said to be collectively exhaustive events.

**Failure modes and effects analysis** A set of guidelines, a process, and a form of identifying and prioritizing potential failures and problems to facilitate process improvement.

**Gantt chart** A type of bar chart used in process/project planning and control to display planned work and finished work about time.

**Goal** The desired result to be achieved in a specified time.

**Green belt** A Six Sigma role associated with an individual who retains his or her regular position within the firm but is trained in the tools, methods, and skills necessary to conduct Six Sigma improvement projects either individually or as part of larger teams.

**Green biology** It is a subject-centered environment and plant sciences, which is an integral part of our ecosystem.

**Green chemistry** It is an area aof chemistry and chemical engineering focused on the designing of products and processes that minimize the use and generation of hazardous substances.

**Green Six Sigma** The qualitative and quantitative assessment of the direct and eventual environmental effects of all processes and products of an organization. The activities involve the systematic usage of infrastructure and manpower, optimum use of technology, and accountability of sustainable business practices.

**Green supply chain management** A methodology to minimize or eliminate wastages including hazardous chemicals, emissions, energy, and solid waste along the supply chain such as product design, material resourcing, and selection, manufacturing process, delivery of the final product, and end-of-life management of the product.

**Green statistics** It is the systematic collection and analysis of data that could lead to assessing environmental impacts of the operations of a firm. These statistics can be qualitative and quantitative.

**GRPI model** A diagnostic management tool to manage a project. Abbreviated for *Goals and responsibilities–processes and procedures–interpersonal relationships.*

**Histogram** A visual display of data organized in a frequency distribution.

**House of quality** A tool for depicting the relationship between quality metrics. The matrix includes all aspects of quality parameters extending beyond the technological solutions to management requirements housed on a particular performance barometer.

**ISO** An organization stands for the promotion of standards in the world to facilitate the international exchange of goods and services and to develop cooperation in the sphere of intellectual, scientific, technological, and economic activity including the transfer of technology to developing countries.

**Kaizen** A Japanese management concept of continuous improvement.

**Kanban** A visual signal for overproduction. It eliminates wastes in a controlled way.

**Lean** A quality improvement concept designed to reduce waste and process cycle time. Lean rules include focusing on the customer "aim", avoiding batching if customers will be delayed, waste elimination, and continuous flow/"pull". Lean is generally not as effective as a stand-alone approach but rather incorporated into Six Sigma, due to Six Sigma's more robust execution method, DMAIC.

**Lean Six Sigma** A quality improvement technique to reduce costs by eliminating product and process waste through a focused approach to eliminating non-value-added activities. See Lean above.

**Life-cycle assessment** It is a technique to assess environmental impacts associated with all the stages of a product's life from raw material extraction through materials processing, manufacture, distribution, use, repair and maintenance, and disposal or recycling.

**Lower specification limit** The smallest allowable value for quality characteristics.

**Machine learning** Machine learning, or learning using empirical data, is one of the fast-developing fields in computer science. They are aimed at solving real-world problems by extracting features for building models, and hence, the application ranges from interactive knowledge acquisition to automated theory formation.

**Master Black Belt (MBB)** A Six Sigma role associated with an individual typically assigned full time to train and mentor Black Belts as well as lead the strategy to ensure improvement projects chartered are the right strategic projects for the organization. They often write and develop training materials, are heavily involved in project definition and selection, and work closely with business leaders called Champions.

**Matrix diagram** A diagram to show the presence or absence of relationships among collected pairs of elements in a problem situation.

**Median** An accuracy measure, which divides the distribution into two equal parts.

**Mean deviation** Also called the average deviation of a set of numbers is defined by the ratio of absolute difference of observations from mean to the number of observations.

**Mode** An accuracy measure that is the most frequently occurring value in a data set.

**Noise factor** An independent variable that is difficult or too expensive to control as part of standard experimental conditions. Hence, they are also called random factors.

**Nominal scale** It is a system of assigning number symbols to events to label them.

**Non-value-added activities** The activities generally do not contribute any value to the process activities and output.

**Occupational health and safety assessment systems** It is an internationally applied British Standard for occupational health and safety management systems.

**Opportunity** A value-added feature of a unit that should meet the specifications proposed by the customer.

**Pareto chart** A quality tool used for understanding the pattern of variations in attribute or categorical type data sets.

**PDCA** Plan, Do Check, Act cycle of improvement created by Walter Stewart in the 30s and enhanced by quality leaders like Edwards Deming and Joseph Juran over the years. The impact of effective use of PDCA concepts in an organization is that the culture evolves to one of high creativity, experimentation, and the defeat of the status quo.

**Poke-Yoke** A powerful method of eliminating waste and controlling the inventories needed for a process. The philosophy allows building quality into the process and simplifies the identification of root causes of a particular problem.

**Population** The entire set of units under study.

**Prioritization matrix** A matrix to evaluate options through a systematic approach of identifying weighing and applying criteria to the options.

**Probability distribution** A description of the possible values of a random variable and their corresponding probabilities of occurrence.

**Project charter** A document stating the purposes of the project.

**Process capability** The determination of whether a process, with normal variation, is capable of meeting customer requirements or measure of the degree a process is/is not meeting customer requirements, compared to the distribution of the process.

**Precision** The abridged range of an estimate of a characteristic.

**Process control** The inspection of the items for analyzing quality problems and improving the performance of the production process.

**Process improvement** A strategy of developing focused solutions to eliminate the root causes of business performance problems.

**Process management** The totality of defined and documented processes monitored on an ongoing basis, which ensures that measures are providing feedback on the flow and function of a process.

**Process performance** A measure of actual results achieved by the ongoing process.

**Project** A chartered activity intended to achieve a stated result in a stated time. A project may vary in size and duration, involving a small group of people or large numbers in different parts of the organization. A project involves many processes, and each process progresses with specific objectives.

**Project management** A dynamic process that utilizes the appropriate resources of the organization in a controlled and structured manner to achieve some clearly defined objectives identifies as strategic needs conducted within a defined set of constraints.

**Project charter** A written declaration of the purpose and expected result of the project.

**Quality function deployment** The technique for documenting overall design logic. QFD is a method for translating customer requirements into an appropriate company program and technical requirements at each phase of the product realization cycle.

**Quartile deviation** Half of the difference between the third quartile and the first quartile. Also called the inter-quartile range.

**Random sampling** The procedure of drawing a sample randomly from a population.

**Random variable** A real-valued function associated with each outcome of a random experiment.

**Randomization** A technique to assign treatments to experimental units so that each unit has an equal chance of being assigned a particular treatment, thus minimizing the effect of variation from uncontrolled noise factors.

**Range** It is the difference between the largest and smallest numbers in the set of observations.

**Redundancies** Activities that are repeated at two points in the process also can be parallel activities that duplicate the same result.

**Reliability engineering** A method to develop the reliability requirements for the product, establish an adequate reliability program, and perform appropriate analyses and tasks to ensure the product will meet its obligations.

**Reliability function (or survival function)** The probability that a device will perform its intended function during a specified period under stated conditions.

**Response variable** The output variable that shows the observed results or value of an experimental treatment. It is sometimes known as the dependant variable.

**Replication** The repetition of the same treatment in several plots, the purpose is to get an estimate of the error variance of the experiment. Each repetition of the experiment is called a replicate. Replication increases the precision of the estimates of the effects in an experiment.

**Reverse logistics** The process of planning, implementing, and controlling the efficient, cost-effective flow of raw materials, in-process inventory, finished goods, and related information from the point of consumption to the point of origin, to recapture value or proper disposal.

**Risk management** A method of managing a project that focuses on identifying and controlling the areas or events that have the potential of creating and causing unwanted changes leading to unwanted results. Because of the complexity of risks, it is impossible to derive a universal process for managing all risks in a project.

**Robustness** The situation where the design of a product/process produces consistent, high-level performance, despite being subjected to a wide range of changing customer and manufacturing conditions.

**Significance level** The probability of rejecting a true hypothesis.

**SMART** A goal statement characteristic. Acronym for S—specific; M—measurable and measured; A—agreed upon by the manager and his/her supervisor; R—realistic, yet stretch; T—time-specific results to be achieved by a defined date.

**Simulation methods** A method of solving decision-making problems by designing, constructing, and manipulating a model of the real system.

**SIPOC diagram** An extended version of the input–process–output (IPO) diagram where the supplier–customer interface is brought into to entire improvement activities of the organization.

**Six Sigma** A disciplined, project-oriented, and methodology for eliminating defects in any process—from manufacturing to transactional and from product to service. Technically speaking, Six Sigma is described as a data-driven approach to reduce defects in a process or cut costs in a process or product, as measured by, "six standard deviations" between the mean and the nearest specification limits.

**Six Sigma marketing** A fact-based data-driven disciplined approach to growing market share by providing targeted product/markets with superior value.

**Special cause variation** The variations which are relatively large in magnitude and viable to identify also called assignable causes of variation.

**Sponsor** Someone, who is accountable for the Six Sigma project and therefore is the appointed guardian of the project on behalf of the organization.

**Standard deviation** The square root of the ratio of the sum of squares of observations taken from the mean to the number of observations and is denoted by $\sigma$.

**Standard error** The standard deviation of the sampling distribution.

**Standard operating procedures** Standard work instructions used anytime during the execution of a project.

**Statistical inference** The process of concluding the nature of some systems based on data subject to random variation.

**Statistical process control** Methods for understanding, monitoring, and improving process performance over time

**Statistical quality control** Methods for understanding, monitoring, and identifying process variation (common cause and special cause) over time, thereby improving the quality of a process.

**Total productive maintenance** It is a system of maintaining and improving the integrity of production and quality systems through the machines, equipment, processes, and employees that add business value to the organization.

**Takt time** The ratio of available production time to the rate of customer demand.

**Total quality management** A strategy for implementing and managing quality improvement activities across organizations.

**Treatment** A specific setting or combination of factor levels for an experimental unit.

**Upper specification limit** The largest allowable value for quality characteristics.

**Value-added activities** The activities, which add value to the customers.

**Value stream** All activities required to bring a product from conception to commercialization.

**Value stream mapping** The mapping of all activities and process steps according to its time of occurrence required bringing a product, service, or capability to the client.

**Variance** The square of the standard deviation and is denoted by $\sigma^2$.

**Voice of customer or voice of the client** A continuous process of collecting customer views on quality and can include customer needs, expectations, satisfaction, and perception.

**Voice of process** The essential channel to establish a communication link between the customer requirements and the process parameters.

**Yield** The total number of units handled correctly through the process steps.